Advances in Industrial Control

Springer-Verlag London Ltd.

Other titles published in this Series:

Adaptive Internal Model Control
Aniruddha Datta

Price-Based Commitment Decisions in the Electricity Market
Eric Allen and Marija Ilic

Compressor Surge and Rotating Stall: Modeling and Control
Jan Tommy Gravdahl and Olav Egeland

Radiotherapy Treatment Planning: New System Approaches
Olivier Haas

Feedback Control Theory for Dynamic Traffic Assignment
Pushkin Kachroo and Kaan Özbay

Autotuning of PID Controllers
Cheng-Ching Yu

Robust Aeroservoelastic Stability Analysis
Rick Lind and Marty Brenner

Performance Assessment of Control Loops: Theory and Applications
Biao Huang and Sirish L. Shah

Data Mining and Knowledge Discovery for Process Monitoring and Control
Xue Z. Wang

Advances in PID Control
Tan Kok Kiong, Wang Quing-Guo and Hang Chang Chieh with Tore J. Hägglund

Advanced Control with Recurrent High-order Neural Networks: Theory and Industrial Applications
George A. Rovithakis and Manolis A. Christodoulou

Structure and Synthesis of PID Controllers
Aniruddha Datta, Ming-Tzu Ho and Shankar P. Bhattacharyya

Data-driven Techniques for Fault Detection and Diagnosis in Chemical Processes
Evan L. Russell, Leo H. Chiang and Richard D. Braatz

Bounded Dynamic Stochastic Systems: Modelling and Control
Hong Wang

Non-linear Model-based Process Control
Rashid M. Ansari and Moses O. Tadé

Identification and Control of Sheet and Film Processes
Andrew P. Featherstone, Jeremy G. VanAntwerp and Richard D. Braatz

Precision Motion Control: Design and Implementation
Tan Kok Kiong, Lee Tong Heng, Dou Huifang and Huang Sunan

Nonlinear Identification and Control: A Neural Network Approach
G.P.Liu

Robert S.H. Istepanian
and James F. Whidborne (Eds)

Digital Controller Implementation and Fragility

A Modern Perspective

With 69 Figures

Springer

Robert S.H. Istepanian, BSc, MSc, PhD
Department of Electronic and Computer Engineering, Brunel University, Uxbridge, Middlesex, UB8 3PH, UK

James F. Whidborne, MA, MSc, PhD
Department of Mechanical Engineering, King's College London, Strand, London, WC2R 2LS, UK

British Library Cataloguing in Publication Data
Digital controller implementation and fragility : a modern perspective. - (Advances in industrial control)
1.Digital control systems
I.Istepanian, Robert H. II.Whidborne, J.F. (James Ferris), 1960-
629.8'312
ISBN 9781447110828

Library of Congress Cataloging-in-Publication Data
Istepanian, Robert S.H.
Digital controller implementation and fragility : a modern perspective / Robert H. Istepanian and James F. Whidborne.
p. cm.
Includes bibliographical references and index.
ISBN 978-1-4471-1082-8 (alk. paper)
1. Digital control systems. 2. Robust control. I. Whidborne, J.F. (James Ferris), 1960- II. Title
TJ223.M53 I88 2001
629.8--dc21 2001020986

ISBN 978-1-4471-1082-8 ISBN 978-1-4471-0265-6 (eBook)
DOI 10.1007/978-1-4471-0265-6

http://www.springer.co.uk

Originally published by Springer-Verlag London Berlin Heidelberg in 2001
Softcover reprint of the hardcover 1st edition 2001

Typesetting: Electronic text files prepared by editors

69/3830-543210 Printed on acid-free paper SPIN 10783749

Advances in Industrial Control

Professor Dr-Ing M. Thoma
Institut für Regelungstechnik
Universität Hannover
Appelstr. 11
30167 Hannover
Germany

Professor H. Kimura
Department of Mathematical Engineering and Information Physics
Faculty of Engineering
The University of Tokyo
7-3-1 Hongo
Bunkyo Ku
Tokyo 113
Japan

Professor A.J. Laub
College of Engineering – Dean's Office
University of California
One Shields Avenue
Davis
California 95616-5294
United States of America

Professor J.B. Moore
Department of Systems Engineering
The Australian National University
Research School of Physical Sciences
GPO Box 4
Canberra
ACT 2601
Australia

Dr M.K. Masten
Texas Instruments
2309 Northcrest
Plano
TX 75075
United States of America

Professor Ton Backx
AspenTech Europe B.V.
De Waal 32
NL-5684 PH Best
The Netherlands

SERIES EDITORS' FOREWORD

The series *Advances in Industrial Control* aims to report and encourage technology transfer in control engineering. The rapid development of control technology has an impact on all areas of the control discipline. New theory, new controllers, actuators, sensors, new industrial processes, computer methods, new applications, new philosophies..., new challenges. Much of this development work resides in industrial reports, feasibility study papers and the reports of advanced collaborative projects. The series offers an opportunity for researchers to present an extended exposition of such new work in all aspects of industrial control for wider and rapid dissemination.

As academic lecturers, when we begin a course on digital control, we usually start with the hardware block diagram for a sampled data system showing D/A and A/D conversion blocks and the controller safely labelled as K(z). We would then very quickly proceed to a design framework in the z-domain. This new volume in the *Advances in Industrial Control* monograph series edited by Robert Istepanian and James Whidborne should make us stop and think a little more carefully about the practical implementation aspects of digital control. They have brought together a wholly international group of researchers to provide excellent contributions on these usually implementation issues which are usually glossed over: *controller fragility; finite word-length; controller noise* and so on. These effects have become more critical with the growing popularity of H_∞ and H_2 controller design methods for high-performance digital controllers. For this reason it is very useful to have available a text which provides a framework for studying these problems.

This book of contributed chapters is nicely crafted together so that the reader obtains the motivation and the experimental evidence before proceeding to a series of chapters making interesting and fundamental contributions to the area. The editors are to be congratulated on producing such a fine volume, which we hope, will be a milestone publication on fragile controllers and the related issue of digital controller implementation.

M.J. Grimble and M.A. Johnson
Industrial Control Centre
Glasgow, Scotland, U.K.

PREFACE

It has long been recognised that the main motivation for feedback in control systems is in order to deal with uncertainty and variations in the plant dynamics and disturbances. Controllers rely on instrumentation and electronics in order to obtain the information about the plant, and since there is some noise and uncertainty in the sensors and actuators, this should also be considered. What is less commonly appreciated is that there also exists noise and uncertainty within the implemented controller. Nowadays most control systems are implemented with a digital computer, hence some noise is introduced in the analogue-to-digital conversion. In addition, rounding errors within the digital computer cause noise to be introduced into the system and the finite-precision representation of the controller parameters causes a reduction in the relative stability and closed-loop performance of the system. The problems resulting from the rounding errors and finite precision of the digital computer implementation is the main topic of this book.

At first glance, it may seem that the uncertainty resulting from finite-precision computing is so small compared to the uncertainty within the plant that it can simply be ignored. However, as numerous examples within this text and elsewhere show, this is not necessarily the case, and that a naive design or an injudicious controller implementation can have startling and severe effects. In a recent paper[1] Keel and Bhattacharyya presented a number of examples of controllers where extremely small perturbations on the controller parameters destabilise the closed-loop system. In the paper the authors described such controllers as being "fragile". Clearly, the implementation of a fragile controller may be problematic due to the finite-precision problem.

Although the effects of finite precision (or finite word-length) computing in digital controllers have been considered since the advent of digital control theory, there has been relatively little work in this area when compared to the numerical analysis community or even the signal processing community. One of the reasons for this is that, until recently, the implemented algorithms for feedback control were comparatively simple and low-order, and any deteriorative effects were easily rectified by manual tuning. However, the significance of finite-precision effects increases enormously with the order of the controller and higher sampling rates, and modern design techniques, such

[1] Keel, L.H. and S.P. Bhattacharyya, "Robust, fragile, or optimal?" *IEEE Trans. Automatic Contr.*, Vol. 42, no. 8, pp. 1098 - 1105, 1997.

as $\mathcal{H}_\infty$-optimisation, result in controllers which are of much higher order and complexity than traditional classical control. Thus consideration of finite precision is essential for controllers designed using advanced techniques. Indeed, the practical success of advanced control theory design methods can only be judged by the degree of take-up in industrial and commercial applications, and since no matter how good the theoretical performance of the controller, poor implementation will result in poor performance.

This book collects together a number of articles by leading researchers who consider state-of-the-art solutions to problems in the implementation of digital controllers and filters. The editors have aimed to produce a volume that will improve the understanding and awareness of the problems involved by both researchers and practitioners in the control field. This will hopefully result in better controller implementations and so improve the performance of control systems.

The book is organised into 14 chapters as follows. In Chapter 1, Istepanian and Whidborne introduce the readers to the area of finite-precision control and relevant controller implementation issues. In Chapter 2, Keel and Bhattacharyya investigate by numerical experiment some of the effects of quantisation and sampling rate on the performance and stability of discrete time control systems. These chapters provide a good starting point for newcomers to the field.

It so happens that there is a large degree of freedom in how the control algorithm is implemented in the digital computer, and the finite-precision effects are highly dependent upon this controller realisation. In Chapter 3, further analysis on the finite-precision and fast sampling rate effects are considered by Ryszard Gessing, and some different controller realisations are investigated. A detailed consideration and analysis of the effects of different computer arithmetics in the digital computer is made by Darrell Williamson in Chapter 4, and a new, low complexity, low sensitivity controller structure is proposed. In Chapter 5 by Stéphane Dussy, the stability of finite-precision digital filters is analysed via a Linear Matrix Inequality (LMI) approach.

The next four chapters consider the problem of optimising the state space realisation of the digital controller in order to minimise the finite-precision effects. In the first of these, Chen and Wu describe a global optimisation approach to minimising the pole-sensitivity by means of simulated annealing. The problem with state space realisations is that they contain a non-minimal number of control parameters. In Chapter 7, Fialho and Georgiou present an LMI based approach to optimising the realisation, and consider methods for obtaining sparse realisations. Gang Li presents an alternative approach to minimising the pole-sensitivity in Chapter 8, along with another method for obtaining sparse realisations. In Chapter 9, Whidborne and Istepanian pose the problem of both minimising the controller complexity and the required parameter word-length as a multi-objective optimisation problem that can be solved using evolutionary computation.

An alternative to obtaining non-fragile controllers by optimising the realisation of an already designed controller is by considering the design of robust control systems subject to controller uncertainty. In Chapter 10, Mäkilä and Paattilammi show how to obtain non-fragile controller designs using modern robust $\mathcal{H}_\infty$ control. Haddad and Corrado, in Chapter 11, develop non-fragile robust controllers via a fixed-structure controller synthesis approach. Yang, Wang and Soh consider the problem of robust non-fragile $\mathcal{H}_2$ controller design in Chapter 12. In Chapter 13, de Oliveira and Skelton consider the design of both the controller and the optimal controller realisation.

To date, almost all the work on finite-precision problems in control has been on linear systems. The final chapter by del Campo, Tarela and Basterretxea analyses the finite-precision effects in a nonlinear control scheme, namely fuzzy logic controllers.

Finally, we would like to express our thanks and gratitude to all the authors for their excellent contributions, without which of course, this volume would not have been possible. We would also like to thank Robert Skelton, Darrell Williamson and Pertti Mäkilä for their various suggestions and advice. The genesis of this book was in a special session arranged by the editors for the 1999 American Control Conference entitled, "New Methodologies for Finite-Precision Digital Controller Structures"; we would like to express thanks to Emmanuel Collins for his help with this session. Robert Istepanian would like to thank the Royal Society, London for their research grant (652053.Q606/AJM) and their support of his work on finite-precision control. We would also like to thank Mike Grimble and Mike Johnson, the series editors, as well as Oliver Jackson, Catherine Drury and the other staff at Springer-Verlag London.

Robert S.H. Istepanian
James F. Whidborne
London, February 2001

CONTENTS

LIST OF CONTRIBUTORS

Koldo Basterretxea
Department of Electronics and Telecommunications
University of the Basque Country
Plaza de la Casilla 3
48012 Bilbao, Vizcaya
Spain

Shankar P. Bhattacharyya
Department of Electrical Engineering
Texas A&M University
College Station, TX 77843
USA

Sheng Chen
Department of Electronics and Computer Science
University of Southampton
Highfield, Southampton SO17 1BJ
United Kingdom
e-mail: `sqc@ecs.soton.ac.uk`

Joseph R. Corrado
Raytheon Systems Company
1151 East Hermans Road
Tucson, AZ 85706-1151
USA
e-mail:
`jcorrado@west.raytheon.com`

Inés del Campo
Department of Electricity and Electronics
University of the Basque Country
48940 Leioa, Vizcaya
Spain

Mauricio C. de Oliveira
Department of Mechanical and Aerospace Engineering
University of California at San Diego
Mail Code 0411, 9500 Gilman Drive,
La Jolla
California, 92093-0411
USA

Stéphane Dussy
EADS Launch Vehicles
66, route de Verneuil
78130 Les Mureaux
France

Ian J. Fialho
Dynacs Engineering - Boeing ISS
2100 Space Park Drive, MS HS44
Houston, TX 77058
USA

Tryphon T. Georgiou
Department of Electrical Engineering
University of Minnesota
Minneapolis, MN 55455
USA

Ryszard Gessing
Institute of Automatic Control
Silesian Technical University
Ul. Akademicka 16, 44-101 Gliwice
Poland
e-mail:
`gessing@ia.gliwice.edu.pl`

Wassim M. Haddad
School of Aerospace Engineering
Georgia Institute of Technology
Atlanta, GA 30332-0150
USA
e-mail:
wm.haddad@aerospace.gatech.edu

Robert S.H. Istepanian
Department of Electronic and Computer Engineering
Brunel University
Uxbridge, Middlesex, UB8 3PH
United Kingdom
e-mail:
Robert.Istepanian@brunel.ac.uk

Lee H. Keel
Centre of Excellence in Information Systems
Tennessee State University
Nashville, TN 37203
USA

Gang Li
School of Electrical and Electronic Engineering
Nanyang Technological University
Nanyang Avenue, Singapore 639798
Singapore

Pertti M. Mäkilä
Automation and Control Institute
Tampere University of Technology
PO Box 692, FIN-33101 Tampere
Finland

Juha Paattilammi
Automation and Control Institute
Tampere University of Technology
PO Box 692, FIN-33101 Tampere
Finland

Robert E. Skelton
Department of Mechanical and Aerospace Engineering
University of California at San Diego
Mail Code 0411, 9500 Gilman Drive,
La Jolla
California, 92093-0411
USA

Yeng Chai Soh
School of Electrical and Electronic Engineering
Nanyang Technological University
Nanyang Avenue, Singapore 639798
Singapore
e-mail: eycsoh@ntu.edu.sg

José M. Tarela
Department of Electricity and Electronics
University of the Basque Country
48940 Leioa, Vizcaya
Spain

James F. Whidborne
Department of Mechanical Engineering
King's College London
Strand, London WC2R 2LS
United Kingdom
e-mail:
james.whidborne@kcl.ac.uk

Darrell Williamson
Faculty of Engineering and Information Technology
The Australian National University
Canberra, ACT 0200
Australia
e-mail:
Darrell.Williamson@anu.edu.au

Jian Liang Wang
School of Electrical and Electronic Engineering
Nanyang Technological University
Nanyang Avenue, Singapore 639798
Singapore
e-mail: ejlwang@ntu.edu.sg

Jun Wu
National Laboratory of Industrial Control Technology
Institute of Advanced Process Control
Zhejiang University
Hangzhou, 310027, P. R. China
e-mail: jwu@iipc.zju.edu.cn

Guang-Hong Yang
School of Electrical and Electronic Engineering
Nanyang Technological University
Nanyang Avenue, Singapore 639798
Singapore
e-mail: egyang@ntu.edu.sg

CHAPTER 1

FINITE-PRECISION COMPUTING FOR DIGITAL CONTROL SYSTEMS: CURRENT STATUS AND FUTURE PARADIGMS

Robert S.H. Istepanian and James F. Whidborne

Abstract. This chapter provides an overview of some of the important issues that need to be considered in the implementation of digital controllers. It focusses on the need to consider the precision of the computing device, and reviews some of the work done in the area. A summary is made of some of the advances being made in developing hardware for use in digital control implementation.

1.1 Introduction

The majority of modern advanced control systems are implemented digitally, usually with some computing device such as a digital signal processor (DSP) or a general purpose microprocessor. The very rapid advances in processor and memory technology has massively decreased the cost of implementing such controllers, and software support mechanisms make the implementation far easier. However, the implementation is far from a trivial matter, and there are a number of issues that need to be considered in deciding on the digital controller software and hardware implementation. These include:

- sampling rate
- analogue-to-digital (A/D) conversion
- computer arithmetic (floating- or fixed-point)
- word-length
- memory requirements
- computational delay
- controller realisation
- anti-windup and bumpless transfer
- concurrency and operator interfacing

Many standard texts on digital control devote some space to many of the above aspects, e.g., [1,2]. The text by Forsythe and Goodall [3] is particularly recommended as an introduction.

The A/D conversion and the storage and computation within the processor cannot be performed with infinite precision. This is because numbers within such devices must be represented using a finite number of bits, the finite word-length (FWL). The resulting finite-precision storage and computation in digital controllers can lead to serious problems. The roundoff errors cause noise to be introduced into the system and the finite-precision representation of the controller parameters causes a reduction in the relative stability and closed-loop performance of the system. Thus, control engineers need to take account of these FWL effects as well as allow for the occurrence of overflow and underflow.

Analysis of the effects of quantisation and rounding in digital controllers has been researched since the early years of digital control [4–8]. Optimal solutions to the problem of minimising the effects of roundoff noise have been reported since the 1970s [9–12]. Scaling to eliminate overflow has also been considered [13]. Early work on the finite precision of the controller parameters includes [14,15]. The text by Morony [16] and the more recent texts by Williamson [17] and Gevers and Li [18] are all seminal.

It has been argued that, with the rapidly decreasing cost and increasing speed of processor hardware, the word-lengths can simply be increased until the FWL effects are negligible. However, it is known that the problems become more marked as (i) the sampling rate increases, and (ii) the controller order increases. Since advanced design techniques can produce controllers of high order, this "brute-force" approach is not appropriate for modern advanced control. In addition, there are frequently engineering constraints on the cost of the implementation. For example, for a mass-produced product such as a portable CD player, the implementation cost in terms of financial cost, chip area and power consumption is critical. Thus it is very important to reduce the word-length, memory requirements and the number of arithmetic operations. In addition, for safety critical applications, there is a requirement for low-complexity hardware and reducing the gate count is essential — this also precludes the use of floating-point arithmetic. A low word-length, low complexity hardware implementation can be described as a "compact" implementation.

Finite-precision operations are neither associative nor distributive. Hence, the result of the controller calculations depends on the order of operations. Since the order of the operations depends upon the controller realisation (or parameterisation), the question arises of how to obtain the controller realisation that minimises the FWL effects. This has been the subject of much research, particularly within the last decade, and the topic is reviewed in the next section, along with a brief look at the idea of a "fragile" controller. In addition, the effects of fast sampling and the δ-operator are introduced. The use of digital control systems in many control engineering applications since the early 1970s has also resulted in strong research on the hardware realisation issues [19]. This topic is reviewed in Section 1.3. The final section outlines some future research paradigms.

1.2 Finite-precision Control and Fragility

1.2.1 Controller Fragility

In the usual design process, the assumption is often made that the controller can be implemented exactly. This assumption is to some extent valid, since clearly, the plant uncertainty is the most significant source of uncertainty in the control system, whilst controllers are implemented with high-precision hardware. However, there will inevitably be some amount of uncertainty in the controller, a fact that is sometimes ignored in advanced robust control design. If the controller is implemented by analogue means, there are some tolerances in the analogue components. More commonly, the controller will be implemented digitally. Subsequently, there will be some rounding of the controller parameters.

If some property of closed-loop system, particularly the stability, is very sensitive to variations or uncertainty in the controller, then the controller can be described as "fragile" [20] or "non-resilient" [21]. Such controllers have to be implemented accurately. It is clear that a controller which is robustly stable to finite-precision effects caused by FWL is also non-fragile.

1.2.2 Non-fragile Controller Realisations

It has been known for a long time [8] that the controller fragility will depend upon the particular realisation of the controller. This can be demonstrated by a simple example. Consider a proportional plus integral (P+I) controller:

$$u(k) = k_P e(k) + k_I \sum_{j=0}^{k} e(j) \tag{1.1}$$

The z-transform $u(z)/e(z) = C(z)$ is given by:

$$C(z) = k_P + k_I/(1 - z^{-1}) \tag{1.2}$$

Letting $a = k_P + k_I$ and $b = k_P$, the realisation can be arranged as:

$$C(z) = \frac{az - b}{z - 1} \tag{1.3}$$

It can also be rearranged to:

$$C(z) = k\frac{z - c}{z - 1} \tag{1.4}$$

where $k = k_P + k_I$ and $c = k_P/(k_P + k_I)$. There are in fact an infinite number of possible controller realisations, this can be easily seen if a state space controller realisation is used. Consider a P+I controller with

$k_P = 0.6$, $k_I = 0.3$ in a simple negative feedback loop with a second-order plant $P(z) = 1/(z^2 + z/2 + z)$. The closed-loop poles are located at $0.35 \pm 0.92j, 0.83$. Consider the variation in the pole at $0.35 + 0.92j$ when subject to perturbations $-0.01 \leq \delta \leq 0.01$ on the values of (k_P, k_I), (a, b) and (c, k) for the realisations given by (1.2), (1.3) and (1.4) respectively. The extremes of the pole for these realisations are shown in Figure 1.1. From this figure, it is clear that the subsequent perturbations on the pole location are dependent on the controller realisation.

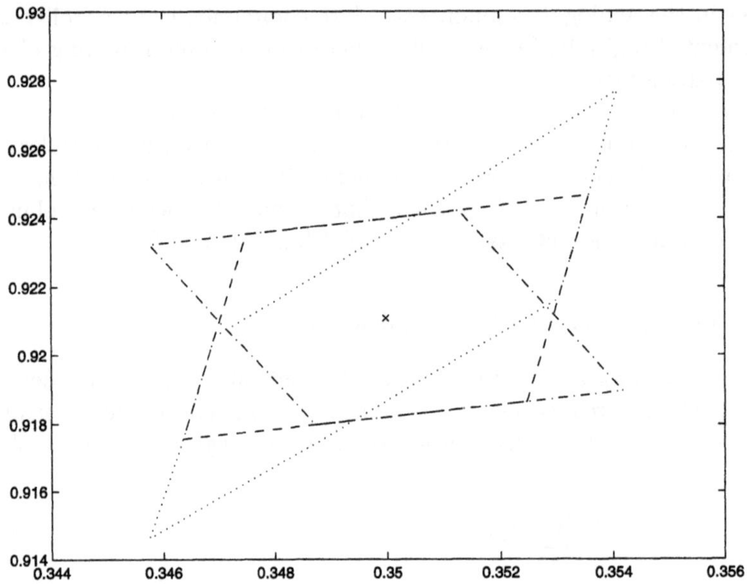

Fig. 1.1. Extremes of the pole, $\times$, at $0.35 + 0.92j$ for perturbations on the P+I controller parameters for realisations $C(z) = k_P + k_I/(1 - z^{-1})$ $(\cdots)$, $C(z) = (az - b)/(z - 1)$ $(- - -)$ and $C(z) = k(z - c)/(z - 1)$ $(- \cdot -)$

The question then arises of what is the best controller realisation so as to minimise the effect of controller parameter errors on the closed-loop system. In order to make the problem tractable, the state space realisation is usually considered. Let $P(z) = C_p(zI - A_p)^{-1}B_p$ be a state space realisation of a discretised plant connected in a feedback loop with a digital controller with a state space realisation $C(z) = C_c(zI - A_c)^{-1}B_c + D_c$. It is well known that, under infinite precision, $C_c(zI - A_c)^{-1}B_c + D_c = C_cT(zI - T^{-1}A_cT)^{-1}T^{-1}B_c + D_c$ where T is a non-singular similarity transformation matrix. Thus the realisations of $C(z)$ are not unique, and in fact, there is an infinite set of equivalent realisations of the controller. However, under finite precision, these realisations are not equivalent. Hence the problem can be posed as one of finding a similarity transformation matrix T that minimises the effect of the FWL.

This is the approach to the problem taken by many researchers. One measure of the fragility (or FWL effect) is to consider the sensitivity of the system transfer function to small changes in the controller parameters set given by $(T^{-1}A_cT, T^{-1}B_c, C_cT, D_c)$. Results minimising the transfer function sensitivity were first obtained for the open-loop case (digital filter) [22,23], and later for the closed-loop case (digital controller) [18,24]. Closed-loop sampled-data systems have also been considered in [25].

An alternative measure is based on the sensitivity of the system eigenvalues to small changes in the controller parameters set. This was solved for the open-loop case by Gevers and Li [18]. The closed-loop case is now considered in more detail. Let $C^0(z) = C_c^0(zI - A_c^0)^{-1}B_c^0 + D_c^0$ be some initial realisation of the designed digital controller described by $(A_c^0, B_c^0, C_C^0, D_C^0)$. Let T be the non-singular similarity transformation matrix such that a realisation $C(T)$ is described by $C(T) = (A_c, B_c, C_c, D_c) = (T^{-1}A_c^0T, T^{-1}B_c^0, C_c^0T, D_c^0)$. It can be shown that the closed-loop system matrix is given by:

$$\begin{aligned}\bar{A} &= \begin{bmatrix} A_p + B_pD_cC_p & B_pC_c \\ B_cC_p & A_c \end{bmatrix} \\ &= \begin{bmatrix} A_p & 0 \\ 0 & 0 \end{bmatrix} + \begin{bmatrix} B_p & 0 \\ 0 & I \end{bmatrix} \begin{bmatrix} D_c & C_c \\ B_c & A_c \end{bmatrix} \begin{bmatrix} C_z & 0 \\ 0 & I \end{bmatrix} \\ &= M_0 + M_1XM_2 = \bar{A}(X)\end{aligned} \tag{1.5}$$

where M_0, M_1 and M_2 are fixed matrices of appropriate dimensions that depend on $P(z)$, I is an identity matrix and $X = \begin{bmatrix} D_c & C_c \\ B_c & A_c \end{bmatrix}$ is the matrix of the parameters of the particular controller realisation $C(T)$.

Suppose $C^0(z)$ has been given to make the closed-loop system stable, then for all i, $|\lambda_i(\bar{A})| < 1$ where $\{\lambda_i\}$ is the set of eigenvalues of $\bar{A}$. The closed-loop eigenvalues of $\bar{A}$ are actually invariant under T and hence independent of the realisation X. However, when realisation (A_c, B_c, C_c, D_c) of $C(z)$ is implemented with finite precision using a fixed-point arithmetic, the controller parameter matrix X is perturbed to $X + \Delta X$, with the maximum perturbation on each element of ΔX dependent on the word-length. Let the maximum perturbation be defined as $\mu(\Delta X)$. In order to be able to choose a suitable arithmetic and word-length, we wish to know what is the smallest value of $\mu(\Delta X)$ that will cause the closed-loop system to go unstable. Thus, a suitable FWL stability measure has been defined [26] as:

$$\mu_0(X) \triangleq \inf\{\mu(\Delta X) : \bar{A}(X + \Delta X) \text{ is unstable}\} \tag{1.6}$$

$\mu_0(X)$ describes the stability robustness of realisation X with FWL effect and is dependent upon the realisation X and hence on T. The problem is then one of finding T and subsequently X such that $\mu_0(X)$ is maximised. Computing the value of $\mu_0(X)$ is numerically very difficult, but since the perturbations on X are actually very small, first-order approximations to $\mu_0(X)$ can be made based on the sensitivity of the system eigenvalues/poles to small changes in the controller realisation parameters.

This pole-sensitivity approach was first considered for the open-loop case in [8], and subsequently by Gevers and Li [18] who solved the problem for state space realisations based on a $1/2$ norm for the open-loop eigenvalue sensitivities. The case of the closed-loop system eigenvalues for state space controller realisations has been considered in [27,26] and sub-optimal solutions proposed (see also Chapter 8). Use of a less conservative $\infty/1$ norm has been proposed [26], but the optimisation problem is not solved in general and requires non-linear programming to find local solutions [28,29]. Global optimisation simulated annealing techniques have also been used [30,31] (see also Chapter 6). The case for a weighted $1/2$ norm for the closed-loop eigenvalue sensitivities has been proposed and completely solved [32].

In deciding the controller structure, it is better to implement the controller using such a realisation that not only yields good stability robustness but also possesses as many trivial parameters (such as $0, \pm 1$) as possible. This reduces the number of parameters and arithmetic operations, and hence reduces the storage requirements and the computation time. Algorithms to obtain such "sparse" realisations can be found in [18,24]. Additional approaches are discussed in later chapters.

Many other approaches to the problem of finding the optimal FWL controllers and realisations have been proposed [33–39]. In addition, the paper by Keel and Bhattacharryya [20] has motivated much study and research into designing "non-fragile" controllers [40–50]. Additional approaches can also be found in the remainder of this book.

1.2.3 Fast sampling and the δ-operator

It is known, that for digital controllers implemented using z-transforms, the effects of rounding become more pronounced as the sampling rate increases. Consider a simple first-order lag with time response e^{-at}. If the sampling period is h, the z-transform is given by $z/(z-e^{-ah})$. The system pole is e^{-ah}, and this gives the ratio of successive samples. Clearly if the sampling rate is increased, the ratio of successive samples tends to unity, that is $\lim_{h\to 0} e^{-ah} = 1$. Thus, the poles move towards the unit circle and so the stability margin decreases with faster sampling. If the poles are very close to the unit circle, small errors in the implementation of the controller parameters can lead to instability and hence the controller is fragile.

This problem can be overcome by means of a δ-operator[1]:

$$\delta = \frac{z-1}{h} \tag{1.7}$$

[1] Note that this is not the standard definition, but is one often used for convenience.

where h is the sampling period. The δ-transform of the first-order lag is thus given by:

$$\frac{h\delta + 1}{h\delta + 1 - e^{-ah}} \tag{1.8}$$

Expanding the exponential function and simplifying gives the transfer function as:

$$\frac{\delta + 1/h}{\delta + a(1 - ah/2! + a^2h^2/3! - \cdots)} \tag{1.9}$$

Thus as the sampling rate increases and h tends to zero, the pole tends to $-a$, which is the continuous time pole.

The FWL characteristics of digital controllers was originally investigated by Middleton and Goodwin [51]. Sensitivity and roundoff noise were considered by Gevers and Li [18]. Further work and comparative study can be found in [52–54]. Explicit relationships between the optimal eigenvalue sensitivity realisations in the z and δ domains have been presented in [55]. Other results on fast sampling have been reported in [56] (also see Chapter 3).

1.3 Hardware Issues and Development of Control System Processing Structures

As described in the earlier section, the DSP hardware revolution has shifted the paradigm of research since the mid-1980s in the area of digital control significantly. Since then, there have been numerous studies that have addressed the implementation issues of area-specific processor structures and purpose-built architectures for control systems dealing with higher integration and improved performance.

The contribution of this section is to describe the development of the area-specific hardware realisations targeted for control system applications and the most successful design systems in this area, and discusses how these advances in electronic design will eventually lead to the realisation of optimal compact computing structures that will benefit from the ongoing theoretical studies outlined above.

The first reported work on the design of a special-purpose digital processor targeted for linear controller design and implementation issues was the specific architectural design of Lang [57] in the mid-1980s. This work triggered the new age of the design and development of specific-area processor architectures targeted for different control applications. These included architectures for high-performance control using DSP chips [58,59]. A specific area that has seen more research work in this control-hardware realisation area is in robotics controller structures. Numerous design and hardware implementation methodologies and VLSI processor designs were addressed in the last two decades [60–63]. However, these studies mainly concentrate on the electronic and processor design issues of these specific robotic applications.

The development of improved "control theories", which demand more processing capabilities and high-performance requirements, lead to a major paradigm shift in the hardware implementation of these structures to off-the-shelf DSP hardware core architectures. These provide superior performance, cost-effective real time for control system processors (CSPs) that can provide optimised electronic gate count in the special core fixed-point processor structure with improved performance and data throughput compared to standard fixed-point DSP architectures [65]. However, these studies do not provide the necessary proof of concept in terms of improved closed-loop control performance and relevant stability issues addressed above. In addition, most of these are restricted to simple linear controller structures. Recent works also address the implementation issues of special-purpose architectures for fuzzy logic controllers [66]. Other soft-computing hardware realisation methods targeted for control system application are expected to emerge in the next few years. The successful mapping of the emerging control design methodologies, targeted for high performance finite-precision controller structures, to hardware architectures require the development of a generic interactive platform that integrate the finite-precision control design framework with the hardware realisation and processor design options. This proposed platform could include an interactive mechanism for both software design elements such as MATLAB®, (for the control design issues) and VHDL for the processor design issues together with a suitable hardware-mapping environment for efficient and shift mapping of the resultant controller structure into a corresponding processor structure that will provide an interactive testing mechanism for the stability and performance issues of the relevant bench mark problem.

The rapprochement mechanism between the theoretical finite-precision control design approaches and the electronic hardware design of the relevant processing architectures is an important area of future research.

1.4 Future Paradigms and Relevant Research Problems

The future remains open for more research in this important area of control engineering. The recent demands for more cost effective and compact control processing systems provide the technological pull for such research.

Recently, several areas have been addressed that highlight such paradigms. These include the application of such compact controller structures with sparse coefficient realisations applied to the areas of telesurgery control systems [67]. This seminal work addresses the need for amalgamation of the finite-precision compact controller studies in biomedical control systems and in particular for different telemedical applications that require mobility in their control system hardware together with the necessary good performance and stability margins. Other recent work addresses the need for using evolutionary methods for the realisation of optimal FWL controller structures [68,69]. These may include for example, the use of different genetic algorithms and other evolutionary optimisation methods for the design of non-fragile control structures.

In summary, the recent advances in both the hardware design requirements and demand for higher performance and critical control applications such as in the areas of military, automotive, consumer electronic, medical and safety critical systems demand further research and in depth analysis of such concepts to fulfil both the high performance, better stability, cost-effective and compactness requirements of these important control applications.

References

1. G.F. Franklin, J.D. Powell, and M. Workman. *Digital Control of Dynamic Systems*, 3rd edition. Addison-Wesley, Menlo Park, CA., 1998.
2. K.A. Åström and B. Wittenmark. *Computer Controlled Systems: Theory and Design*, 3rd edition. Prentice-Hall, Upper Saddle River, NJ., 1997.
3. W. Forsythe and R.M. Goodall. *Digital Control — Fundamentals, Theory and Practice.* Macmillan, Basingstoke, U.K., 1991.
4. J.E. Bertram. The effects of quantization in sampled-feedback systems. *Trans. AIEE*, 77:177–182, 1958.
5. J.B. Slaughter. Quantization errors in digital control systems. *IEEE Trans. Autom. Control*, 9:70–74, 1964.
6. J.B. Knowles and R. Edwards. Effect of a finite-word-length computer in a sampled-data feedback system. *Proc. IEE*, 112(6):1197–1207, 1965.
7. E.E. Curry. The analysis of round-off and truncation errors in a hybrid control system. *IEEE Trans. Autom. Control*, 12:601–604, 1967.
8. P.E. Mantey. Eigenvalue sensitivity and state-variable selection. *IEEE Trans. Autom. Control*, 13(3):263–269, 1968.
9. C.T. Mullis and R.A. Roberts. Synthesis of minimum round off noise fixed-point digital filters. *IEEE Trans. Circuits & Syst.*, 23:551–562, 1976.
10. S.Y. Hwang. Minimum uncorrelated unit noise in state-space digital filtering. *IEEE Trans. Accoustics, Speech & Sig. Proc.*, 25:273–281, 1977.
11. P. Morony, A.S. Willsky, and P.K. Houpt. Roundoff noise and scaling in the digital implementation of control compensators. *IEEE Trans. Accoustics, Speech & Sig. Proc.*, 31(6):1464–1474, 1983.
12. M.E. Ahmed and P.R. Belanger. Scaling and roundoff in fixed-point implementation of control algorithms. *IEEE Trans. Ind. Electr.*, 31(3):228–234, 1984.
13. M. Steinbuch, G. Schoostra, and H.T. Goh. Closed-loop scaling of fixed-point digital control. *IEEE Trans. Control Syst. Technology*, 2:312–317, 1994.
14. P. Morony, A.S. Willsky, and P.K. Houpt. The digital implementation of control compensators: the coefficient word length issue. *IEEE Trans. Autom. Control*, 25:621–630, 1980.
15. A.J.M. Van Wingerden and W.L. de Koning. The influence of finite word-length on digital optimal-control. *IEEE Trans. Autom. Control*, 29(5):385–391, 1984.
16. P. Morony. *Issues in the Implementation of Digital Feedback Compensators.* Number 5 in Signal Processing, Optimization, and Control Series. MIT Press, Cambridge, MA, 1983.

17. D. Williamson. *Digital Control and Implementation: Finite Wordlength Considerations.* Systems and Control Engineering. Prentice Hall, Englewood Cliffs, NJ, 1991.
18. M. Gevers and G. Li. *Parametrizations in Control, Estimations and Filtering Problems: Accuracy Aspects.* Springer-Verlag, Berlin, 1993.
19. Nagle, H.T. and Carrolf, C. 'Hardware realization of sampled data controllers', Proc. IFAC Symp. Aut. Contr. in Space, Armenia, pp.23-27, Aug.1974.
20. L.H. Keel and S.P. Bhattacharryya. Robust, fragile, or optimal? *IEEE Trans. Autom. Control*, 42(8):1098–1105, 1997.
21. W.M. Haddad and J.R. Corrado. Robust resilient dynamic controllers for systems with parametric uncertainty and controller gain variations. In *Proc. 1998 Amer. Contr. Conf.*, pages 2837–2841, Philadelphia, PA., 1998.
22. L. Thiele. Design of sensitivity and round-off noise optimal state-space discrete systems. *Int. J. Circuit Theory Appl.*, 12:39–46, 1984.
23. L. Thiele. On the sensitivity of linear state space systems. *IEEE Trans. Circuits & Syst.*, 33(5):502–510, 1986.
24. G. Li, B.D.O. Anderson, M. Gevers, and J.E. Perkins. Optimal FWL design of state space digital systems with weighted sensitivity minimization and sparseness consideration. *IEEE Trans. Circuits & Syst. II*, 39(5):365–377, 1992.
25. A.G. Madievski, B.D.O. Anderson, and M. Gevers. Optimum realizations of sampled-data controllers for FWL sensitivity minimization. *Automatica*, 31(3):367–379, 1995.
26. R.H. Istepanian, G. Li, J. Wu, and J. Chu. Analysis of sensitivity measures of finite-precision digital controller structures with closed-loop stability bounds. *IEE Proc. Control Theory and Appl.*, 145(5):472–478, 1998.
27. G. Li. On the structure of digital controllers with finite word length consideration. *IEEE Trans. Autom. Control*, 43(5):689–693, 1998.
28. S. Chen, J. Wu, R.H. Istepanian, and J. Chu. Optimizing stability bounds of finite-precision PID controller structures. *IEEE Trans. Autom. Control*, 44(11):2149–2153, 1999.
29. J. Wu, R.H. Istepanian, and S. Chen. Stability issues of finite-precision controller structures for sampled-data systems. *Int. J. Control*, 72(15):1331–1342, 1999.
30. R.H. Istepanian, S. Chen, J. Wu, and J.F. Whidborne. Optimal finite-precision controller realization of sampled-data systems. *Int. J. Syst. Sci.*, 31(4):429–438, 2000.
31. J. Wu, S. Chen, G. Li, and J. Chu. Optimal finite-precision state-estimate feedback controller realizations of discrete-time systems. *IEEE Trans. Autom. Control*, 45(8):1550 – 1554, 2000.
32. J.F. Whidborne, R.S.H. Istepanian, and J. Wu. Reduction of controller fragility by pole sensitivity minimization. *IEEE Trans. Autom. Control*, 46(2):320–325, 2001.
33. D. Williamson and K. Kadiman. Optimal finite wordlength linear quadratic regulation. *IEEE Trans. Autom. Control*, 34(12):1218–1228, 1989.
34. D. Williamson and R.E. Skelton. Optimal Q-markov cover for finite word-length implementation. *Math. Syst. Theory*, 22(4):255–273, 1989.
35. K. Liu, R.E. Skelton, and K.M. Grigoriadis. Optimal controllers for finite word length implementation. *IEEE Trans. Autom. Control*, 37(9):1294–1304, 1992.

36. I.J. Fialho and T.T. Georgiou. On stability and performance of sampled-data systems subject to word-length constraint. *IEEE Trans. Autom. Control*, 39(12):2476–2481, 1994.
37. M.A. Rotea and D. Williamson. Optimal realizations of finite word-length digital-filters and controllers. *IEEE Trans. Circuits & Syst. II*, 42(2):61–72, 1995.
38. J.F. Whidborne, J. Wu, and R.H. Istepanian. Finite word length stability issues in an ℓ_1 framework. *Int. J. Control*, 73(2):166–176, 2000.
39. R. D'Andrea, and R.S.H. Istepanian. Design of full state feedback finite precision controllers'. *Int. J. of Robust and Non-Linear Control*, 2001 (to appear).
40. P.M. Mäkilä. Comments on "Robust, fragile, or optimal?". *IEEE Trans. Autom. Control*, 43(9):1265–1268, 1998.
41. P. Dorato. Non-fragile controller design: an overview. In *Proc. 1998 Amer. Contr. Conf.*, pages 2829–2831, Philadelphia, PA., 1998.
42. D. Kaesbauer and J. Ackermann. How to escape from the fragility trap. In *Proc. 1998 Amer. Contr. Conf.*, pages 2832–2836, Philadelphia, PA., 1998.
43. A. Jadbabaie, C.T. Abdallah, D. Famularo, and P. Dorato. Robust, non-fragile and optimal controller design via linear matrix inequalities. In *Proc. 1998 Amer. Contr. Conf.*, pages 2842–2846, Philadelphia, PA., 1998.
44. D. Famularo, P. Dorato, C.T. Abdallah, W.M. Haddad, and A. Jadbabaie. Robust non-fragile LQ controllers: the static state feedback case. *Int. J. Control*, 73(2):159–165, 2000.
45. W.M. Haddad and J.R. Corrado. Robust resilient dynamic controllers for systems with parametric uncertainty and controller gain variations. *Int. J. Control*, 73(15):1405–1423, 2000.
46. J. Paattilammi and P.M. Mäkilä. Fragility and robustness : A case study on paper machine headbox control. *IEEE Control Syst. Magaz.*, 20:13–22, 2000.
47. G.H. Yang, J.L. Wang, and Y.C. Soh. Guaranteed cost control for discrete-time linear systems under controller gain perturbations. *Linear Alg. Appl.*, 312(1-3):161–180, 2000.
48. G.H. Yang, J.L. Wang, and C. Lin. $\mathcal{H}_\infty$ control for linear systems with additive controller gain variations. *Int. J. Control*, 73(16):1500–1506, 2000.
49. J.S. Yee, G.H. Yang, and J.L. Wang. Non-fragile $\mathcal{H}_\infty$ flight controller design for large bank-angle tracking manoeuvres. *Proc. IMechE, Part I: J. Syst. & Contr.*, 214(3):157–172, 2000.
50. G.H. Yang and J.L. Wang. Robust nonfragile Kalman filtering for uncertain linear systems with estimator gain uncertainty. *IEEE Trans. Autom. Control*, 46(2):343–348, 2001.
51. R.H. Middleton and G.C. Goodwin. Improved finite word-length characteristics in digital-control using delta-operators. *IEEE Trans. Autom. Control*, 31(11):1015–1021, 1986.
52. S. Chen, J. Wu, R.H. Istepanian, J. Chu, and J.F. Whidborne. Optimizing stability bounds of finite-precision controller structures for sampled-data systems in the δ-operator domain. *IEE Proc. Control Theory and Appl.*, 146(6):517–526, 1999.
53. J. Wu, S. Chen, G. Li, R.H. Istepanian, and J. Chu. Shift and delta operator realisations for digital controllers with finite word length considerations. *IEE Proc.-Control Theory Appl.*, 147(6):664–672, 2000.

54. S. Chen, R.H. Istepanian, J. Wu, and J. Chu. Comparative study on optimizing closed-loop stability bounds of finite-precision controller structures with shift and delta operators. *Syst. Control Lett.*, 40(3):153–163, 2000.
55. T.L. Song, E.G. Collins, and R.H. Istepanian. Improved closed-loop stability for fixed-point controller realizations using the delta operator. *Int. J. Robust Nonlinear Control*, 11(1):41–57, 2001.
56. R. Gessing. Word length of pulse transfer function for small sampling periods. *IEEE Trans. Autom. Control*, 44(9):1760–1764, 1999.
57. J.H. Lang. On the design of a special-purpose digital control processor, *IEEE Trans. Autom. Control*, 29(3):195–201, 1984.
58. S. Battilotti and G. Ulivi. An architecture for high performance control using digital signal processor chips, *IEEE Control Syst. Magazine*, 10(6):20-23, 1990.
59. S.C. Chen. Application of one-chip signal processor in digital controller implementation, *IEEE Control Syst. Magazine*, 2:16-22, 1982.
60. K. Shimabukuro, M. Kameyama and T. Higuchi. Design of a multiple-valued VLSI processor for digital control, *IEICE Trans. Inf. Systems*, E-75(5):709-719, 1992.
61. Kitticakoonkit, S., Kampyama, M. and Higuchi, T. Design of a matrix multiply-addition VLSI processor for robot inverse dynamics computation, *IEICE Trans. Inf. Systms*, E-74(11):3819–3827, 1994.
62. E. Kappos and D.J. Kinniment. Application-specific processor architectures for embedded control: Case studies. *Microprocessors and Microsystems*, 20(4):225–232, 1996.
63. P. Sadayappan, Y.L.C. Ling, K.W. Olson, and D.E. Orin. A restructurable VLSI robotics vector processor architecture for real-time control. *IEEE Trans. Robot. Autom.*, 5(5):583–599, 1989.
64. M.K. Masten and I. Panahi. Digital signal processors for modern control systems. *Control Eng. Practice*, 5(4):449–458, 1997.
65. S. Jones, R. Goodall, and M. Gooch. Targeted processor architectures for high-performance controller implementation. *Control Eng. Practice*, 6(7):867–878, 1998.
66. V. Samoladas and L. Petrou. Special-purpose architectures for fuzzy-logic controllers, *Microprocessing and Microprogramming*, 40(4):275–289, 1994.
67. R.S.H. Istepanian, J. Wu, and J.F. Whidborne. Controller realizations of a teleoperated dual-wrist assembly system with finite word length considerations. *IEEE Trans. Control Syst. Technology*, 9(4):624–628, 2001.
68. J.F. Whidborne and R.S.H. Istepanian. Optimal finite-precision PID controller structures using genetic algorithms. In *Proc. 14th IFAC World Congress*, volume F, pages 181–186, Beijing, 1999.
69. J.F. Whidborne and R.H. Istepanian. A genetic algorithm approach to designing finite-precision controller structures. *IEE Proc. Control Theory and Appl.*, 2000 (submitted).

CHAPTER 2
STABILITY MARGINS AND DIGITAL IMPLEMENTATION OF CONTROLLERS

Lee H. Keel and Shankar P. Bhattacharyya

Abstract. This chapter is motivated by recent results on the fragility of optimal controllers and investigates the robustness and related behaviour of control systems containing a digital implementation of a designed continuous time controller, mainly through examples. We determine the coefficient stability margin of the digital controller as the sampling time is varied. We also study the stability and inter-sample response of the hybrid system as the sampling period varies. The main general trend indicated by the examples are a) that the stability margin of the digital controller often decreases as the sampling frequency increases and b) that the actual hybrid system may exhibit unstable inter-sample behaviour even though the discrete time stability margin of the digital controller may be large especially at low sampling frequencies. These examples raise a number of issues related to digital implementation of controllers.

2.1 Introduction

A recent paper [1] has shown that controllers designed in the continuous time domain, by several optimal and robust control approaches, can give rise to fragile designs in the sense that the controller itself possesses a vanishingly small amount of tolerance to coefficient perturbations. This result challenges the fundamental notion that has driven modern and post-modern control theory since 1960, namely that good or even just acceptable control systems can be designed through optimisation. An additional layer of complication is the issue of controller order which cannot be fixed if one uses the available theory. The final important consideration is the issue of digital implementation of a designed continuous controller.

This last issue initially motivated us to ask: what happens when these optimal or fragile controllers are implemented digitally? Due to the spectacular developments in computer technology, control systems are now mainly implemented digitally in practice. Even in the case of simple PID type of controllers, digital implementation through microprocessors is the common practice. In pursuing this question we realised that there are some basic design issues related to stability margins, sampling time, inter-sample behaviour of the hybrid system simulation and the effects of finite word-length (FWL) that were not clear to us. This prompted us to investigate some of these interrelationships through examples. The chapter presents the results of these numerical experiments.

In this chapter, we examine this digital implementation from the point of view of stability and performance of the closed-loop system. The analytical solution of this problem is very difficult because the relationships that we want to study, namely that between stability margin, sampling frequencies and the behaviour of a digitally implemented system is highly nonlinear. Thus, in an attempt to develop some insights, we examine these via examples.

Our examination starts with the discrete time equivalent of the continuous time controller under various sampling periods. The discretisation method we used here is Tustin's method, which is based on trapezoidal approximation of the integration operation. For each sampling period, the parametric stability margin of the discrete time equivalent controller is computed, by the methods developed in [2]. This shows how the choice of sampling period affects the parametric stability margin. Then the unit step responses of the continuous and discrete systems are compared to examine the accuracy of discrete equivalent system with respect to the sampling period T. In each case, the step response of the hybrid configuration (a discrete controller plus a continuous plant) is also computed.

In a real-world digital control implementation, the output signals of the plant are converted into binary numbers by an analogue-to-digital (A/D) converter. These binary numbers are then fed to a computer algorithm to generate the output of the controller again in binary number form. A digital-to-analogue (D/A) converter is converting this into analogue signals. During this conversion process, the accuracy of the signals largely depend upon the quantisation steps that are limited by finite word-lengths of A/D and D/A devices. We also observe the effect of this quantisation process in the examples. Finally, we give some discussion of observed trends based on the simulation results.

2.2 Digital Implementation

A typical digital control configuration is shown in Figure 2.1.

A reference signal r is usually generated internally in the computer. When r is generated externally, the signal must be fed to the computer through an A/D block. It is noted that the A/D and D/A devices are built with finite word-length. It means that if the operating voltage of an n-bit A/D is $[-V, V]$, the quantisation step cannot be smaller than $\frac{2V}{2^n}$. This is called "quantisation effect" in later examples.

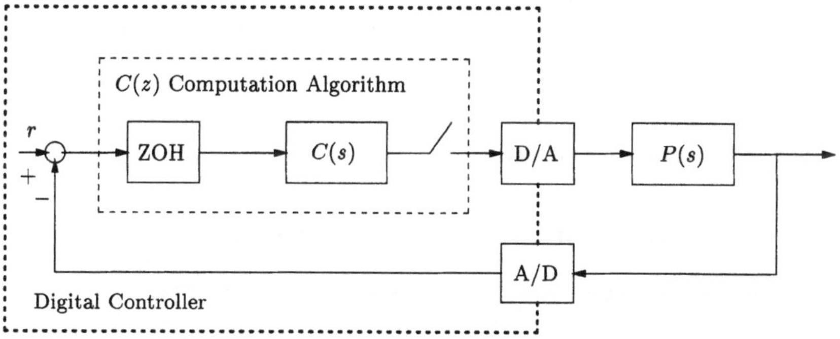

Fig. 2.1. A typical digital control configuration

2.3 Simulation Setup

In our simulation, we consider four different configurations. The simulation tool we used here are both MATLAB® v.4/SIMULINK® v.1 and MATLAB® v.5/SIMULINK® v.2. All simulations have been carried out in both 233MHz Pentium/Windows 95 and 200MHz SUN UltraSparc 2 Station/Solaris 2.5. The discretisation method we employed in each example here is Tustin's method.

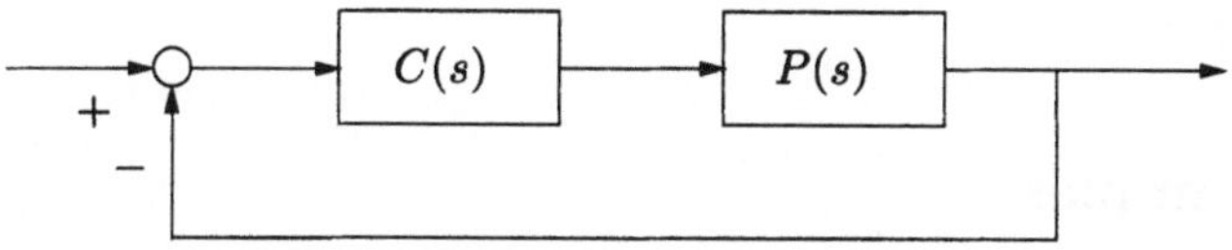

Fig. 2.2. A continuous controller/continuous plant feedback configuration

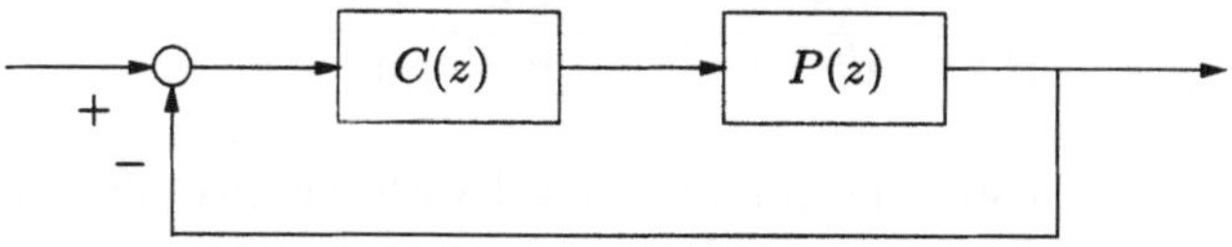

Fig. 2.3. A discrete controller/discrete plant feedback configuration

The configuration shown in Figure 2.2 simply simulates the ideal continuous time system. Figure 2.3 shows the discrete equivalent of the configuration given in Figure 2.2. Since no inter-sample behaviour and quantisation effects

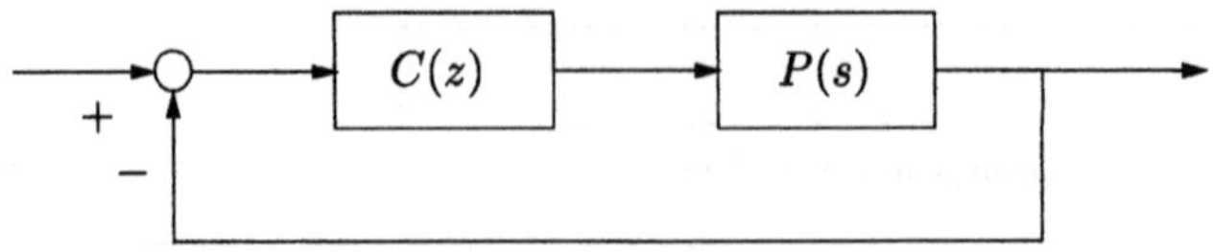

Fig. 2.4. A hybrid system configuration

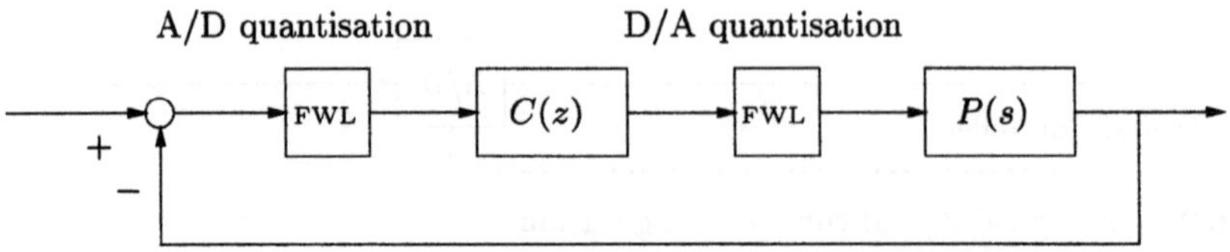

Fig. 2.5. A hybrid system configuration with FWL quantisation

play a role, the simulation result will be close to the theoretical values only within the numerical accuracy limitations of the computer as long as the discretisation of $C(s)$ and $P(s)$ are accurate enough. Figure 2.4 shows the hybrid configuration. In this configuration, the inter-sample behaviour is in effect. As a result, the simulation is closer to the real-world digital control than the previous two cases shown in Figures 2.2 and 2.3. In Figure 2.5, both inter-sample behaviour and FWL quantisation are in effect. This may be the closest real-world digital control simulation structure one can construct in the computer.

2.4 Examples

Example 2.1. (Example 1 of [1])

$$P(s) = \frac{s-1}{s^2 - s - 2}$$

$$C(s) = \frac{q_6^0 s^6 + q_5^0 s^5 + q_4^0 s^4 + q_3^0 s^3 + q_2^0 s^2 + q_1^0 s + q_0^0}{p_6^0 s^6 + p_5^0 s^5 + p_4^0 s^4 + p_3^0 s^3 + p_2^0 s^2 + p_1^0 s + p_0^0}$$

The plant and controller are discretised with various different sampling periods. For each discrete equivalent closed-loop system, we computed the parametric stability margin of the digital controller and these are listed below (see Table 2.2).

The unit step responses of the closed-loop systems given in Figures 2.2, 2.3, and 2.4 are obtained for each case. Figure 2.6 shows the responses of these three types of systems with $T = 0.5$ sec and $T = 0.05$ sec.

Table 2.1. Controller coefficients and closed-loop poles (Example 2.1)

numerator	denominator	closed-loop poles
$q_6^0 = 379$	$p_6^0 = 3$	$-0.4666 \pm j14.2298$
$q_5^0 = 39383$	$p_5^0 = -328$	$-5.5333 \pm j11.3290$
$q_4^0 = 192306$	$p_4^0 = -38048$	-1.0002
$q_3^0 = 382993$	$p_3^0 = -179760$	$-0.9999 \pm j0.00021$
$q_2^0 = 383284$	$p_2^0 = -314330$	-0.9997
$q_1^0 = 192175$	$p_1^0 = -239911$	
$q_0^0 = 38582$	$p_0^0 = -67626$	

Table 2.2. Sampling periods vs. stability margins

T	$\hat{\rho}$
0.5	$7.006326574597361 \times 10^{-4}$
0.1	$1.305146761417940 \times 10^{-4}$
0.05	$8.500629640436455 \times 10^{-6}$
0.03	$1.139921321461374 \times 10^{-6}$
0.01	$4.483408705860278 \times 10^{-8}$
0.005	$7.956037771058994 \times 10^{-11}$
0.001	unstable

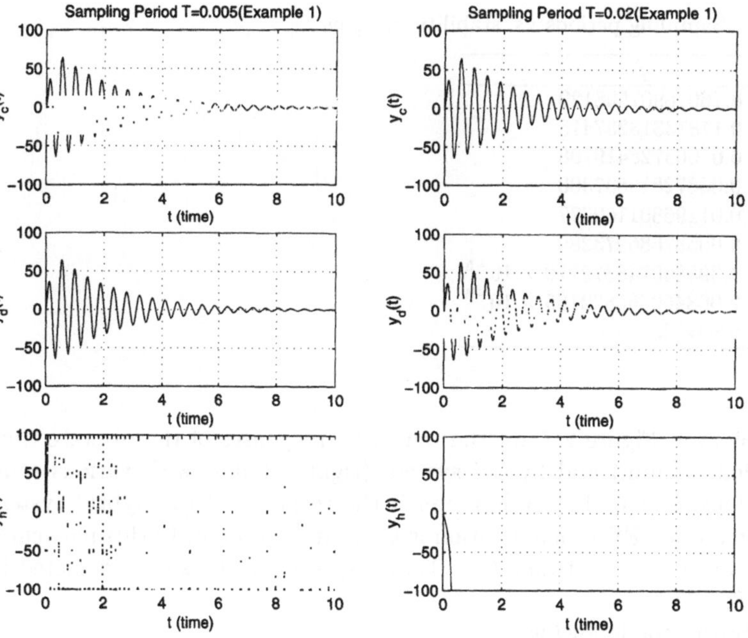

Fig. 2.6. Top: Continuous system, Mid: Discrete system, Bottom: Hybrid system (Example 2.1)

We observe the following:

1. It appears that the parametric stability margin of the equivalent discrete system becomes smaller as the sampling period T reduces.
2. For all T above, it is found that the response of the hybrid system is unbounded (see the bottom figures) even though the continuous and discrete equivalent systems show stable behaviour. It suggests that the inter-sample behaviour of the system is not properly captured even when T is very small. Moreover, even though a large discrete time stability margin may be obtained, the actual hybrid system may be unstable especially at low sampling frequencies.

Example 2.2. (Example 2 of [1])

$$P(s) = \frac{s-1}{s^2-s-2}$$

$$C(s) = \frac{11.44974739s + 11.24264066}{s - 7.03553383}$$

The roots of the closed-loop system are -1.4142 and $-1.0000 \pm j0.9999$. The parametric stability margins of discrete time equivalent controller, with different Ts, are tabulated below (see Table 2.3).

Table 2.3. Sampling periods vs. stability margins

T	$\hat{\rho}$
0.5	0.76078907508129
0.1	0.17813313287418
0.05	0.07005728419798
0.03	0.03872810837300
0.01	0.01196801162777
0.005	0.00587686973289
0.001	$4.393860748379719 \times 10^{-5}$
0.00001	$2.008460365822165 \times 10^{-6}$
0.000001	unstable

The plots in Figure 2.7 are the unit step responses of the discrete time system (left column) and hybrid system (right column) with various sampling periods T. We observe how much the response of the hybrid system deteriorates when FWL quantisation is taken into account. In this particular example, we set $T = 0.1$ and $T = 0.05$, respectively. These are depicted in Figure 2.8.

We observe the following:

1. As in Example 2.1, it appears that the parametric stability margin of the discrete equivalent controller is reduced as the sampling period is reduced.

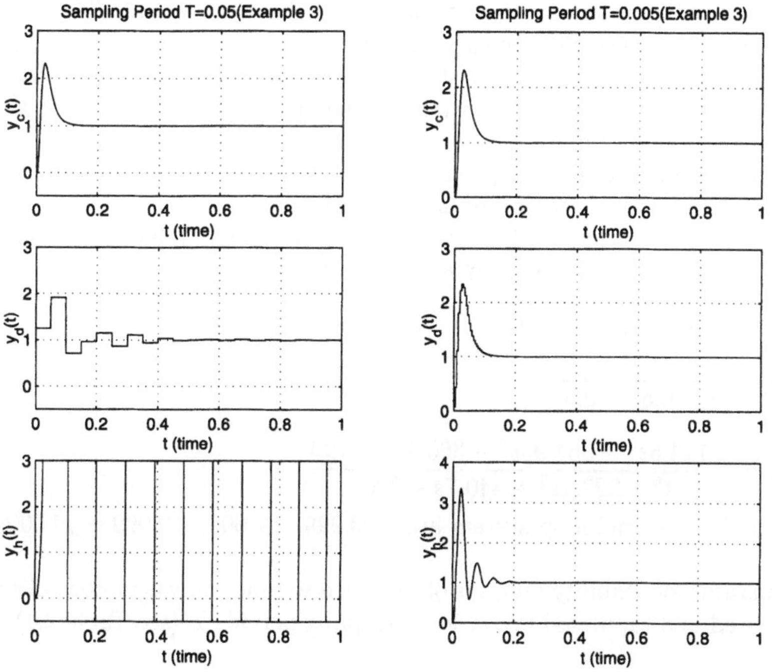

Fig. 2.7. Top: Continuous system, Mid: Discrete system, Bottom: Hybrid system (Example 2.2)

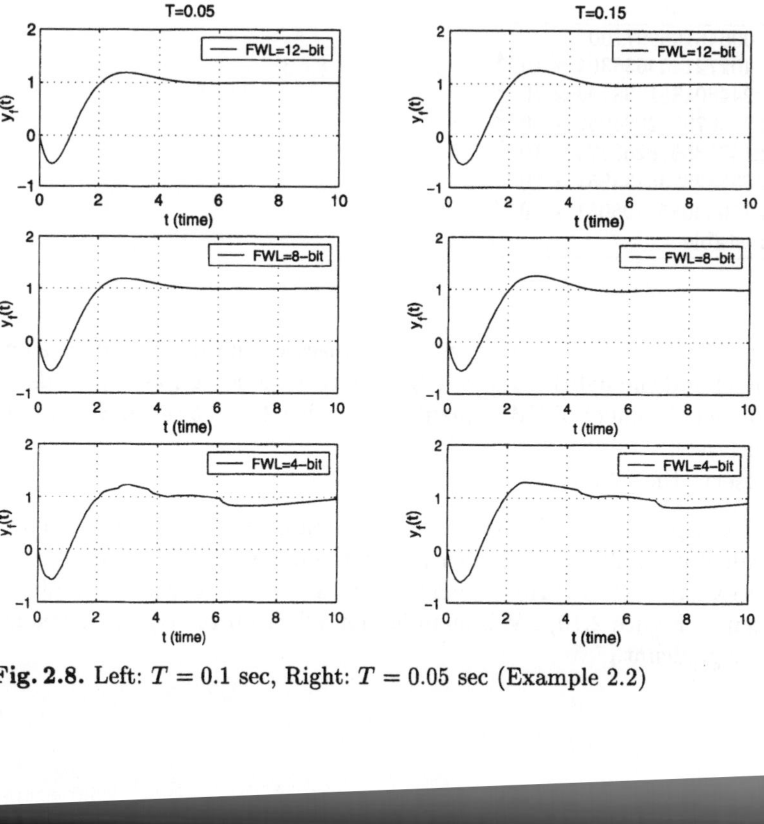

Fig. 2.8. Left: $T = 0.1$ sec, Right: $T = 0.05$ sec (Example 2.2)

2. The discrete equivalent systems show stable behaviour with all Ts. However, the hybrid system becomes unstable when $T = 0.5$ sec. In other words, the system becomes unstable if the sampling rate is not fast enough, even if its discrete equivalent system possesses relatively large parametric stability margin around the controller coefficients.
3. For smaller Ts, the hybrid system behaves stably even if it has smaller parametric stability margin around the controller.
4. As expected (see Figure 2.8), the response deterioration due to a rough quantisation is compensated by a smaller sampling period.

Example 2.3. (Example 3 of [1])

$$P(s) = \frac{s-1}{s^2+0.5s-0.5}$$

$$C(s) = \frac{-124.5s^3-364.95s^2-360.45s-120}{s^3+227.1s^2+440.7s+220}$$

The roots of the closed-loop system are -99.999, -1.000, $-0.999 \pm j0.000$, -0.100.

The parametric stability margins of the discrete time equivalent controller are computed and tabulated for various sampling periods T (see Table 2.4).

Table 2.4. Sampling periods vs. stability margins

T	$\hat{\rho}$
0.5	0.00646195997100
0.1	$3.402721454636304 \times 10^{-4}$
0.05	$8.244801851338506 \times 10^{-5}$
0.03	$2.749079002084932 \times 10^{-5}$
0.01	$2.137538978200774 \times 10^{-6}$
0.005	$3.651828235223951 \times 10^{-7}$
0.001	$3.770540851795051 \times 10^{-9}$
0.0004	unstable

Figure 2.9 illustrates the unit step responses of the discrete equivalent system (left) and the hybrid system (right) with various sampling periods T. We also depict the effect of FWL quantisation with $T = 0.5$ sec and $T = 0.05$ sec, respectively.

We observe the following:

1. As in Examples 2.1 and 2.2, the parametric stability margin of the discrete equivalent controller gets smaller as the sampling period is reduced.
2. In this example, the hybrid system behaves well under all Ts we examined.
3. As seen in Figure 2.10, FWL quantisation effect on the output response is not significant.

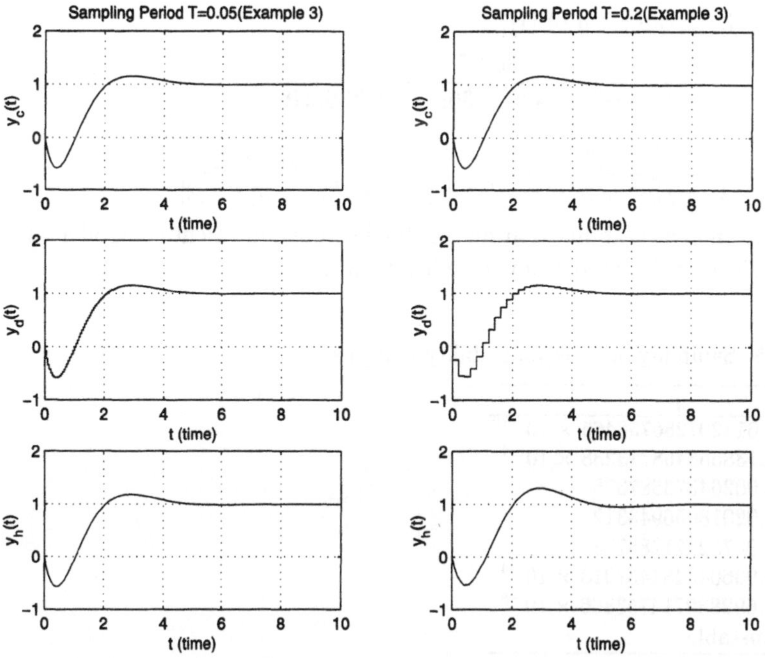

Fig. 2.9. Left: Discrete equivalent system, Right: Hybrid system (Example 2.3)

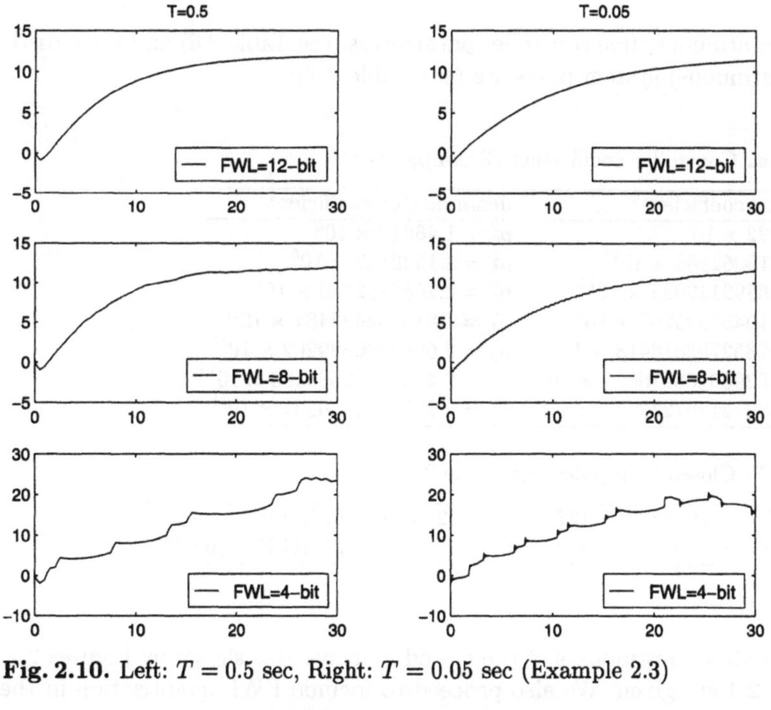

Fig. 2.10. Left: $T = 0.5$ sec, Right: $T = 0.05$ sec (Example 2.3)

Example 2.4. (Example 4 of [1])

$$P(s) = \frac{-36.27}{s^3 + 45.69s^2 - 4480.9636s - 204735.2269}$$

$$C(s) = \frac{q_6^0 s^6 + q_5^0 s^5 + q_4^0 s^4 + q_3^0 s^3 + q_2^0 s^2 + q_1^0 s + q_0^0}{s^7 + p_6^0 s^6 + p_5^0 s^5 + p_4^0 s^4 + p_3^0 s^3 + p_2^0 s^2 + p_1^0 s + p_0^0}$$

The parametric stability margins of the discrete equivalent controller with different Ts are tabulated below (see Table 2.5).

Table 2.5. Sampling periods vs. stability margins

T	$\hat{\rho}$
0.5	$1.614207286758465 \times 10^{-10}$
0.1	$1.138558458713238 \times 10^{-4}$
0.05	0.00264373687375
0.03	0.02018056943312
0.01	0.00752321285733
0.005	$2.936043484437013 \times 10^{-4}$
0.001	$2.692882714753866 \times 10^{-8}$
0.0005	unstable

The continuous time controller parameters (see Table 2.6) and the closed-loop (continuous) system poles are (see Table 2.7):

Table 2.6. Controller coefficients (Example 2.4)

numerator coefficient	denominator coefficient
$q_6^0 = -5.22 \times 10^8$	$p_6^0 = 1.46817 \times 10^3$
$q_5^0 = -1.19062980 \times 10^{11}$	$p_5^0 = 8.1539147 \times 10^5$
$q_4^0 = -1.0892119025 \times 10^{13}$	$p_4^0 = 2.2686802480 \times 10^8$
$q_3^0 = -5.10462225207 \times 10^{14}$	$p_3^0 = 1.818763428484 \times 10^{10}$
$q_2^0 = -1.2852702818418 \times 10^{16}$	$p_2^0 = 5.6984090389202 \times 10^{11}$
$q_1^0 = -1.62953268976593 \times 10^{17}$	$p_1^0 = 6.28454292585598 \times 10^{12}$
$q_0^0 = -7.93721797233977 \times 10^{17}$	$p_0^0 = 6.2277404850231 \times 10^{11}$

Table 2.7. Closed-loop poles (Example 2.4)

−557.7809	−424.0627	$-127.9871 \pm j32.7509$	−677.9148
−66.93999	−45.69000	−42.99716	$-26.31167 \pm j0.09297$

The unit step responses of the closed-loop systems shown by Figures 2.2, 2.3, and 2.4 are given. We also proceed to include FWL quantisation in the hybrid configuration with $T = 0.005$ sec (left of Figure 2.12).

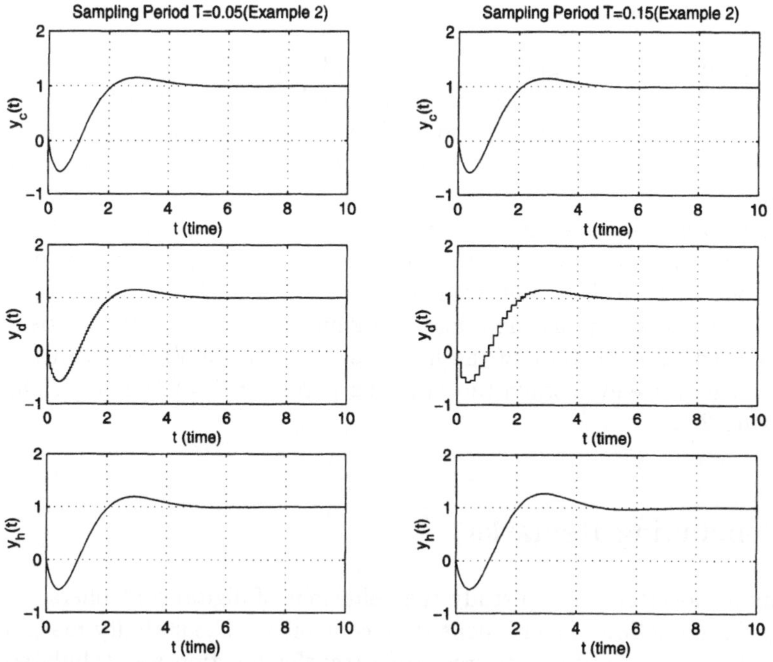

Fig. 2.11. Top: Continuous system, Mid: Discrete system, Bottom: Hybrid system (Example 2.4)

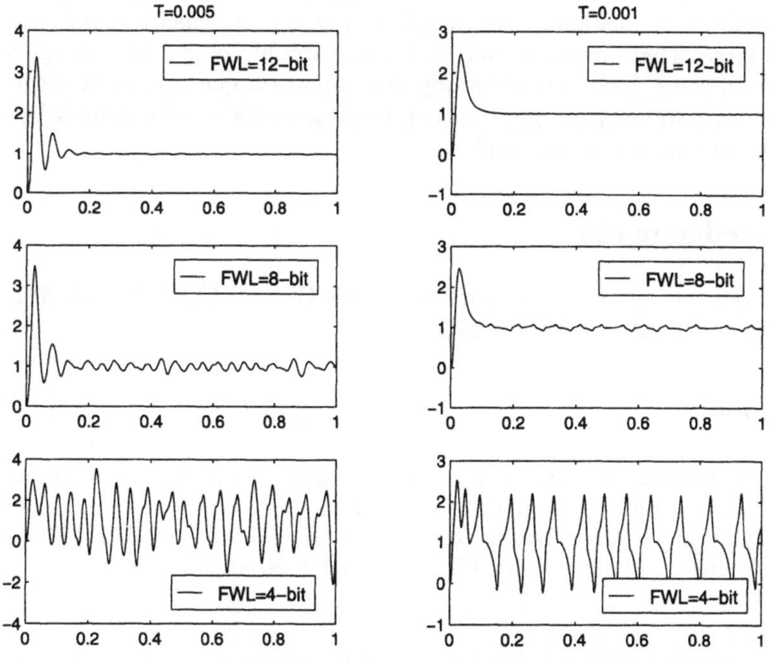

Fig. 2.12. Left: $T = 0.005$ sec, Right: $T = 0.001$ sec (Example 2.4)

We observe the following:

1. This example shows a slightly different trend from those of previous examples. The parametric stability margin of the discrete equivalent systems are small for large sampling periods. Then, it follows the trend seen in the previous examples when T ranges from 0.05 to an arbitrary small number.
2. For a slightly larger sampling period (*i.e.*, $T = 0.05$ in Figure 2.11) the hybrid system shows unstable behaviour. However, it shows stable behaviour for a much smaller value of T (*i.e.*, $T = 0.005$).
3. It shows that the response deteriorates significantly when 4-bit quantisation is used. The response of the hybrid system with FWL quantisation does not improve even when the sampling period is $T = 0.001$ sec (right of Figure 2.12).

2.5 Concluding Remarks

This chapter has attempted to study the behaviour of digital controllers obtained by discretisation of continuous time controllers. The stability margin in the coefficient space of the discrete time transfer function was tabulated against the sampling time. Roughly speaking, we observed that the stability margin decreases with increasing sampling frequency and that the hybrid system often shows unstable inter-sample response even when the stability margin is large. These issues as well as the relationship of stability margins to truncation errors bears further study. An important question is: how can one introduce a stability margin that captures the unstable inter-sample behaviour of the true hybrid system?

Acknowledgements

This research was supported in part by NASA Grant NCC-5228 and NSF Grants HRD-9706268 and ECS-9417004.

References

1. Keel, L.H., Bhattacharyya S.P. (1997) Robust, Fragile, or Optimal? IEEE Transactions on Automatic Control. **42**(8), 1098–1105.
2. Bhattacharyya, S.P, Chapellat, H, Keel, L.H. (1995) Robust Control: The Parametric Approach. Prentice Hall PTR, Upper Saddle River, NJ.

CHAPTER 3

FINITE WORD-LENGTH EFFECTS IN SYSTEMS WITH FAST SAMPLING

Ryszard Gessing

Abstract. Discrete time systems composed of sampler, zero-order hold and continuous time plants are investigated. The well-known Åström's *et al.* [1] theorem in a reformulated version is first proved and then utilised for estimating the length of a digital word needed for recording the pulse transfer function parameters, in the case of small sampling periods. It is shown that only non-zero poles of the plant transfer function cause an increase of the needed word-length; the zero poles and the zeros of the latter transfer function have no influence on this length. Next, the consequences for the control system design and regulator implementation are discussed. Finally, the identification of discrete time models under fast sampling and limited measurements accuracy is described. It is then shown that the identification accuracy of the shift operator and delta operator models is comparable. Considerations are illustrated by examples and simulations.

3.1 Introduction

These days most control systems contain regulators implemented with microprocessors. In this case both the parameters and the signals of the regulator are recorded using the arithmetic with finite word-length (FWL). In connection with this, two kinds of errors are distinguished: first, the errors due to the FWL implementation of the regulator coefficients and second, the errors due to the roundoff of the regulator signals. Additional errors of the roundoff type result from the FWL realisation of the analogue-to-digital (A/D) and digital-to-analogue (D/A) converters.

The FWL effects were researched mainly in connection with realisation of digital filters. By comparison, the FWL effects have received much less attention in control literature. One exception is the book of Gevers and Li [4] where in two chapters the control problems with accounting FWL effects are summarised and some new results are described. In [4] the attention is focused on the choice of the regulator model structure realised with FWL, for which both the above mentioned errors take the minimal values.

Performance requirements often command that a reasonably small sampling period be used. It is known that in this case the use of FWL leads to poor properties of the pulse transfer function (TF) models $H(z)$ based on z-transform (*i.e.*, the shift operator models). In connection with this, Middleton and Goodwin [8,11], to improve description properties of discrete time (DT) systems, use the delta operator models. They stress that the latter models have better numerical properties than the shift operator models. Another approach is based on application of the bilinear transformation resulting from the Tustin formula [4,12]. In this approach the TF $H(z)$ by means of the formula $H^*(w) = H(\frac{2+wh}{2-wh})$, with h being a sampling period, is transformed to the TF $H^*(w)$, which is less sensitive to parameter quantisation.

However, in many cases, the design approach based on the shift operator models also gives satisfactory results. In this approach the continuous time plant together with zero-order hold is described by the TF and the FWL effects are not taken into account. The approach is justified if the digital word of A/D and D/A converters (which should correspond to the plant output measurement accuracy), as well as of the microprocessor regulator, is sufficiently long. But one should remember that, in the case of small sampling periods, some sufficiently long digital word must also be used to record the TF parameters of the plant. In the case of the latter model, frequently a longer word-length must be used since, as will be shown, this length for small sampling periods is increasing with the model order. The latter property was also noticed in early digital filter literature [9].

Therefore, in the case of small sampling periods, it can be interesting to answer the following questions: which should be the design calculation precision for the given plant and sampling period? Or: whether the programs with single precision calculation can be used in the design? Or: whether it is possible to identify the long words recording the plant coefficients, using the input and output measurements with real, limited, say 0.1% accuracy?

Following these questions, in the present chapter the attention is focused on the plant model representing the DT system composed of sampler, zero-order hold and a linear continuous time plant in series. Only the direct model structure [4] is researched, since its coefficients together with the coefficients of the controller appear in the closed-loop system description. In Section 3.2 the researched model is described. The Åström's *et al.* theorem [1] in a reformulated version is proved in Section 3.3. In Section 3.4 the method is proposed for estimating the word-length, needed for recording the coefficients of the TF, which assures the prescribed model accuracy. The examples illustrating the influence of the FWL on the accuracy of the considered models are described in Section 3.5. Section 3.6 contains some remarks concerning regulator design and implementation. Finally in Section 3.7, some problems with model identification under fast sampling and limited measurement accuracy are discussed. The contents of the chapter present results mainly from the author's previous papers [5,6].

3.2 The Case of Small Sampling Periods

Let us consider the system composed of the sampler, zero-order hold and a linear continuous time system G in series, as shown in Figure 3.1. The system G is assumed to have a rational TF,

$$G(s) = \frac{C(s)}{D(s)} = \frac{c_0 s^m + c_1 s^{m-1} + \ldots + c_m}{d_0 s^n + d_1 s^{n-1} + \ldots + d_n} \tag{3.1}$$

where $C(s)$ and $D(s)$ are polynomials with degrees m and n, respectively, and $m < n$. The considered DT system is characterised by the TF $H(z)$, which can be calculated from the formula:

$$H(z) = (1 - z^{-1})\mathcal{Z}[y_1(kh)] \tag{3.2}$$

where $\mathcal{Z}$ denotes the z-transform, $y_1(kh) = y_1(t)|_{t=kh}$, $k = 0, 1, 2, \cdots$ is the discrete time, $y_1(t)$ is the step response of the continuous time system G, h denotes the sampling period and $(1 - z^{-1})$ is the inverse of the step function z-transform. Let s_i and z_i, $i = 1, 2, \ldots, n$ be the poles of the transfer functions $G(s)$ and $H(z)$, respectively. Then we have:

$$z_i = e^{s_i h}, \quad i = 1, 2, \ldots, n \tag{3.3}$$

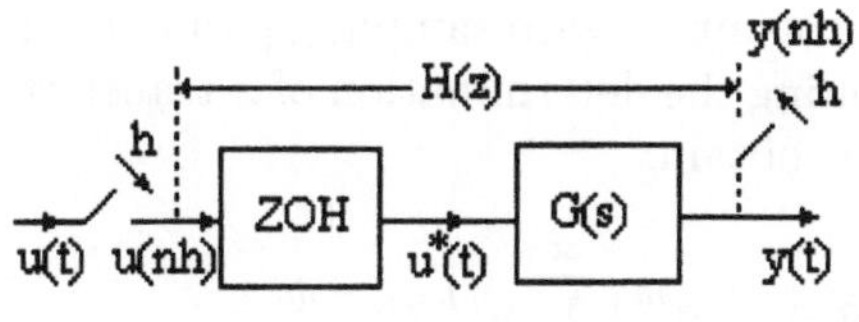

Fig. 3.1. Discrete time system

It is worthwhile to stress that by means of the substitution $t = kh$ we change the scale of the time from the real time t with the unit "1" to the discrete time k with the unit "h". In connection with this, the loci of the roots z_i do not correspond to the real transient responses. It can be shown that the latter responses correspond to the roots:

$$z_i^* = z_i^{1/h} = e^{s_i}, \quad i = 1, 2, \cdots, n \tag{3.4}$$

in the discrete time case, or to the roots s_i, $i = 1, 2, \ldots, n$ in the continuous time case.

The limiting case when h tends to zero must be treated with caution. First, in this case all the poles z_i tend to 1, which means that these limits contain no information about transient responses. Second, if we treat z as a complex variable independent of h and different from 1 then for the considered case we obtain:

$$\lim_{h\to 0} H(z) = 0 \tag{3.5}$$

The property (3.5) is expressed in this manner so that the coefficients of the numerator of $H(z)$ tend to zero when $h \to 0$.

The case $z \neq 1$ corresponds to unsteady state appearing for finite values of the discrete time k. The limit (3.5) results from the fact that for any finite k, $y_1(kh) \to 0$ if $h \to 0$. The limit (3.5) has no practical sense and results from using the time scale with the unit "h", which tends to zero. On the other hand, all practitioners know that when $h \to 0$ then the discrete time system works as a continuous time one. This property results also from the unified approach of [8,11]. In order to show this we must return to the real time scale and continuous time description by substituting in $H(z)$ the dependence:

$$z = e^{sh} \tag{3.6}$$

and with h tending to zero.

Let us consider the dependence:

$$Y(z) = H(z)U(z) \tag{3.7}$$

where $Y(z) = \mathcal{Z}[y(kh)]$ and $U(z) = \mathcal{Z}[u(kh)]$ are the z-transforms of the output y and input u, sampled with sampling period h. Multiplying both sides of (3.7) by h, accounting the determination of z-transforms of $Y(z)$, $U(z)$ and substituting (3.6) we obtain:

$$\sum_{k=0}^{\infty} y(kh)e^{-shk}h = H(e^{sh}) \sum_{k=0}^{\infty} u(kh)e^{-shk}h \tag{3.8}$$

Note that:

$$\lim_{h\to 0} \sum_{k=0}^{\infty} y(kh)e^{-shk}h = \int_0^{\infty} y(t)e^{-st}dt = \mathcal{L}[y(t)] = \bar{Y}(s) \tag{3.9}$$

and similarly:

$$\lim_{h\to 0} \sum_{k=0}^{\infty} u(kh)e^{-shk}h = \mathcal{L}[u(t)] = \bar{U}(s) \tag{3.10}$$

where $kh = t$, $h = dt$ $\bar{Y}(s)$ and $\bar{U}(s)$ determine the Laplace transforms of $y(t)$ and $u(t)$, respectively, while $\mathcal{L}$ denotes the symbol of Laplace transform. Tending with h to zero in (3.8) and accounting (3.9) and (3.10) we obtain:

$$G(s) = \lim_{h\to 0} H(z)|_{z=e^{sh}} = \lim_{h\to 0} H(e^{sh}) \tag{3.11}$$

The formula (3.11) has a basic meaning, but usually, it does not appear in the books devoted to DT systems. The exception is the book of Kuzin [10] in which the modified version of this formula appears (corresponding to the case in which the zero-order hold is replaced with the ideal sampler). The formula (3.11) will be used for completing the proof of the reformulated Åström's *et al.* theorem [1], as shown below.

3.3 A Reformulated Åström's Theorem

The original formulation of the theorem [1] speaks about the limits to which the zeros of $H(z)$ tend when $t \to 0$. In this formulation there is no information about the accuracy of the DT TF resulting from the theorem for small h. Additionally, the proof of the theorem is made under the silent assumption that $w \neq 0$ (in the original notation), which means that the steady state is not taken into account in considerations.

In the reformulated theorem below a correction $o(h^n)$ is added, which for small h establishes the error order of the approximation of $H(z)$ resulting from the theorem. The proof of the reformulated theorem takes into account the steady and unsteady state of the system response and is supported by basic property (3.11). Some steps of the proof are utilised in the further considerations concerning the word-length.

Theorem 3.1. *Let $G(s)$ be a rational function:*

$$G(s) = c\frac{(s - s_1^{'})(s - s_2^{'})\cdots(s - s_m^{'})}{(s - s_1)(s - s_2)\cdots(s - s_n)} \tag{3.12}$$

and $H(z)$ the corresponding TF. Assume that $m < n$. Then for sampling period h tending to zero the transfer function $H(z)$ takes the form:

$$H(z) = c\frac{h^{n-m}B_{n-m}(z)(z - e^{s_1^{'}h})\cdots(z - e^{s_m^{'}h}) + o(h^n)}{(n-m)!(z - e^{s_1 h})\cdots(z - e^{s_n h})} \tag{3.13}$$

where:

$$B_j(z) = b_1^j z^{j-1} + b_2^j z^{j-2} + \cdots + b_j^j \tag{3.14}$$

$$b_l^j = \sum_{i=1}^{l}(-1)^{l-i} i^j \binom{j+1}{l-i}, \quad l = 1, 2, \cdots, j \tag{3.15}$$

and:

$$\lim_{h \to 0} \frac{o(h^n)}{h^n} = 0$$

Proof. Using (3.14), (3.15) one can prove that $B_k(1) = k!$. Taking into account that for any finite s:

$$\lim_{h\to 0} e^{sh} = 1, \quad e^{sh} - e^{s_i h} = (s - s_i)h + o(h)$$

we obtain:

$$\lim_{h\to 0} H(z)|_{z=e^{sh}} = \lim_{h\to 0} H(e^{sh}) = G(s) \tag{3.16}$$

then (3.11) is fulfilled. In order to show that (3.13) really represents the TF $H(z)$, we investigate its properties for steady and unsteady state.

For steady state, *i.e.*, for $z = 1$, since $(z - e^{s_i h}) = z - 1 - s_i h + o(h)$, then under the assumption that $s_i \neq 0,\ i = 1, 2, \ldots, n$, from (3.13) we obtain:

$$\lim_{h\to 0} H(1) = c\frac{(-s_1')(-s_2')\cdots(-s_m')}{(-s_1)(-s_2)\cdots(-s_n)} = G(0) = k_o \tag{3.17}$$

If only one pole is equal to zero, e.g., $s_1 = 0$, then we obtain from (3.13):

$$\lim_{h\to 0}\lim_{z\to 1} \frac{z-1}{h} H(z) = c\frac{(-s_1')(-s_2')\cdots(-s_m')}{(-s_2)(-s_3)\cdots(-s_n)} = \lim_{s\to 0} sG(s) = k_0' \tag{3.18}$$

For only two zero poles, e.g., $s_1 = 0,\ s_2 = 0$ from (3.13) we obtain:

$$\lim_{h\to 0}\lim_{z\to 1} \frac{(z-1)^2}{h^2} H(z) = c\frac{(-s_1')(-s_2')\cdots(-s_m')}{(-s_3)(-s_4)\cdots(-s_n)} = \lim_{s\to 0} s^2 G(s) = k_0'' \tag{3.19}$$

Let $y_1(t)$ be the step response of the system $G(s)$, then $\mathcal{L}[y_1(t)] = Y_1(s) = (1/s)G(s)$, where $\mathcal{L}[\cdot]$ denotes the Laplace transform. Let $y_1^*(t) = y_1(th)$ and $Y_1^*(s) = \mathcal{L}[y_1^*(t)]$. Using the theorem about the change of time scale we have:

$$Y_1^*(s) = \mathcal{L}[y_1(th)] = (1/h)Y_1\ (s/h) = (1/s)G(s/h) = (1/s)G^*(s)$$

where:

$$G^*(s) = G(s/h)$$

Taking into account (3.12) we obtain:

$$G^*(s) = c\left(\frac{h}{s}\right)^{n-m} \frac{(1 - s_1' h/s)\cdots(1 - s_m' h/s)}{(1 - s_1 h/s)\cdots(1 - s_n h/s)} \tag{3.20}$$

Thus, for unsteady states when $s \neq 0 (z \neq 1)$ we have:

$$\lim_{h\to 0} h^{m-n} G^*(s) = \frac{c}{s^{n-m}} \tag{3.21}$$

The DT description corresponding to (3.21) takes the form [1]:

$$\lim_{h\to 0} h^{m-n} H(z) = c\frac{B_{n-m}(z)}{(n-m)!(z-1)^{n-m}} \tag{3.22}$$

On the other hand the formula (3.22) results directly from (3.13). □

3.4 Estimation of the Word-length

For small sampling periods, in order to obtain an appropriate model accuracy, the DT TF coefficients must be recorded using a sufficiently long digital word in decimal code. Here, the word-length that assures the prescribed accuracy of the system gain will be estimated. The case of floating-point arithmetic will be considered.

Let us denote:

$$H(z) = \frac{B(z)}{A(z)} = \frac{b_1 z^{n-1} + b_2 z^{n-2} + \cdots + b_n}{a_0 z^n + a_1 z^{n-1} + \cdots + a_n} \tag{3.23}$$

where:

$$\begin{aligned} a_0 &= 1 \\ a_1 &= -z_1 - z_2 \cdots - z_n \\ a_2 &= z_1 z_2 + z_1 z_3 \cdots + z_{n-1} z_n \\ &\dots\dots\dots\dots \\ a_n &= (-1)^n z_1 z_2 \cdots z_n \end{aligned} \tag{3.24}$$

The roots $z_i \to 1$ when $h \to 0$, and the denominator of (3.13) tends to the polynomial $(z-1)^n$. Therefore, for sufficiently small h, from accounting (3.3) in (3.24), we obtain:

$$a_i = (1)^i \binom{n}{i} + \Delta a_i \tag{3.25}$$

where $(-1)^i \binom{n}{i}$ determines the coefficients of the polynomial $(z-1)^n$ and $|\Delta a_i|$ are small numbers tending to zero when $h \to 0$. Note that the delta operator approach [8,11] simply leaves Δa_i, as these are parts of (3.25) that contain the information about the system parameters. Due to this the delta operator approach has some better properties for small h.

The gain k_0 of the system in steady state is determined by:

$$k_0 = \frac{\bar{b}}{\bar{a}}, \qquad \bar{b} = \sum_{i=1}^{n} b_i, \qquad \bar{a} = \sum_{i=0}^{n} a_i \tag{3.26}$$

or by (3.17).

Denote:

$$b_i^1 = \frac{b_i}{k_0}, \quad \bar{b}^1 = \sum_{i=0}^{n} b_i^1 \tag{3.27}$$

The coefficients a_i, $i = 0, 1, \ldots, n$, b_i^1, $i = 1, 2, \ldots, n$ describe the DT TF with gain equal to 1. Three cases will be considered below.

Case a : $m = 0,\ \ s_i \neq 0,\ \ i = 1, 2, \ldots, m.$

For sufficiently small h, from (3.25), (3.13) and (3.17) it results that:

$$\bar{a} = \sum_{i=0}^{n} \Delta a_i = \bar{b}^1 \approx \frac{c}{k_0} h^n = (-s_1)(-s_2)\ldots(-s_n)h^n = v \tag{3.28}$$

The value v may be treated as the estimate of $\bar{a}$ and/or $\bar{b}^1$ and may be used for estimating the word-length needed for recording the coefficients a_i and b_i^1.

We assume that the same word-length is used for recording the coefficients a_i and b_i^1; in this case the errors of the coefficients b_i^1 resulting from quantisation are significantly smaller than those of the coefficients a_i. This will be shown further on. Therefore, the errors of b_i^1 may be neglected and the sum $\bar{b}^1$ may be treated as accurate. The record errors of $a_i, i = 1, 2, \ldots, n$ are summed up. If we would like to obtain the gain $\bar{b}^1/\bar{a}$ with at least 1% accuracy, then the sum $\bar{a} \approx v$ should be determined with the same accuracy. This will be obtained when the quantity 10^{-l} resulting from the FWL fulfils the inequality:

$$10^{-l} < \frac{v}{n \cdot 100} < 10^{-(l-1)} \tag{3.29}$$

where l determines the position of the last, least meaning digit after the point and $v/(n \cdot 100)$ determines the record error of the parameters a_i, which assures 1% accuracy of the sum $\bar{a}$ and of the gain. Since a_i have the shape (3.25) and for $n \leq 4$ the integers $\binom{n}{i}$ occupy one decimal digit then the FWL N^* for recording the parameters a_i, b_i^1 (which assures at least 1% accuracy of the gain) is determined by:

$$N^* = l + 1 \tag{3.30}$$

For higher order systems e.g., for $5 \leq n \leq 8$ some of the numbers $\binom{n}{i}$ need two decimal digits for recording and we have:

$$N^* = l + 2 \tag{3.31}$$

Another situation is in the case of coefficients b_i^1. From formula (3.13) it results that in the considered case ($m = 0$) the numerator of the TF $H(z)$ is determined approximately by the polynomial:

$$\frac{ch^n}{k_0 n!} B_n(z) \tag{3.32}$$

This means that the coefficients b_i^1 are determined by the coefficients of the polynomial $B_n(z)$ (formula (3.15)) multiplied by $(ch^n/k_0 n!)$. Since the signs of these coefficients are positive, the value $\bar{b}^1 = \sum_{i=1}^{n} b_i^1$ is really determined by sum, and the word-length for recording b_i^1 and assuring 1% accuracy of $\bar{b}^1$ is significantly smaller. Therefore, if we use the word-length assuring 1% accuracy of $\bar{a}$ also for recording b_i^1, (in floating-point arithmetic) then the record errors of b_i^1 may be neglected.

Case b : $m \neq 0,\ \ s_i \neq 0,\ \ i = 1, 2, \ldots, n.$

Since:

$$1 - e^{s_i' h} = -s_i' h + o(h) \qquad i = 1, 2, \ldots, m \tag{3.33}$$

then from the formula (3.25) and (3.13) written for $z = 1$, as well as from (3.17), it results:

$$\bar{a} = \sum_{i=0}^{n} \Delta a_i = \bar{b}^1 \approx h^{n-m} \frac{c}{k_o} (-s_1')(-s_2') \ldots (-s_m') h^m = \\ (-s_1)(-s_2) \ldots (-s_n) h^n = v \tag{3.34}$$

Comparing (3.28) with (3.34) we see that v is independent of the zeros of $G(s)$.

Similarly, as in Case a, for $m \neq 0$ and $m < n$, it may be shown that if the same word-length is used for recording a_i and b_i^1, then the recording errors of b_i^1 may be neglected. Therefore, the formulas (3.29)–(3.31) may now also be used for determining the required word-length. This means that the zeros of $G(s)$ do not influence the word-length, assuring the prescribed accuracy of the gain.

Case c : $m = 0,\ \ s_1 = 0,\ \ s_i \neq 0$ for $2 \leq i \leq n.$

Denote:

$$Q(z) = \frac{1}{z-1} A(z) = q_0 z^{n-1} + q_1 z^{n-2} + \ldots + q_{n-1} \tag{3.35}$$

For sufficiently small h using (3.18) and (3.13) we obtain:

$$\bar{q} = \sum_{i=0}^{n-1} q_i = \frac{\bar{b}}{k_0' h} \approx \frac{c}{k_0'} h^{n-1} = (-s_2)(-s_3 \ldots (-s_n) h^{n-1} = v' \tag{3.36}$$

The word-length may be determined similarly as in Case a, using formulas (3.29)–(3.31), in which v is replaced by v'. Comparing (3.36) with (3.28) we see that the zero poles of $G(s)$ do not influence the required word-length. Taking into account the numerator of (3.13), it may be shown that the latter statement is valid when m is smaller than the number of non-zero poles.

The above considerations concern the decimal code without taking into account the coefficient sign.

Let L be the number of bits needed for recording the transfer function coefficients together with the sign. From considerations similar to those concerning the formulas (3.29)–(3.31) it results that for Cases a and b we have:

$$L \gtrsim 1 + \log_2 \binom{n}{i}_{mx} + \log_2(n \cdot 100) + [-\log_2 \prod_{i=1}^{i=n} (-s_i h)] \tag{3.37}$$

where $\binom{n}{i}_{mx} = \max_i \binom{n}{i}$. The "1" appearing on the right hand side of (3.37) corresponds to the sign; the third and fourth components result from $\log_2(n \cdot 100/v)$ with accounting (3.28). For Case c the inequality (3.37) should be appropriately modified.

Comparing the condition (3.37) with that of Kaiser [9] (the latter written for the bits, with an appropriate correction of the signs of s_i) it can be noticed that the condition (3.37) gives the word-length greater by $\log_2(n \cdot 100)$ than that of Kaiser. Since Kaiser's condition assures that the model remains stable, it means that the condition (3.37) assures also that the model remains stable. From the many step responses, calculated by the author for different examples, it seems that the system gain is most sensitive to the coefficients quantisation.

3.5 Examples

Below, for given $G(s)$ and h, the results of calculations of the gain accuracy corresponding to different word-lengths are presented in four examples. First, the value v is calculated using (3.28) and from (3.29), (3.30) the length N^* is determined. Next, the gain and the gain error are calculated for several different word-lengths N. In the examples below, N^* is calculated so that it assures the gain recording accuracy of 1% in the first three examples and of 2% in the fourth example.

Example 3.1. Given $n = 2$, $m = 0$, $c = 10$, $s_1 = -1$, $s_2 = -10$, $h = 0.015$. From calculations $k_0 = 1$, $v = 2.25 \times 10^{-3}, \bar{b}^1 = 2.07 \times 10^{-3}$ $N^* = 6$. For the word-length N equal to 6, 5, 4, 3, we obtain the gain k_0 equal to $1.00183, 0.98752$, 1.03650, -1.87×10^{13}, and the percentage gain error equal to 0.183, -1.247, 3.65, -1.87×10^{15}, respectively.

Example 3.2. Given $n = 3$, $m = 0$, $c = 1 + 4\pi^2$, $s_1 = -1$, $s_2 = -1 + j2\pi$, $s_3 = -1 - j2\pi$, $h = 0.01$. From calculations $k_0 = 1, v = 4.0478 \times 10^{-5}$, $\bar{b}^1 = 3.986 \times 10^{-5}$, $N^* = 8$. For the word-length N equal to 8, 7, 6, 5 we obtain the gain k_0 equal to 0.99907, 0.9966, 0,9966, 0.3986, and the percentage error equal to -0.092, -0.342, -0.342, -60.14, respectively.

Example 3.3. Given $n = 3$, $m = 1$, $c = 2 + 8\pi^2$, $s_1 = 0$, $s_2 = -1 + j2\pi$, $s_3 = -1 - j2\pi, s_1' = -0.5$, $h = 0.01$. From calculations $k_0' = 1, v' = 4.0478 \times 10^{-3}$, $(\bar{b} \,/\, (k_0' h)) = 4.006 \times 10^{-3}$, $N^* = 6$. For word-length N equal to 6, 5, 4, 3 we obtain the gain k_0' equal to 0.9990, 1, 1, 7.5×10^{12}, and the percentage error equal to -0.097, 0, 0, 7.5×10^{14}, respectively.

Example 3.4. Given $n = 2$, $m = 1$, $c = -20$, $s_1 = -1$, $s_2 = -10$, $s_1' = 0.5$, $h = 0.01$. From calculations $k_0 = 1, v = 10^{-3}$, $\bar{b}^1 = 0.947 \times 10^{-3}$, $N^* = 6$. For word-length N equal to 6, 5, 4, 3 we obtain the gain k_0 equal to 1.0074, 1.0555, 0.899, 0.100 and the percentage error equal to 0.744, 5.55, -10, -90, respectively.

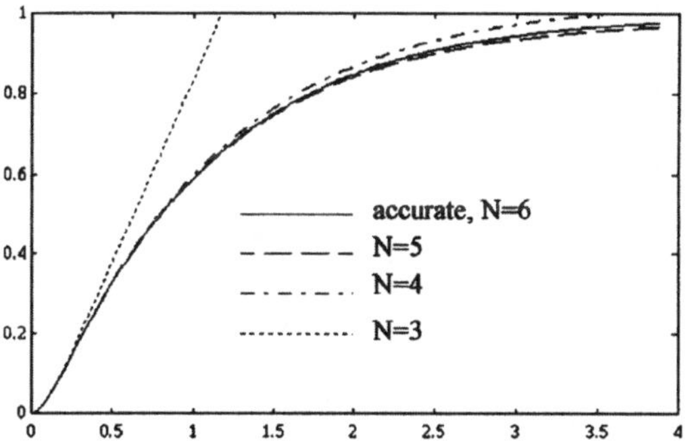

Fig. 3.2. Step responses of Example 3.1

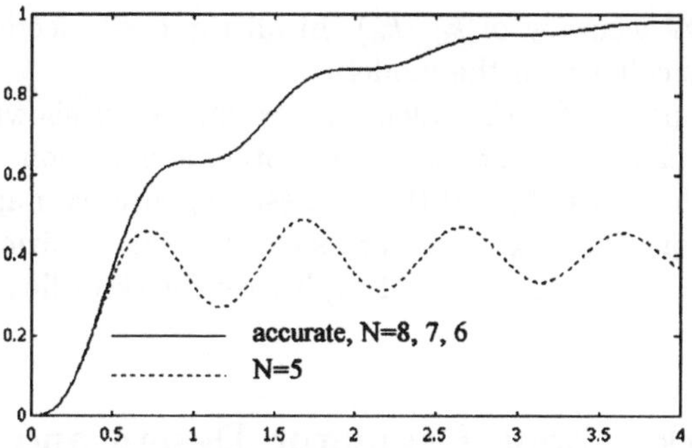

Fig. 3.3. Step responses of Example 3.2

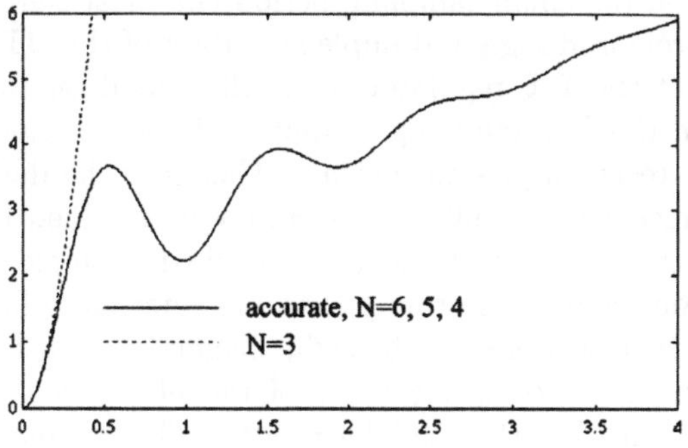

Fig. 3.4. Step responses of Example 3.3

It is seen that for all the examples the quantities v, (v') are close to that of $\bar{b}^1$, $(\bar{b}/(k_0'h))$. In all the cases N^* determines the word-length well, assuring

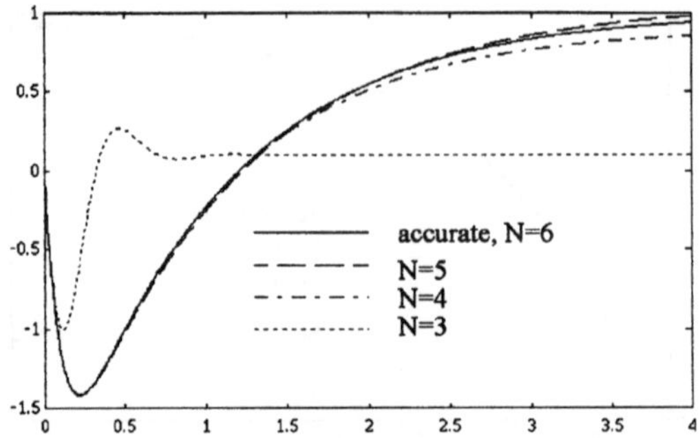

Fig. 3.5. Step responses of Example 3.4

the assumed 1% accuracy of $k_0, (k_0')$. In all the cases decrease of N^* by 3 causes complete collapse of the model.

The step responses for the calculated examples are shown in Figures 3.2 to 3.5, respectively. From these and many other step responses calculated for different examples it results that the greatest response error appears in steady state; in transient time the errors are smaller. This confirms the view that the gain is most sensitive to word-length used for recording the parameters a_i, b_i.

3.6 Remarks about Regulator Design and Implementation

The long digital word needed for recording the coefficients of the discrete time models when the small sampling period (*i.e.*, fast sampling) is applied may play some role in design and implementation of the DT regulator.

The design of the DT regulator is usually based on the DT model of the plant. When the fast sampling is applied the latter model needs a long digital word for recording its coefficients. Therefore the design calculations must be performed with a sufficient accuracy determined by the adequate word-length of the mantissa. From the described examples it is seen that, even for the second and third-order plants and sufficiently fast sampling, the needed word-length occupies six to eight decimal digits. Since the needed word-length depends also on the order of the plant then for higher order plants a bigger word-length could be needed. This means that sometimes the standard word-length of 24-bit mantissa (seven decimal digits) may be insufficient. This brings into question the possibility of application of the shift operator model (3.23) for fast sampling. In these cases either the delta operators models [8,11], or the models resulting from bilinear transformation [12] may be used. Another approach based on some appropriately modified continuous time methods may be also applied [7].

Another problem in which the FWL may play some role is the implementation of a DT regulator using microprocessors. There arises the question of how many bits should microprocessors have to implement the DT regulator. One can note that in the case of DT approximation of PID continuous time regulators, the required word-length results rather from the accuracy of the description of the signals, since the zero pole of the integrator does not influence the required word-length (which results from the above considerations). But if we would like to apply the DT approximation of higher order regulators (e.g., linear quadratic regulators [3]) then considerations similar to those presented above could help in estimating the required word-length.

3.7 Problems with Model Identification

The DT models described by the TF (3.23) and based on the z-transform are also called the shift operator (SO) models. The long digital word needed for recording the coefficients of the SO model may create some problems during performing identification on the basis of measurements of u and y. It is known that the delta operator (DO) models need a shorter digital word for recording their coefficients [8,11]. Therefore there is a common conviction that for small h the DO models are better than the SO ones. This view also concerns model identification and is justified if too short a digital word is used in information processing (too short a mantissa), as in [8]. However from an application point of view, more realistic is the assumption that the main source of errors results from inaccurate measurements and the errors resulting from information processing are negligible. It will be shown further on under this assumption, the superiority of DO models related with the accuracy of identification disappears.

3.7.1 Shift and Delta Operator Model

The SO model (3.23) may be written in the form:

$$y(kh) = \Phi_k^T \Theta \tag{3.38}$$

where:

$$\Phi_k^T = [-y(kh-h), \ldots, -y(kh-nh), u(kh-h), \ldots, u(kh-nh)] \tag{3.39}$$

is the vector of measurements of output y and input u while:

$$\Theta^T = [a_1, \ldots, a_n, b_1, \ldots, b_n] \tag{3.40}$$

is the vector of estimated coefficients.

Using the model (3.38) and the measurements of y and u the vector Θ may be determined in real time by means of the recursive least square (RLS) method [2]. In the case of fast sampling, a long digital word (e.g., six decimal digits) is needed for recording the coefficients appearing in Θ. But the measurements are performed with some limited accuracy; usually a relative accuracy of measurement is 0.1% – 1%, which corresponds to three digits of the mantissa in decimal code. Of course, it is impossible to estimate the model coefficients with accuracy to six decimal digits if the measurements accuracy corresponds only to three digits. At first glance, it seems that this difficulty could be overcome by identifying the delta operator (DO) model.

The TF of DO model is determined by [11]:

$$\bar{H}(\gamma) = H(z)_{|z=1+h\gamma} \tag{3.41}$$

and takes the form:

$$\bar{H}(\gamma) = \frac{\bar{b}_1\gamma^{n-1} + \bar{b}_2\gamma^{n-2} + \ldots + \bar{b}_n}{\gamma_n + \bar{a}_1\gamma^{n-1} + \ldots + \bar{a}_n} \tag{3.42}$$

where the coefficients $\bar{a}_i$, $\bar{b}_i$ result from the coefficients a_i, b_i. The shift z and delta γ operators are related by:

$$\gamma = \frac{z-1}{h} \tag{3.43}$$

The DO model (3.42) may be described in the form:

$$\Delta^n y(kh - nh) = \bar{\Phi}^T_{k-n}\bar{\Theta} \tag{3.44}$$

where:

$$\begin{aligned}\Phi^T_{k-n} = [&-\Delta^{n-1}y(kh-nh), \ldots, -y(kh-nh),\\ &\Delta^{n-1}u(kh-nh), \ldots, u(kh-nh)]\end{aligned} \tag{3.45}$$

$$\bar{\Theta}^T = [\bar{a}_1, \ldots, \bar{a}_n, \bar{b}_1, \ldots, \bar{b}_n] \tag{3.46}$$

To justify the form of (3.44), note that the term $\gamma^n y$ determines the multiple difference $\Delta^n y(kh)$, where:

$$\Delta^{j+1}y(kh) = [\Delta^j y(kh+h) - \Delta^j y(kh)]/h, \quad j = 1, 2, \ldots, n-1$$

and:

$$\Delta^1 y(kh) = [y(kh+h) - y(kh)]/h$$

But the difference $\Delta y(kh)$ cannot be determined using actual and past measurements *i.e.*, $y(kh), y(kh-h), y(kh-2h), \ldots$ Therefore all the differences like $\Delta^j y(kh)$ are back shifted by nh in (3.44), (3.45) and replaced with $\Delta^j y(kh-nh)$. Owing to this, the latter differences may be calculated in any current time kh using the actual and past measurements $y(lh), u(lh), l \leq k$.

As will be shown, the limited accuracy of measurements also creates problems in identification of the DO models for small h. This may be explained by the fact that under fast sampling, the increase of the variables u and y over the period h becomes comparable with the measurement errors, so causing an increase of the noise level in the measurement vector $\bar{\Phi}^T_{k-n}$.

3.7.2 Results of Simulations

In the following simulations the limited accuracy of measurements of $u(kh)$, $y(kh)$ was obtained by recording them with a digital word having N digits in floating-point arithmetic. The N-digit mantissa was created by accounting the first N, most significant digits of the MATLAB® mantissa. In simulations $N = 3$ was used, which corresponds to the relative accuracy ranging from $0.1\% - 1\%$.

The identification experiment was simulated using MATLAB®/SIMULINK® for the system shown in Figure 3.1 in which:

$$G(s) = \frac{1}{s^2 + 3s + 2} \tag{3.47}$$

The estimates of DT TF $H(z)$ were calculated using the RLS algorithm with forgetting factor λ [2] and measurements with limited accuracy. The system was excited by input u obtained from a white noise generator; the same input u was used in all the simulations described below.

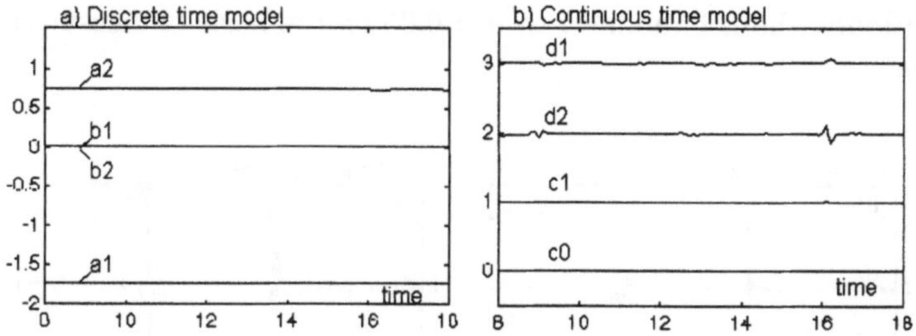

Fig. 3.6. Estimates for $h = 0.1$, $\lambda = 0.96$, $N = 3$

In Figure 3.6a the estimates of b_1, b_2, a_1, a_2 of DT TF $H(z)$ for $h = 0.1$, $N = 3$, $\lambda = 0.96$ are plotted as functions of time. From these plots, in the form of horizontal lines, it is not possible to determine the estimates' accuracy. Really, for small sampling periods, some long digital words are needed for recording the coefficients of $H(z)$ and the information about these coefficients are contained in successive, less meaningful digits of these words; therefore even inaccurate estimates of coefficients of $H(z)$ may have the shape of horizontal lines. To obtain a visual presentation of the accuracy, the estimates of c_0, c_1, d_1, d_2 of the continuous time plant $G(s)$ are calculated with double precision accuracy, in each period h, using b_1, b_2, a_1, a_2 and the MATLAB® functions `tf2ss`, `d2c`, `ss2tf`. The estimates of c_0, c_1, d_1, d_2, corresponding to those from Figure 3.6a, are shown in Figure 3.6b; they are almost accurate, only d_2 has some slight fluctuations.

In Figures 3.7a and 3.7b the estimates of c_0, c_1, d_1, d_2 of the continuous time plant are shown; they were calculated as previously from those of b_1,

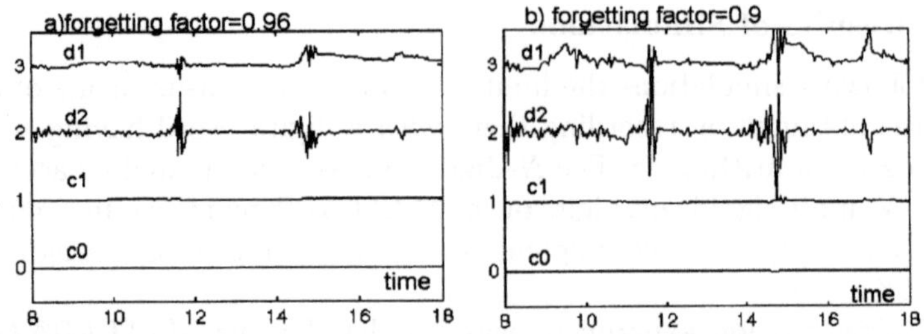

Fig. 3.7. Estimates of the continuous time plant calculated from those of SO model, $h = 0.05$, $N = 3$

b_2, a_1, a_2, for $h = 0.05$, $N = 3$, $\lambda = 0.96$ and $\lambda = 0.9$, respectively; the estimates of b_1, b_2, a_1, a_2 are not shown since they, as previously, have the form of horizontal lines. It is seen that the estimates of d_1, d_2 are yet more inaccurate – they have some big fluctuations – bigger if λ is smaller. This means that the measurement accuracy determined by $N = 3$ is not sufficient for identifying the DT model with $h = 0.05$ on the basis of measurements of u and y.

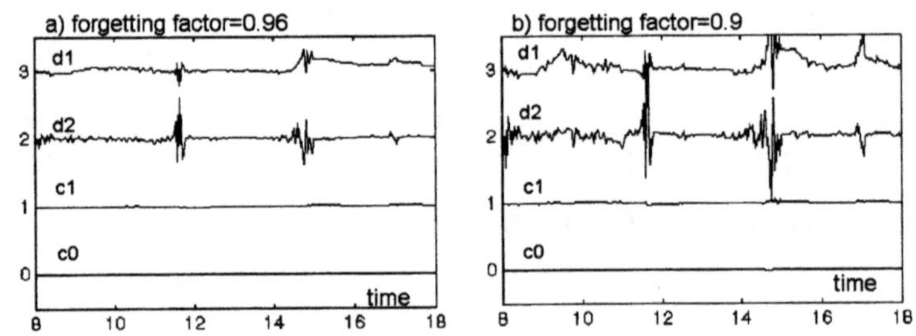

Fig. 3.8. Estimates of the continuous time plant calculated from those of the DO model, $h = 0.05$, $N = 3$

In Figures 3.8a and 3.8b the estimates c_0, c_1, d_1, d_2 of the continuous time plant $G(s)$ are shown; they were calculated similarly as previously from the estimates of $\bar{b}_1$, $\bar{b}_2$, $\bar{a}_1$, $\bar{a}_2$ of the DO model $\bar{H}(\gamma)$, for $h = 0.05$, $N = 3$, $\lambda = 0.96$ and $\lambda = 0.9$, respectively. Comparing the plots of Figure 3.7 and Figure 3.8 we see that they are similar. This and other performed simulations show that the accuracy of estimation of SO and DO models, under the assumptions made, is comparable. This observation is in contradiction to the view that for small sampling periods the DO models may be identified more accurately than the SO models [8]. Note that the latter statement is also true but under a different assumption: that the main source of errors results from too short a mantissa used for calculations (information processing). The assumptions made in this chapter say that the main source of errors is the limited measurements accuracy and the information processing is relatively accurate; these assumptions are well justified from the application point of view.

To improve the accuracy of identification in the case of fast sampling and limited measurements accuracy, either some additional measurements should be used, or the information about the known parameters should be utilised [6].

3.8 Conclusions and Notes

For small sampling periods h, a sufficiently long digital word is required to record the coefficients of $H(z)$. The smaller the ratio of h to the time constants of the system, the greater the word-length that must be used. Only non-zero poles of $G(s)$ cause the increase of the word-length, as the zero poles and the zeros of $G(s)$ have no influence on it. The value v, which results directly from $G(s)$ and h, gives a good basis for estimating this length.

The poles of $H(z)$ determine the characteristic polynomial of the discrete time system. Then the same word-length is needed in the case of companion state space description.

From comparison of our condition (3.29) with that of Kaiser [9] it results that our condition assures also non-destabilisation of the model.

The design calculations of the system with fast sampling must be performed with some adequate accuracy, making it possible to record the long digital words representing the shift operator (SO) model coefficients. One can note that a shorter digital word needed for recording the delta operator (DO) model coefficients is not sufficient during the simulations when the time responses are calculated. Therefore one can suppose that during the design calculations in which the simulations are performed, the advantage of the DO models related with a shorter digital word may disappear.

The identification of the SO operator model, under fast sampling and limited measurement accuracy, may be impossible. Really, it is impossible to identify the model coefficients with the accuracy, e.g., six decimal digits (needed for recording the model coefficients) if the measurement accuracy corresponds to only three digits (relative accuracy 0.1%-1%).

Under a realistic assumption that the main source of errors comes from inaccuracies in measurements and the errors resulting from information processing are negligible, the identification accuracy of the SO and DO model under fast sampling is comparable. This observation is in contradiction to the common view concerning the superiority of the DO models. Note that the latter view is valid, but under the different assumption: that the main source of errors results from information processing.

References

1. Åström, K.J., P. Hagander and J. Sternby. (1984) Zeros of Sampled Systems. *Automatica*, vol.30, No 1, pp.31–38
2. Åström. K.J. and B. Wittenmark. (1989) *Adaptive Control*. Addison-Wesley, Reading, MA
3. Dorato, P., C. Abdallach and V. Cerone. (1995) *Linear-Quadratic Control, An Introduction*. Prentice Hall, Englewood Cliffs, NJ
4. Gevers, M. and G. Li. (1993) *Parametrizations in Control, Estimation and Filtering Problems*. Springer Verlag, London
5. Gessing, R. (1999) Word Length of Pulse Transfer Function for Small Sampling Periods. *IEEE Trans. Automat. Control*, vol. 44, no. 9, pp. 1760-1764
6. Gessing, R. (1999) Towards Increase of Identification Accuracy for Small Sampling Periods. *CD-Rom Proc. European Control Conf. ECC'99*
7. Gessing, R. (1995) Comments on "A Modification and the Tustin Approximation" with a Concluding Proposition. *IEEE Trans. Automat. Control*, vol. 40, no. 5, pp. 942–944
8. Goodwin, G.C., R.H. Middleton and H.V. Poor. (1992) High Speed Digital Signal Processing and Control. *Proc. IEEE*, vol. 80, no. 2, pp. 240–259
9. Kaiser, J.F. (1965) Some Practical Considerations in the Realization of Linear Digital Filters. *Proc. 3-d Annu. Allerton Conf. Circuit System Theory*, pp.621–633
10. Kuzin, L.T. (1962) *Calculation and Design of Discrete Control Systems*, Moscow (in Russian)
11. Middleton, R.H. and G.C. Goodwin. (1990) *Digital Control and Estimation – A Unified Approach*. Englewood Cliffs, NJ
12. Rabiner, L.R. and B. Gold. (1975) *Theory and Application of Digital Signal Processing*, Prentice-Hall, Englewood Cliffs, NJ

CHAPTER 4

IMPLEMENTATION OF A CLASS OF LOW COMPLEXITY, LOW SENSITIVITY DIGITAL CONTROLLERS USING ADAPTIVE FIXED-POINT ARITHMETIC

Darrell Williamson

Abstract. The performance of a real time implementation of a digital control algorithm especially under a high sampling rate is affected by arithmetic errors caused by finite word-length arithmetic. The effects of these numerically induced errors can be measured both in terms of the sensitivity of the closed-loop performance due to changes in the coefficients of the control algorithm, and the effect of internal "noise" resulting from roundoff errors on the controller states. The least sensitive structures generally have the highest complexity in terms of the number of arithmetic operations required to compute the next controller output, while the most sensitive structures are often the least complex. A new class of low complexity algorithms is presented, which have only moderate complexity. In addition, it is shown how a software modification can eliminate the occurrence of internal arithmetic overflow, which are usually present in a conventional fixed-point implementation.

4.1 Introduction

The design of any control system must take into account the lack of information. In particular, we need to consider uncertainties in the model used to represent the physical system, in the characteristics of signals that are likely to disturb the system, in the accuracy of the measurements, in the control actuators, and uncertainty associated with the generation of the control signal itself. Errors that occur in the sensors and actuators can be attributed in part to errors in analogue circuits that are used to precondition (*i.e.*, filter and scale) analogue signals prior to measurement or control. In the worst case, either as a result of component failure or the presence of extraordinary large disturbances, signals can overload the analogue conditioning circuits.

Analogue circuits (and control systems) are realised using both active and passive components. Disadvantages of (purely) passive circuits (that is, those realised with only inductors L, capacitors C and resistors R) occur as a result of the restricted class of transfer functions that can be realised and the fact that large inductance values may be required. The advantage of passive circuits results from the inherent robustness in the realisation in that small changes in the component values of R, C and L only cause small changes to occur in the system transfer function. In particular, since a passive circuit has no internal energy source, any perturbation of a passive circuit remains passive.

An active analogue circuit uses both operational amplifiers and $\{R, L, C\}$ components. Two advantages of active circuits are: any transfer function can be realised using an active circuit and any (or all) inductors can be omitted. Disadvantages of active circuits are: the inherent internal noise of operational amplifiers and the (possible) lack of robustness of the circuit realisation in that small changes in dc bias of the operational amplifier and/or component values $\{R, L, C\}$ can lead to large changes in the controller transfer function. In terms of minimising the sensitivity of the controller transfer function to changes in circuit parameters, the internal structure of active analogue circuits is very important.

Digital control systems require the use of analogue-to-digital converters (ADCs), digital-to-analogue converters (DACs), and digital circuits. Digital circuits are implemented using either general purpose digital computers and/or special purpose digital signal processor (DSP) technology. The choice of both the ADCs and DACs must take into account the effects of possible signal saturation and signal quantisation. The digital input from an ADC and the digital output to a DAC can only assume one of a finite number of possible quantised values. In a *linear* ADC/DAC configuration, the quantisation levels are uniformly distributed from the minimum to the maximum level whereas a *compounding* ADC/DAC configuration provides a *nonlinear distribution* of signal quantisation levels. An advantage of a compounding ADC/DAC over a linear ADC/DAC configuration is that a greater dynamic range can be achieved for the same number of quantisation levels but at the expense of having a relatively large step in quantisation level for a large signal value.

The possible lack of robustness in the realisation of an analogue control system has a duality with the possible lack of robustness in the implementation of a discrete algorithm for digital control implementation. Specifically: (i) the inherent operational amplifier noise in active analogue circuits is analogous to a statistical representation of arithmetic quantisation noise in a fixed-point digital algorithm, (ii) the effect of changes in analogue circuit component values is analogous to the effect of changes in coefficient values in a digital algorithm, and (iii) the dc offset in operational amplifiers is analogous to the error that arises in truncation quantisation or biased rounding quantisation. Consequently, just as the internal structure of the active circuit for realising an analogue control system is important to sensitivity minimisation so too is the internal structure of the numerical algorithm for realising a digital control system. Research on wave (or lattice) digital filters beginning with the work of Fettweis [1] is important in this regard. As reflected in the references, most of the key results for control applications on analysis and design in the presence of finite-precision arithmetic have their origins in the field of signal processing.

A number in a digital algorithm is represented as a binary representation consisting of a string of bits or ***words*** where the ***word-length*** is the number of bits in the word. The most common arithmetic representations that are supported by standard digital processors are ***fixed-point*** and ***floating-point*** arithmetic. The ***dynamic range*** of an arithmetic representation is the total numerical range of numbers that can be represented, while the ***resolution*** of a arithmetic representation is the difference between consecutively represented numerical values. For fixed-point arithmetic, the resolution is constant, but in floating-point arithmetic, the resolution increases as the magnitude of the number increases. As a consequence, for the same number of bits, a floating-point format has a greater dynamic range than a fixed-point format. This property is analogous to the difference between a compounding and a linear ADC/DAC configuration.

When two binary words are added or multiplied, the result may possibly exceed the allowable dynamic range. If this range is exceeded, then ***overflow*** is said to have occurred, and a numerical error results. However, even if overflow has not occurred, the numerical result may not be able to be exactly represented within the available precision. In this case, a numerical error also results, and we say that ***underflow*** has occurred. If either overflow or underflow occurs after an addition or multiplication, then a numerical result must be selected from the allowable values that can be represented. This selection process is referred to as ***(numerical) quantisation.***

The floating-point representation of a number w is defined by two words: the ***mantissa*** m and the ***exponent*** exp. One possible floating-point representation is to have the mantissa m as a ***sign magnitude*** fixed-point fraction, and the exponent exp as a ***sign magnitude*** integer. In this case, it is usual to ***normalise*** the mantissa m such that $0.5 \leq |m| < 1$. Arithmetic operations between floating-point numbers are more complicated than fixed-point, and take more time to complete. Floating-point addition or subtraction first requires an alignment of the exponents before the mantissa can be added or subtracted. The product of two floating-point numbers is formed by a fixed-point multiplication of the two mantissa, and an addition of the exponents. No alignment of exponents is required, but a normalisation of the mantissa of the product may be necessary. If overflow occurs because the dynamic range of the exponent is exceeded, then saturation arithmetic is implemented whereby $|x + y|$ or $|xy|$ is fixed at the maximum allowable value consistent with the correct sign. Unlike fixed-point arithmetic, ***underflow*** quantisation can also occur in ***floating-point addition.*** Specifically, if $Q_{RO}[m]$ denotes the fixed-point ***roundoff*** quantisation of the (normalised) ***mantissa*** m of the finite word-length (FWL) sum $x+y$, and $FL[x+y]$ denotes the floating-point quantisation of $x + y$, then:

$$FL[x + y] = Q_{RO}[m] \times 2^{exp} \tag{4.1}$$

Applications in both signal processing and control require the calculation of inner products of the form:

$$r = \sum_{j=0}^{n} \alpha_j \gamma_j \tag{4.2}$$

In floating-point arithmetic, arithmetic quantisation occurs after every multiplication and addition. Hence, when the inner product r in (4.2) is computed for (say) $n = 1$ using floating-point arithmetic, the FWL result $\hat{r}$ is given by:

$$\hat{r} = FL[FL[\alpha_0\gamma_0] + FL[\alpha_1\gamma_1]] \tag{4.3}$$

and the complex nature of the quantisation process makes it difficult to both control and tightly bound the computation error $r - \hat{r}$. On the other hand, since no quantisation need occur after either a multiplication or addition in fixed-point arithmetic, it is possible to compute the inner product without any computation error.

Both fixed- and floating-point implementations of recursive algorithms can suffer from the effects of both underflow and overflow. Underflow can either manifest itself in the form of "random like" noise at the controller output or, in the worst case, in the form of limit cycles. However, overflow can be the most devastating form of numerical error. In particular, an overflow in the controller output $y(k_1)$ at discrete time instant $k = k_1$ can result in significant errors in all future values $y(k)$ for $k > k_1$. Overflow is most likely to occur in a fixed-point implementation, and so in spite of the speed and cost disadvantages, floating-point arithmetic is often preferred over fixed-point arithmetic for a digital control implementation.

This chapter focuses on the analysis of a new Q-parameterised class of low complexity low sensitive linear recursive control structures with a view to presenting a form of fixed-point (or block floating-point) implementation that avoids overflow. In Section 4.2, we begin by introducing a fixed-point implementation for a controller in the standard input/output form. We then show how the FWL performance with respect to both coefficient error and internal signal quantisation error can be analysed.

In Section 4.3, we introduce and examine the FWL characteristics of the Q-parameterised control structures. The analysis in terms of coefficient sensitivity applies to both a fixed- and a floating-point implementation. However, the sensitivity to internal signal roundoff error is developed only for a fixed-point implementation.

Finally, in Section 4.4, we show how dynamic scaling can be employed in a fixed-point implementation of a recursive control algorithm to both further improve numerical accuracy while reducing the possibility of overflow to the level that would occur in a floating-point implementation.

4.2 Digital Feedback Controller

An Nth-order (single-input single-output) linear algorithm:

$$\begin{aligned} y(k) &= a_1 y(k-1) + \ldots + a_N y(k-N) \\ &\quad + b_0 u(k) + b_1 u(k-1) + \ldots + b_N u(k-N) \end{aligned} \tag{4.4}$$

represents a digital controller with frequency response function $H_c(z)$, where:

$$H_c(z) = b_c(z)/a_c(z) \tag{4.5}$$

and:

$$b_c(z) = \sum_{i=0}^{N} b_i z^{N-i} \; ; \; a_c(z) = z^N - \sum_{i=1}^{N} a_i z^{N-i} \tag{4.6}$$

This algorithm is defined by at most $2N$ coefficients and requires at most $2N$ multiplications to compute the current controller output value $y(k)$ from past values of y, and present and past values of the controller input signal u. The most general state space equation:

$$\begin{aligned} \boldsymbol{x}(k+1) &= \boldsymbol{F}\boldsymbol{x}(k) + \boldsymbol{g}u(k) \quad ; \quad \boldsymbol{x}(k) \in \mathbb{R}^N \\ y(k) &= \boldsymbol{h}^T\boldsymbol{x}(k) + du(k) \end{aligned} \tag{4.7}$$

of (4.4) when $H_c(z) = d + \boldsymbol{h}^T(z\boldsymbol{I} - \boldsymbol{F})^{-1}\boldsymbol{g}$ is described by $N^2 + 2N + 1$ coefficients and requires $N^2 + 2N + 1$ multiplications, which is generally impractical for any real time implementation.

The design of structures that analyse and minimise coefficient sensitivity is frequently based on early work in signal processing [2,9]. Optimal (open-loop) fixed-point structures, which minimise the output roundoff noise sensitivity, have been reported [4,11,20], and an extension for the design of controllers in feedback is also available [15]. Results on optimising the design of Kalman filters [8] and linear quadratic regulators [13,17] have also appeared. However all these optimal structures have high complexity in that they require the order of N^2 multiplications to compute each controller output value.

A *low complexity* structure is one in which the number of multiplications (and coefficients) grow *linearly* with the order N. For example, a cascade or parallel structure of $M = N/2$ (N even) second-order sections can be implemented with $4N$ multiplications. As shown in [15], a lattice structure in state space form can also be implemented in $4N$ multiplications. Another structure:

$$\begin{aligned} r(k) &= u(k) + a_1 r(k-1) + \ldots + a_N r(k-N) \\ y(k) &= b_0 r(k) + b_1 r(k-1) + \ldots + b_N r(k-N) \end{aligned} \tag{4.8}$$

like (4.4) also only requires $2N$ multiplications.

When a controller $H_c(z)$ is realised using a general-purpose digital computer or a programmable DSP chip, the results of the computation at each discrete time step k must be stored in binary form in registers of finite length. In this respect, the representation (4.8) *may* be preferred to the representation (4.4) since in representation (4.8) the current and past values of the controller input u do not have to be stored. This property is also true for more general state space representations (4.7).

In an FWL implementation, the controller coefficients $\{a_p, b_q\}$ (or the components of the matrices $\{\boldsymbol{F}, \boldsymbol{g}, \boldsymbol{h}, d\}$), as well as the results of the computation at each discrete time step k must be stored in binary form in registers of finite length. The relative *significance* of the FWL depends very much on the numerical sensitivity of the control algorithm. Controllers implemented with a high sample rate can be particularly sensitive to numerical errors, and sometimes small arithmetic errors can lead to a significant loss in performance and even (numerically induced) instability if suitable precautions are not taken.

4.2.1 DSP Implementation

The essential difference between a DSP chip and a microprocessor is that a DSP chip has features designed to support high-performance, repetitive, numerically intensive tasks. In contrast, general-purpose processors or microcontrollers are either not specialised for a specific kind of application (in the case of general-purpose processors), or they are designed for logic-oriented control applications (in the case of microcontrollers).

A DSP implementation can be based on either a fixed- or floating-point format. Features that accelerate performance in DSP applications include: single-cycle multiply-accumulate capability; the availability in high-performance DSPs of two multipliers that enable two multiply-accumulate operations per instruction cycle; and specialised addressing modes (such as pre- and post-modification of address pointers, circular addressing, and bit-reversed addressing). Most DSPs provide various configurations of on-chip memory and peripherals tailored for DSP applications, and generally feature multiple-access memory architectures that enable the completion of several accesses to memory in a single instruction cycle, and specialised execution control. DSP processors are also known for their irregular instruction sets, which generally allow several operations to be encoded in a single instruction. For example, a processor that uses 32-bit instructions may encode two additions, two multiplications, and four 16-bit data moves into a single instruction.

Fixed-point Format. The fixed-point method for representing a number w with a word-length of n $(= 1 + p + t)$ bits assigns 1 bit for the sign, p bits for the integer, and t bits for the fraction. We then say that w has a $[1 + p, t]$

bit representation, and write:

$$w \approx [1 + p, t] \tag{4.9}$$

Each particular word w has the binary form:

$$a_p a_{p-1} ... a_1 a_0 \ \Delta \ a_{-1} a_{-2} ... a_{-t} \tag{4.10}$$

where a_p is the sign bit with 0 for a positive and 1 for a negative number. The string $a_{p-1}a_{p-2}...a_1a_0$ represents the magnitude of the integer part, and the string $a_{-1}a_{-2}...a_{-t}$ represents the (unsigned) fractional part. The value of a number w in fixed-point arithmetic is normally represented in the *two's complement format* in which the value of a positive number (*i.e.*, $a_p = 0$) is given by:

$$2^{p-1}a_{p-1} + 2^{p-2}a_{p-2} + ... + 2a_1 + a_0 + 2^{-1}a_{-1} + ... + 2^{-t}a_{-t} \tag{4.11}$$

For a *negative* number (*i.e.*, $a_p = 1$), the two's complement number w has the value determined by:

$$-(2^{p-1}\bar{a}_{p-1} + 2^{p-2}\bar{a}_{p-2} + ... + 2\bar{a}_1 + \bar{a}_0 + 2^{-1}\bar{a}_{-1} + ... + 2^{-t}\bar{a}_{-t}) - 1 - 2^{-t} \tag{4.12}$$

where $\bar{a}_i$ denotes the complement of the bit a_i.

When two fixed-point numbers having a $[1 + p, t]$ bit representation are added, the fractional representation of the sum also only requires t fractional bits. However, the product of two such numbers generally requires $2t$ bits for an exact representation of the fraction, and $2p$ bits for the integer. If less than $2p$ bits are available, then *overflow* may occur in which case *saturation arithmetic* should be implemented.

If less than $2t$ bits are available to represent the fraction, then *underflow* can occur in which case, a *quantisation* $Q[w]$ of w must be implemented. The two possible quantisation characteristics are *roundoff* and *truncation.* Ignoring the least significant bit(s) in a two's complement representation is equivalent to implementing *truncation* $Q_T[w]$. The *truncation quantisation error* $w - Q_T[w]$ will therefore be *non-negative* for both positive *and* negative values of w. We therefore say that truncation quantisation introduces a *bias.*

If a number w is quantised using *roundoff quantisation*, then the error $e \triangleq w - Q_{RO}[w] \geq 0$ always has the *same sign* as w; that is, $w(w - Q_{RO}[w]) \geq 0$. The *roundoff* operation Q_{RO} selects $Q_{RO}[w]$ to be the nearest representation to w. However when w is exactly halfway between two consecutive values of $Q_{RO}[w]$, the decision as to whether to round up or round down must be selected in advance. If either *rounding up* or *rounding down* is always implemented, then a *small bias* will be added to the signal. However, even though this bias is much smaller than that which is introduced when *truncation* is implemented, sometimes even a small bias can cause problems in particular control and signal processing applications.

An improved rounding procedure that overcomes this bias problem is known as *convergent rounding.* In this case, when a number that is to be rounded lies at the midpoint range between two allowable number representations, then the number is rounded either higher or lower according to the following algorithm: if the bit that will become the least significant bit of the rounded number is 1, then the number is rounded up, and if the least significant bit of the rounded number is 0, then the number is rounded down. Convergent rounding is not supported in hardware by all digital signal processors. However the analogue devices ASP-21xx and Motorola DSP5600x families both support convergent rounding.

When using fixed-point arithmetic, an inner product (4.2) can be determined without any computational error, provided an accumulator of sufficient length is available. Fortunately, many commercially available fixed-point DSP chips do indeed have this feature. For example, the *Analog Devices* $ADSP-21xx$ series provides 10 to 50 MIPS 16-bit fixed-point operation with a 40-bit accumulator. The *Lucent Technologies* $DSP16xxx$ series, which provides 100 MIPS 16-bit fixed-point operation, has eight 40-bit accumulators. The *Motorola* $DSP563xx$, which operates at 66, 80 or 100 MHz, has a 24-bit fixed-point multiplier and two 56-bit accumulators. Then (for example), using a $DSP563xx$ chip, we can have $t = 24$ fractional bits for all coefficients $\{\alpha_j\}$ and all signals $\{\gamma_j\}$. After taking into account the sign bit, the maximum number ($= n$) of inner products that can be computed in (4.2) *without* overflow is given by $n = 2^{56-48-1} = 128$.

Fixed-point Algorithm. Suppose a given fixed-point DSP chip has a B-bit multiplier, and suppose without any loss of generality that the controller input signal u (arising from an analogue-to-digital conversion) has been (analogue) scaled such that $|u(k)| \leq 1$ for all k so that $u(k)$ has a $[1, B-1]$-bit representation. Then one possible fixed-point implementation of (4.4) is given by:

$$\begin{aligned}\hat{y}(k) &= \hat{a}_1 Q[\hat{y}(k-1)] + \ \dots \ + \hat{a}_N Q[\hat{y}(k-N)] \\ &\quad + \hat{b}_0 u(k) + \hat{b}_1 u(k-1) + \ \dots \ + \hat{b}_N u(k-N)\end{aligned} \tag{4.13}$$

In this implementation, we suppose the quantised signal $Q[\hat{y}(k)]$ has a $[1 + p_1, t_1]$ bit representation, the quantised coefficients $\hat{a}_n$ have a $[1 + p_2, t_2]$ bit representation and the quantised coefficients $\hat{b}_m$ have a $[1 + p_3, t_3]$ bit representation. Since a B-bit multiplier is available, it is not necessary that $p_1 = p_2 = p_3$ but only that:

$$B = 1 + p_1 + t_1 = 1 + p_2 + t_2 = 1 + p_3 + t_3$$

The choice of $\{p_1, p_2, p_3\}$ (and consequently $\{t_1, t_2, t_3\}$) can then reflect the need for greater or less numerical resolution and corresponding less or greater dynamic range in the signal values $Q[\hat{y}(k)]$ and the coefficient values $\{\hat{a}_n, \hat{b}_m\}$.

The result $\hat{y}(k)$ in (4.13) can be computed as the sum of two separate inner products: $\sum_{j=1}^{N} \hat{a}_j Q[\hat{y}(k-j)$, and $\sum_{j=0}^{N} \hat{b}_j u(k-j)$. However, in order to avoid both overflow and underflow, each term $\hat{a}_j Q[\hat{y}(k-j)]$ requires a $[1+p_1+p_2, t_1+t_2]$ bit representation, while the accumulated result $\sum_{j=1}^{N} \hat{a}_j Q[\hat{y}(k-j)]$ generally requires a $[1+n_{ab}+p_1+p_2, t_1+t_2]$ representation where the minimum value of n_{ab} satisfies the condition $N \leq 2^{n_{ab}} < N+1$. Similarly, each term $\hat{b}_j u(k-j)$ requires a $[1+p_3, B-1+t_3]$ bit representation, and the inner product $\sum_{j=0}^{N} \hat{b}_j u(k-j)$ generally requires a $[1+n_{ab}+1+p_3, B-1+t_3]$ bit representation.

Also, if $t_1+t_2 \neq B-1+t_3$, an alignment of the two inner products is required before addition, and an accumulator of size $n_{ab}+L_{ab}$ is required for the result $\hat{y}(k)$ where:

$$L_{ab} = \max\{1+p_3+t_1+t_2, p_1+p_2+B+t_3\}$$

Finally, $\hat{y}(k)$ is then *quantised* (e.g., using convergent rounding) to a $[1+p_1, t_1]$ bit representation $Q[\hat{y}(k)]$ to be used for subsequent calculations, and results in the (roundoff) quantisation error $\epsilon_y(k)$ given by

$$\epsilon_y(k) \triangleq \hat{y}(k) - Q[\hat{y}(k)] \tag{4.14}$$

with

$$|\epsilon_y(k)| \leq 2^{-(t_1+1)} \tag{4.15}$$

A possible fixed-point implementation of (4.8) is given by:

$$\begin{aligned}\hat{r}(k) &= u(k) + \hat{a}_1 Q[\hat{r}(k-1)] + \dots + \hat{a}_N Q[\hat{r}(k-N)] \\ \hat{y}(k) &= \hat{b}_0 Q[\hat{r}(k)] + \hat{b}_1 Q[\hat{r}(k-1)] + \dots + b_N Q[\hat{r}(k-N)]\end{aligned} \tag{4.16}$$

where the quantised signals $\{Q[\hat{r}(k)], Q[\hat{y}(k)]\}$ have a $[1+p_1, t_1]$ bit representation, and the quantised coefficients $\{\hat{a}_n, \hat{b}_m\}$ have a $[1+p_2, t_2]$ bit representation. Since a B-bit multiplier is available, it is also not necessary that $p_1 = p_2$ but only that $B = 1+p_1+t_1 = 1+p_2+t_2$ where again, the choice of $\{p_1, p_2\}$ (and consequently $\{t_1, t_2\}$) can then reflect the need for greater or less numerical resolution and corresponding less or greater dynamic range in the signal values $Q[\hat{r}(k)]$ and the coefficient values $\{\hat{a}_n, \hat{b}_m\}$.

We make the following remarks: (i) in (4.16), unlike in the implementation (4.13), if only a B-bit multiplier is available, then both sets of coefficients $\{\hat{a}_n\}$ and $\{\hat{b}_m\}$ need to have the same representation; (ii) even though the ideal (*i.e.*, infinite-precision) representations (4.4) and (4.8) are equivalent, the FWL implementations (4.13) and (4.16) because of the nonlinear (quantisation) function $Q[\cdot]$ are *different*; (iii) in terms of numerical accuracy, one FWL implementation is not always better than the other for all input signals $\{u\}$.

Numerical Accuracy with Coefficient Quantisation. Assume that there are *no signal errors*; that is, the word-length of the signal $\hat{y}$ is arbitrarily large so that in (4.14), $\epsilon_y(k) = 0$ for all k, or equivalently:

$$Q[\hat{y}(k)] = \hat{y}(k) \tag{4.17}$$

and define the controller output numerical error:

$$E_y(k) \triangleq y(k) - \hat{y}(k) \tag{4.18}$$

Then it follows from (4.4), (4.13) and (4.14) that:

$$E_y(k) = \sum_{i=1}^{N} \hat{a}_i E_y(k-i) + \sum_{i=0}^{N} (\Delta b_i) u(k-i) + \sum_{i=1}^{N} (\Delta a_i) y(k-i) \tag{4.19}$$

with the coefficient errors:

$$\Delta a_n \triangleq a_n - \hat{a}_n \ ; \ \Delta b_m \triangleq b_m - \hat{b}_m \tag{4.20}$$

More simply, in terms of the z-transforms $\mathcal{Z}\{E_y\}$ of the output error signal E_y and $\mathcal{Z}\{u\}$ of the input signal u:

$$\mathcal{Z}\{E_y\} = [H_c(z) - \hat{H}_c(z)]\mathcal{Z}\{u\}$$

where:

$$\hat{H}_c(z) = \hat{b}_c(z)/\hat{a}_c(z) \tag{4.21}$$

with:

$$\hat{b}_c(z) = \sum_{i=0}^{N} \hat{b}_i z^{N-i} \ ; \ \hat{a}_c(z) = z^N - \sum_{i=1}^{N} \hat{a}_i z^{N-i} \tag{4.22}$$

We can now establish the general result.

Theorem 4.1. *Suppose the finite coefficient word-length controller $\hat{H}_c(z)$ defined by (4.21), (4.22) is used to approximate the ideal controller $H_c(z)$ in (4.5), (4.6). Then the z-transform $\mathcal{Z}\{E_y\}$ of the resulting controller output error E_y due to controller coefficient error is given in terms of the error transfer function $\mathcal{G}_E(z)$ and the controller input z-transform $\mathcal{Z}\{u\}$ by:*

$$\mathcal{Z}\{E_y\} = \mathcal{G}_E(z)\mathcal{Z}\{u\} \tag{4.23}$$

where $\mathcal{G}_E(z)$ is given by:

$$\mathcal{G}_E(z) = \frac{[\sum_{i=1}^{N} (\Delta a_i) z^{N-i}]\hat{H}_c(z) + \sum_{i=0}^{N} (\Delta b_i) z^{N-i}}{a_c(z)} \tag{4.24}$$

Furthermore, for a unity feedback closed-loop control system with open-loop transfer function $\hat{H}_c(z)H_p(z)$, where $H_p(z)$ is the open-loop plant transfer function, external reference signal $\{r_{ref}\}$, the z-transform $\mathcal{Z}\{E_z\}$ of the plant output error E_z is given by:

$$\mathcal{Z}\{E_z\} = \mathcal{G}_E(z)\frac{H_p(z)}{1+\hat{H}_c(z)H_p(z)}\mathcal{Z}\{r_{ref}\} \tag{4.25}$$

The expression in (4.25) allows the consequences of coefficient error in implementing a controller in a feedback loop to be investigated. As is often the case in feedback control as compared to signal processing applications, some *frequency weighting* must be applied to the controller error transfer function $\mathcal{G}_E(z)$.

Similarly, it follows from (4.8) and (4.16) when $Q[\hat{r}(k)] = \hat{r}(k)$ that the z-transform $\mathcal{Z}\{E_y\}$ of the error signal $\{E_y\}$ is again given by (4.23). Therefore in terms of the numerical error due only to *coefficient error*, the implementations (4.13) and (4.16) are *equivalent.* This analysis of the effects of coefficient error applies equally to both a fixed- and a floating-point implementation.

Numerical Accuracy with Signal Quantisation. Suppose there are no coefficient errors; that is, $\{\hat{a}_n = a_n, \hat{b}_n = b_n\}$ for all n. Then it follows from (4.4), (4.13) and (4.14) that:

$$E_y(k) = \sum_{i=1}^{N} a_i E_y(k-i) + \sum_{i=1}^{N} a_i \epsilon_y(k-i) \tag{4.26}$$

Equivalently, in terms of z-transforms $\mathcal{Z}\{E_y\}, \mathcal{Z}\{\epsilon_y\}$ and the error transfer function $\mathcal{H}_1(z)$, we have:

$$\frac{\mathcal{Z}\{E_y\}}{\mathcal{Z}\{\epsilon_y\}} = \mathcal{H}_1(z) \triangleq \frac{\sum_{i=1}^{N} a_i z^{N-i}}{z^N - \sum_{i=1}^{N} a_i z^{N-i}} \tag{4.27}$$

which does not depend on the coefficients $\{b_m\}$.

Similarly, if we define the quantisation error:

$$\epsilon_r(k) \triangleq \hat{r}(k) - Q[\hat{r}(k)] \tag{4.28}$$

and the numerical error:

$$E_r(k) \triangleq r(k) - \hat{r}(k) \tag{4.29}$$

it follows from (4.8), (4.16), (4.28) and (4.18) that:

$$\begin{aligned} E_r(k) &= \sum_{i=1}^{N} a_i E_r(k-i) + \sum_{i=1}^{N} a_i \epsilon_r(k-i) \\ E_y(k) &= \sum_{i=0}^{N} b_i E_r(k-i) + \sum_{i=0}^{N} b_i \epsilon_r(k-i) \end{aligned} \tag{4.30}$$

Equivalently, in terms of the z-transforms $\{\mathcal{Z}\{E_y\}, \mathcal{Z}\{\epsilon_r\}\}$ and the error transfer function $\mathcal{H}_2(z)$, we have:

$$\frac{\mathcal{Z}\{E_y\}}{\mathcal{Z}\{\epsilon_r\}} = \mathcal{H}_2(z) \triangleq \frac{\sum_{i=0}^{N} b_i z^{N-i}}{z^N - \sum_{i=1}^{N} a_i z^{N-i}} \tag{4.31}$$

which (unlike $\mathcal{H}_1(z)$ in (4.27)) depends on both sets of coefficients $\{a_n, b_m\}$.

Since the (roundoff) error signals $\{\epsilon_y\}$ and $\{\epsilon_r\}$ in the two different implementations are different, it is not possible to directly compare the accuracy of the two implementations except by simulation. However, we can provide some analysis. In particular, since $\{\epsilon_y\}$ and $\{\epsilon_r\}$ are both bounded by $2^{-(t_1+1)}$ as in (4.15), then when $\{\epsilon_y(k) = \epsilon_r(k) = 0; k \leq 0\}$, an ***upper bound*** on the ***magnitude*** of the output error signal $\{E_y\}$ is given by:

$$|E_y(k)| \leq ||\mathcal{H}||_1 . 2^{-(t_1+1)}$$

where $\mathcal{H} = \mathcal{H}_1$ in (4.27) for the implementation (4.13) and $\mathcal{H} = \mathcal{H}_2$ (4.31) for the implementation (4.16).

In applications, it often follows that upper bounds on $|E_y(k)|$ obtained in this way are far too conservative, and it is therefore more common to base the FWL performance due to signal quantisation in terms of a "noise model" whereby the roundoff signals $\{\epsilon_y\}$ and $\{\epsilon_r\}$ are assumed to be a zero mean white noise signals which are uniformly distributed over the interval $[-2^{t_1}, 2^{t_1}]$ with variance $\frac{1}{12}2^{-2t_1}$. The resulting ***output error variance*** σ_y is then given by:

$$\sigma_y = ||\mathcal{H}||_2 . \frac{1}{12} 2^{-2t_1}$$

where $\mathcal{H} = \mathcal{H}_1$ for (4.13) and $\mathcal{H} = \mathcal{H}_2$ for (4.16). Therefore in terms of the numerical error due to ***internal signal quantisation***, the implementations (4.13) and (4.16) are ***not*** equivalent. The equivalent analysis for the effects of internal signal quantisation due to a floating-point implementation [21] is more complicated and provides more conservative predictions for the consequences of numerical errors.

4.3 *Q*-Parameterized Controller

The single-input single-output Nth-order digital controller $H_c(z)$ in (4.5) has N zeros $\{q_m\}$ given by the roots of $b_c(z) = 0$ and N poles $\{p_m\}$ given by the roots of $a_c(z) = 0$. The significance or otherwise of a change Δq_m in the location of a particular controller zero q_m or a change Δp_m in the location of a particular controller pole p_m depends on the closed-loop system performance. An extreme case of pole sensitivity occurs when a change Δp_m results in an otherwise stable controller pole becoming unstable. Small variations in the poles (through changes in the coefficients from $\{a_n\}$ to $\{\hat{a}_n\}$) will generally have a greater impact on the numerical error compared with changes in the zeros (through changes in the coefficients from $\{b_m\}$ to $\{\hat{b}_m\}$). Accordingly, much of the research activity is focused on pole sensitivity. Nevertheless, with regard to performance, zero sensitivity can also be important.

The variation $\Delta\lambda_n$ in the root λ_n of a polynomial equation:

$$f(z) = \prod_{j=1}^{N}(z - \lambda_j) = 0 \tag{4.32}$$

where:

$$f(z) = z^N - \sum_{j=1}^{N} f_j z^{N-j} \tag{4.33}$$

due to a change Δf_m in the coefficient f_m can be measured in terms of the *root coefficient sensitivity* S_{nm} by:

$$\Delta\lambda_n \approx S_{nm}.\Delta f_m \; ; \; S_{nm} \triangleq \frac{\partial \lambda_n}{\partial f_m}$$

After equating the differentials $f^{(n)}(z)$ of $f(z)$ in (4.32), (4.33) at $z = 0$, it follows that for all distinct roots λ_j:

$$S_{nm} = \frac{\lambda_n^{N-m}}{D_n} \; ; \; D_n = \prod_{j \neq n}(\lambda_n - \lambda_j) \tag{4.34}$$

Hence $|S_{nm}| \approx 0$ for all m when $|\lambda_n| \approx 0$.

In terms of the numerical realisation of a digital controller, a variation $\Delta f_m \triangleq f_m - \hat{f}_m$ occurs because a controller parameter f_m must be realised in FWL with resolution $2^{-(t_2+1)}$ as $\hat{f}_m$. For FWL coefficient selection, the resolution (and not the dynamic range per se) is more important, and is an issue for both fixed- and floating-point implementations. Once the FWL selection $\hat{f}_m$ is made, the coefficient error Δf_m is *fixed* and *known*. Nevertheless, coefficient sensitivity has also been analysed [9] based on a statistical representation of the error (essentially, it is assumed that each coefficient f_m is replaced by an independent white noise source with mean $\hat{f}_m$ and variance σ_f where $\sigma_f = \frac{1}{12}2^{-2t_2}$).

Another approach aimed at reducing the effects of coefficient sensitivity has been based on the representation of a polynomial in the shift operator z in terms of another operator (e.g., delay-replaced operator [3,6], delta operator [16,12], alternative discrete time operators [23]). The next result is an extension of the earlier result [23].

Theorem 4.2. *Suppose $f(z)$ in (4.32) is expressed in the form:*

$$f(z) = (z - q_1)^N - \sum_{j=1}^{N} f_{j1}(z - q_1)^{N-j} \tag{4.35}$$

Then for all distinct roots λ_j of $f(z) = 0$:

$$S_{jk}(q_1) \triangleq \frac{\partial \lambda_j}{\partial f_{k1}} = \frac{(\lambda_j - q_1)^{N-k}}{D_j} \ ; \ 1 \leq k \leq N \tag{4.36}$$

where $D_j = \prod_{k \neq j}(\lambda_j - \lambda_k)$.

This result may be derived by equating the differentials of both expressions for $f(z)$ with respect to f_{k1} and then evaluating at $z = \lambda_j$. The significance of the parameter q_1 is that $|S_{jk}(q_1)| \approx 0$ for all $k \neq N$ when $|\lambda_j - q_1| \approx 0$.

There are many ways for obtaining expressions for the parameters $\{f_{k1}\}$ of $f(z)$ in (4.33) in terms of the parameters $\{f_k\}$ of $f(z)$ in (4.36), including evaluating $f(z)$ at distinct values of z. The explicit expression:

$$f_{k1} = \sum_{j=1}^{N} f_j \binom{j}{k} q_1^{j-k} \tag{4.37}$$

may be derived by equating expressions for the N derivatives $\{f^{(n)}(q_1)\}$.

Optimal Root Sensitivity. Since $S_{jN}(q_1) = D_j^{-1}$ is independent of q_1, an *optimal* choice q_1^* for q_1 can be obtained [23] by minimising the *largest root sensitivity* (excluding the $\{S_{jN}(q_1)\}$); that is, we can define:

$$q_1^* \triangleq \arg\min_{q_1}\{\max_{j,k \neq N} |S_{jk}(q_1)|\} \tag{4.38}$$

where:

$$\max_{j,k \neq N} |S_{jk}(q_1)| = \max_{j} \begin{cases} |\lambda_j - q_1|^{N-1} D_j^{-1} & ; \ |\lambda_j - q_1| \geq 1 \\ |\lambda_j - q_1| D_j^{-1} & ; \ |\lambda_j - q_1| \leq 1 \end{cases}$$

4.3.1 Improved Coefficient Sensitivity Characteristics

A significant improvement in root sensitivity due to changes in parameters is possible by means of the following extension of Theorem 4.2.

Theorem 4.3. *Suppose $f(z)$ in (4.32) is expressed in the Q-parameterised form:*

$$f(z) = \prod_{i=1}^{N}(z - q_i) - \sum_{n=1}^{N} \alpha_n \prod_{i=1}^{N-n} (z - q_i) \tag{4.39}$$

Then for all distinct roots λ_j *of* $f(z) = 0$*:*

$$S_{jk}(\boldsymbol{q}) \triangleq \frac{\partial \lambda_j}{\partial \alpha_k} = \frac{\prod_{i=1}^{N-k}(\lambda_j - q_i)}{D_j} \tag{4.40}$$

where $\boldsymbol{q}^T = [q_1, q_2, \dots, q_N]$ *and* $D_j = \prod_{k \neq j}(\lambda_j - \lambda_k)$*.*

We shall refer to the $\{q_i\}$ as the *Q-parameters* of the Q-parameterization (4.39) [19].

There are a variety of ways for deriving expressions for the coefficients $\{\alpha_k\}$ in (4.39) in terms of the coefficients $\{f_k\}$ in (4.33) and the Q-parameters $\{q_i\}$ in (4.39). For example, in the special case when all the coefficients $\{q_i\}$ are *distinct*, then α_N is given directly from $f(q_1)$, while α_{N-k} for $k = 1, 2, \dots, N-1$ is given recursively in terms of $\{f(q_{k+1}), \alpha_N, \alpha_{N-1}, \dots, \alpha_{N-k+1}\}$. More generally, if the coefficients $\{q_i; 1 \le i \le m\}$ are *distinct*, then α_{N-k} for $k = 1, 2, \dots, m-1$ is given recursively in terms of $\{f(q_{k+1}), \alpha_N, \alpha_{N-1}, \dots, \alpha_{N-k+1}\}$. Then if $q_{N-m} = q_{N-i}$ for some $1 \le i \le m-1$, the coefficient α_{M-m} is given from the derivative $f'(q_{N+1-m})$.

Note that the Nth-order polynomial f in (4.39) is over-parameterised in that $2N$ parameters $\{q_j, \alpha_j; 1 \le j \le N\}$ define the N roots of $f(z) = 0$. In any FWL representation $\hat{f}$ of f we can therefore assume without loss of generality that the N parameters $\{q_j; 1 \le j \le N\}$ have no error; that is, we can assume:

$$\hat{f}(z) = \prod_{i=1}^{N}(z - q_i) - \sum_{n=1}^{N} \hat{\alpha}_k \prod_{i=1}^{N-n}(z - q_i) \tag{4.41}$$

Consequently, in Theorem 4.3, we are not concerned with the sensitivity of the roots with respect to the Q-parameters $\{q_i\}$. We have the following Corollary.

Corollary 4.1. *Let* $\boldsymbol{q}^T = [q_1, q_2, \dots, q_N]$ *and* $\{S_{jk}(\boldsymbol{q})\}$ *be defined as in (4.40). Then:*

(i) $S_{jk}(\boldsymbol{q})$ *is independent of the Nth component* q_N *of* $\boldsymbol{q}$ *for all* k*.*

(ii) The sensitivity measure $S_{jk}(\boldsymbol{q})$ *in (4.40) for* $N \ge 2$ *is equal to the sensitivity measure* $S_{jk}(q_1)$ *in (4.36) when* $\boldsymbol{q}^T = [q_1, q_1, q_1 \dots, 0]$*.*

(iii) $S_{jN}(\boldsymbol{q}) = D_j^{-1}$ *is independent of* $\boldsymbol{q}$*, and* $S_{j,N-k}(\boldsymbol{q})$ *for* $k = 1, 2, \dots, N-1$ *depends only on* $\{q_i; 1 \le i \le k\}$*.*

Low Root Sensitivity. There are at least three ways of achieving low root sensitivity by selecting the Q-parameters $\{q_i\}$. (i) If we follow on from (4.38), we could define:

$$\boldsymbol{q}^* \triangleq \arg\min_{\boldsymbol{q}}\{\max_{j,k \neq N} |S_{jk}(\boldsymbol{q})|\} \tag{4.42}$$

However, it can be seen from property (ii) in Corollary 4.1 that $\boldsymbol{q}^*$ is non-uniquely given by $\boldsymbol{q}^* = [q_1^*, q_2^*, \ \dots \ , 0]^T$. Hence, one possible optimisation process is to first define q_1^* by (4.42), and then $\{q_j^*;\ 2 \le j \le N-1\}$ can be used to successively minimise the jth largest sensitivity for $j = 2, 3, \ \dots \ , N-1$.

(ii) A second possible method to achieve low root sensitivity, using property (iii) in Corollary 4.1, is to successively define optimal coefficients $\{q_i^*\}$ as follows:

$$q_1^* \triangleq \arg\min_{q_1}\{\max_j S_{j,N-1}(\boldsymbol{q})\}$$

$$q_2^* \triangleq \arg\min_{q_2}\{\max_j S_{j,N-2}(\boldsymbol{q}) \ ; \ q_1 = q_1^*\}$$

$$\cdots \quad \cdot \quad \cdots$$

$$q_{N-1}^* \triangleq \arg\min_{q_{N-1}}\{\max_j S_{j,N-1}(\boldsymbol{q}) \ ; \ q_1 = q_1^*, q_2 = q_2^*, \ \dots \ , q_{N-2} = q_{N-2}^*\}$$

(It can be shown that method (i) is equivalent to method (ii) if $|q_i^* - \lambda_i| < 1$ for all i.)

(iii) Finally, a third method to achieve low root sensitivity is to choose $q_i^* = \text{Re}\{\lambda_i\}$ to reduce the sensitivity of the $N-1$ most critical roots (say) $\{\lambda_i;\ 1 \le i \le N-1\}$ (which are generally those closest to the unit circle), and where $\text{Re}\{\cdot\}$ denotes the real part.

Example 4.1. Suppose $f(z) = (z - 0.3)(z - 0.7)(z - 0.9)$, which corresponds in (4.32) to $\{\lambda_1 = 0.3, \lambda_2 = 0.7, \lambda_3 = 0.9\}$. Then with:

$$f(z) = (z - q_1)^3 - f_{11}(z - q_1)^2 - f_{21}(z - q_1) - f_{31}$$

and:

$$f(z) = (z - q_1)(z - q_2)(z - q_3) - \alpha_1(z - q_1)(z - q_2) - \alpha_2(z - q_1) - \alpha_3$$

we have from (4.36) and (4.40) that:

$$S_{j1}(q_1) = D_j^{-1}(\lambda_j - q_1)^2 \ ; \ S_{j2}(q_1) = D_j^{-1}(\lambda_j - q_1) \ ;$$

$$S_{j1}([q_1, q_2]) = D_j^{-1}(\lambda_j - q_1)(\lambda_j - q_2) \ ; \ S_{j2}([q_1, q_2]) = D_j^{-1}(\lambda_j - q_1)$$

with $S_{j3}(q_1) = S_{j3}([q_1, q_2]) = D_j^{-1}$ and $\{|D_1| = 0.24, |D_2| = 0.08, |D_3| = 0.12\}$. (Note that since all sensitivities are independent of q_n for $n = 3$, we have omitted explicit reference to q_3.)

(a) In terms of the root sensitivities $\{S_{ij}(q_1)\}$, we can proceed as follows:

$$q_1^* = \arg\min_{q_1}\{\max_j\{|S_{j1}(q_1)|, |S_{j2}(q_1)|\}\}$$

$$= \arg\min_{q_1}\max\{\frac{|0.3 - q_1|}{0.24}, \frac{|0.7 - q_1|}{0.08}, \frac{|0.9 - q_1|}{0.12}\} = 0.7$$

with $\{|S_{11}(0.7)| = 0.67, |S_{12}(0.7)| = 1.67, |S_{21}(0.7)| = |S_{22}(0.7)| = 0, |S_{31}(0.7)| = 0.33, |S_{32}(0.7)| = 1.67\} = 0.23$.

(b) In terms of the root sensitivities $\{S_{ij}(q_1, q_2)\}$, we can proceed as follows:

(i) With $q_1^* = 0.7$, we have $\{|S_{12}(0.7, q_2)| = |S_{32}(0.7, q_2)| = 1.67, |S_{21}(0.7, q_2)| = |S_{22}(0.7, q_2)| = 0\}$ and

$$|S_{11}(0.7, q_2)| = \frac{0.4(0.3 - q_2)}{0.24} \; ; \; |S_{31}(0.7, q_2)| = \frac{0.2(0.9 - q_2)}{0.12}$$

(Observe that $|S_{11}(0.7)| = |S_{11}(0.7, 0.7)| = 0.67$ and $S_{31}(0.7)| = |S_{31}(0.7, 0.7)| = 0.33$.)

The optimal coefficient q_2^* can then be determined according to:

$$q_2^* = \arg\min_{q_2} \max\{|S_{11}(0.7, q_2)|, |S_{31}(0.7, q_2)|\} = 0.6$$

with $|S_{11}(0.7, 0.6)| = |S_{31}(0.7, 0.6)| = 0.50$.

(ii) Since $\{|q_i^* - \lambda_1| < 1, |q_i^* - \lambda_2| < 1\}$ for all i, method (i) and method (ii) achieve the same result.

(iii) Suppose we select $\{\lambda_3 = 0.7, \lambda_2 = 0.9\}$ as the most critical roots. Then $\{q_1 = 0.9, q_2 = 0.7, q_3 = 0\}$ implies:

$$|S_{31}(0.7, 0.9)| = |S_{32}(0.7, 0.9) = 0 \; ; \; |S_{21}(0.7, 0.9)| = 0$$

with:

$$|S_{22}(0.7, 0.9)| = 2.5 \; ; \; |S_{11}(0.7, 0.9)| = 1, \; |S_{12}(0.7, 0.9)| = 2.5$$

In this case, in terms of the Q-parameterisation in (4.39), we have:

$$(z - 0.3)(z - 0.7)(z - 0.9) = (z - 0.9)(z - 0.7)(z) - \alpha_1(z - 0.9)(z - 0.7)$$

with $\{\alpha_1 = 0.3, \alpha_2 = \alpha_3 = 0\}$.

Complex Roots. As indicated in Example 4.1, one or more root sensitivities can be made equal to zero provided one or more of the Q-parameters $\{q_i\}$ are equal to one or more of the real roots λ_j. It also follows that root sensitivities S_{ij} for a complex root λ_j can be made small with $q_i = \mathrm{Re}\{\lambda_j\}$ provided $\mathrm{Im}\{\lambda_j\}$ is small, where $\mathrm{Im}\{\cdot\}$ denotes the imaginary part. However since all the q_i are real, a root sensitivity with respect to a complex root cannot be made zero. In order to achieve this outcome, we need to modify the Q-parameterisation as follows.

Express $f(z)$ in (4.32) in the form:

$$f(z) = [z^2 - q_1 z - q_2] \left\{ \prod_{i=3}^{N} (z - q_i) - \sum_{n=1}^{N-2} \alpha_n \prod_{i=1}^{N-n} (z - q_i) \right\}$$
$$-\alpha_{N-1}(z - q_1) + \alpha_N q_2 \qquad (4.43)$$

which like the representation (4.39) involves N Q-parameters $\{q_i\}$ and N coefficients $\{\alpha_n\}$. The following result can be established.

Lemma 4.1. *Suppose:*

$$f(z) = [z^2 - r_1 z - r_2] \prod_{i=3}^{N} (z - \lambda_i) \tag{4.44}$$

is represented in the form (4.43). Then when $\{q_1 = r_1, q_2 = r_2\}$

$$\frac{\partial r_1}{\partial \alpha_n} = \frac{\partial r_2}{\partial \alpha_n} = 0$$

for $n = 1, 2, \ldots, N-2$.

Equations can be developed that can be solved sequentially for $\{\alpha_j;\ j = 1, 2, \ldots, N\}$ in terms of $\{r_1, r_2, \lambda_i\}$ and the Q-parameters $\{q_i\}$.

Example 4.2. (i) The polynomial:

$$f(z) = (z^2 + 0.81)(z - 0.8)(z - 0.1)$$

can be written in the Q-parameterised form (4.44) for $N = 4$ as follows:

$$[z^2 - q_1 z - q_2]\{(z - q_3)(z - q_4) - \alpha_1(z - q_3) - \alpha_2\} - \alpha_3(z - q_1) + \alpha_4 q_2$$

For example, when $\{q_1 = 0, q_2 = -0.81, q_3 = 0.8, q_4 = 0\}$, we have

$$[z^2 + 0.81]\{(z - 0.8)(z) - (0.3)(z - 0.8)\} = [z^2 + 0.81]\{(z - 0.8)(z - 0.3)\}$$

which corresponds to $\{\alpha_2 = \alpha_3 = \alpha_4 = 0\}$. This outcome is not surprising. However, in FWL implementations, we shall see that in general it is not possible to choose a parameter q_i equal to a root λ_j .

A more interesting use of the Q-parameterisation is when $\{q_1 = 0, q_2 = -1, q_3 = 1, q_4 = 0\}$. Now, we have:

$$f(z) = [z^2 + 1]\{(z - 1)(z) - \alpha_1(z - 1) - \alpha_2\} - \alpha_3(z) + \alpha_4$$

where:

$$\begin{aligned} 1 + \alpha_1 = 1.1 \ &;\ 1 + \alpha_1 - \alpha_2 = 1.05 \\ 1 + \alpha_1 + \alpha_3 = 0.891 \ &;\ \alpha_1 - \alpha_2 + \alpha_4 = 0.1944 \end{aligned}$$

These equations can be solved sequentially to give $\{\alpha_1 = 0.1, \alpha_2 = 0.05, \alpha_3 = -0.219, \alpha_4 = 0.1444\}$.

(ii) The polynomial:

$$f(z) = (z^2 + 0.99)(z - 0.99)^2 \tag{4.45}$$

can be written in the Q-parameterised form (4.44) for $N = 4$ as follows:

$$f(z) = [z^2 + 1]\{(z - 1)^2 - \alpha_{11}(z - 1) - \alpha_{21}\} - \alpha_{31}(z) + \alpha_{41} \tag{4.46}$$

where:

$$\alpha_{11} = -0.02,\ \alpha_{21} = 0.0099,\ \alpha_{31} = -0.0198,\ \alpha_{41} = 0.000199 \tag{4.47}$$

Note that the roots of:

$$f_\alpha(q) \triangleq q^4 + \alpha_1 q^3 + \alpha_2 q^2 + \alpha_3 q + \alpha_4 \tag{4.48}$$

with $\{\alpha_i = \alpha_{i1}\}$ are given by $\{-0.2928, 0.0101, 0.1313 \pm j0.2237\}$.

The delta operator parameterisation of $f(z)$ in (4.45) is given by:

$$f(z) = (z-1)^4 - \alpha_{12}(z-1)^3 - \alpha_{22}(z-1)^2 - \alpha_{32}(z-1) - \alpha_{42} \tag{4.49}$$

where:

$$\alpha_{12} = -2.02,\ \alpha_{22} = -2.0301,\ \alpha_{32} = -0.04,\ \alpha_{42} = 0.000199 \tag{4.50}$$

Now, the roots of $f_\alpha(q)$ in (4.48) with $\{\alpha_i = \alpha_{i2}\}$ are given by $\{-0.0243, 0.0041, -0.9999 \pm j0.9950\}$.

As is well known, the *delta operator* (where $q_i = 1$ for all i) maps roots of $f(z) = 0$ near $\mathrm{Re}\{z\} = \rho$ to roots of $f_\alpha(q) = 0$ near $\mathrm{Re}\{z\} = \rho - 1$. Unfortunately, while this operator is known to reduce the root sensitivity when *all* roots of $f(z)$ are near $z = +1$, it does not necessarily improve (and may decrease) the overall root sensitivity characteristics when one or more roots are *not* near $z = +1$. The following criteria is justified.

Criteria for Low Root Sensitivity. A Q-parameterisation (4.39), (4.43) has low coefficient sensitivity when all roots r_i of $f_\alpha(q) = 0$ where:

$$f_\alpha(q) \triangleq q^N + \sum_{i=1}^{N} \alpha_i q^{N-i}$$

are such that $|r_i| \approx 0$.

Based on this criteria, it follows that the sensitivity of a pair of roots $\{\rho^{\pm j\theta}\}$ of $f(z) = 0$ near the unit circle (*i.e.*, $|\rho| \approx 1$) may be improved using the Q-parameterisation (4.43) by choosing $\{q_1 \approx 2\cos\theta, q_2 = -1\}$. As we shall see in Section 4.3.2, extensions of the Q-parameterisation (4.43) are possible to accommodate the case when more than one pair of roots are near $|z| \approx 1$.

4.3.2 Q-Operator Canonical Form

The usefulness of Theorem 4.3 lies in the fact that any Nth-order (controller) transfer function $H_c(z)$ can be expressed in terms of $3N$ parameters: the $2N$

coefficients $\{\alpha_n, \beta_m\}$ and the N *Q-parameters* $\{q_i\}$ in the Q-parameterised form:

$$H_c(z) = b_c(z)/a_c(z) \tag{4.51}$$

where:

$$a_c(z) = \prod_{i=1}^{N}(z - q_i) - \sum_{n=1}^{N} \alpha_n \prod_{i=1}^{N-n} (z - q_i) \; ; \; b_c(z) = \sum_{n=0}^{N} \beta_n \prod_{i=1}^{N-n} (z - q_i) \tag{4.52}$$

Now, the Q-parameters $\{q_i\}$ can be used to either optimise the root sensitivities $\{S^b_{ij}; j \neq N\}$ of the zeros $\{q_i\}$ with respect to the coefficients $\{\beta_j\}$ or the root sensitivities $\{S^a_{k\ell}; \ell \neq N\}$ of the poles $\{p_k\}$ with respect to the coefficients $\{\alpha_\ell\}$ of $H_c(z)$. However, Q-parameters $\{q_i = q_i^*\}$ which optimise with respect to pole sensitivity will not normally also optimise with respect to zero sensitivity. Alternatively, one could seek *optimal* Q-parameters $\{q_i = q_i^*\}$ which optimise with respect to the *transfer function sensitivity*, in terms of the N largest values of the combined sensitivities $\{S^b_{ij}, S^a_{k\ell};\ j, \ell \neq N\}$.

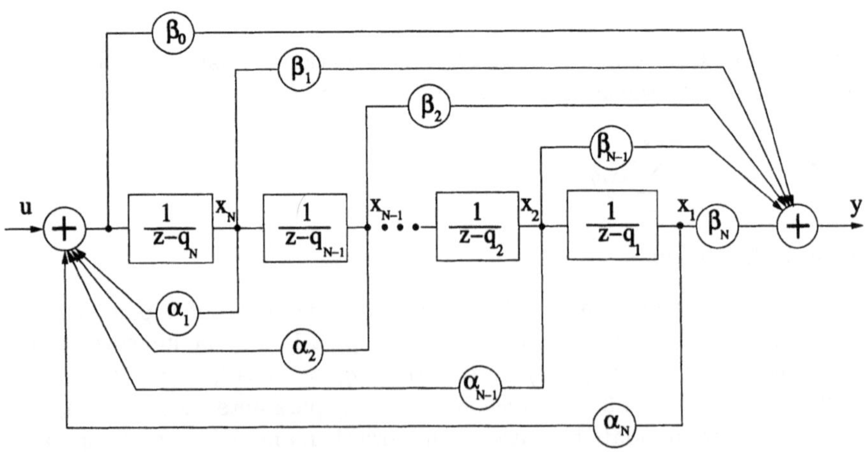

Fig. 4.1. Block-diagram of Q-parameterised structure

Recursive Representation. The block-diagram for the Q-parameterised structure (4.51), (4.52) is illustrated in Figure 4.1. It therefore follows that one recursive (state space) representation for the controller $H_c(z)$ in (4.51),

(4.52) for $q_N = 0$ is given by:

$$x_i(k+1) = x_{i+1}(k) + q_i x_i(k) \; ; \; 1 \le i \le N-1$$

$$x_N(k+1) = \sum_{j=1}^{N} \alpha_{N+1-j} x_j(k) + u(k)$$

$$y(k) = \beta_0 u(k) + \sum_{j=1}^{N} h_{N+1-j} x_j(k) \; ; \; h_j \triangleq \beta_j - \beta_0 \alpha_j \tag{4.53}$$

Equivalently, we can define the *Q-operator canonical form* $\{\boldsymbol{F} = \boldsymbol{\mathcal{F}}, \boldsymbol{g}, \boldsymbol{h}, d\}$ of the controller $H_c(z)$ by (4.7) with $d = \beta_0$, and:

$$\boldsymbol{\mathcal{F}} = \begin{bmatrix} q_1 & 1 & 0 & . & . & 0 \\ 0 & q_2 & 1 & . & . & 0 \\ 0 & 0 & q_3 & 1 & . & 0 \\ . & . & . & . & . & . \\ 0 & . & . & 0 & q_{N-1} & 1 \\ \alpha_N & \alpha_{N-1} & \alpha_{N-2} & . & . & \alpha_1 \end{bmatrix} ; \boldsymbol{g} = \begin{bmatrix} 0 \\ 0 \\ 0 \\ . \\ 0 \\ 1 \end{bmatrix} ; \boldsymbol{h} = \begin{bmatrix} h_N \\ h_{N-1} \\ h_{N-2} \\ . \\ h_2 \\ h_1 \end{bmatrix} \tag{4.54}$$

A more general Q-operator form for $\boldsymbol{\mathcal{F}}$ which can provide better eigenvalue sensitivity for a pair of complex eigenvalues near the roots of $z^2 - q_1 z - q_2 = 0$, is given by:

$$\boldsymbol{\mathcal{F}} = \begin{bmatrix} 0 & 1 & 0 & . & . & 0 \\ q_2 & q_1 & 1 & . & . & 0 \\ 0 & 0 & q_3 & 1 & . & 0 \\ . & . & . & . & . & . \\ 0 & . & . & 0 & q_{N-1} & 1 \\ \alpha_N & \alpha_{N-1} & \alpha_{N-2} & . & . & \alpha_1 \end{bmatrix} \tag{4.55}$$

If eigenvalue sensitivity is to be improved with respect to two pairs of complex eigenvalues near the roots $(z^2 - q_1 z - q_2)(z^2 - q_3 z - q_4) = 0$, then one should choose:

$$\boldsymbol{\mathcal{F}} = \begin{bmatrix} 0 & 1 & 0 & . & . & . & . & 0 \\ q_2 & q_1 & 1 & . & . & . & . & 0 \\ 0 & 0 & 0 & 1 & . & . & . & 0 \\ 0 & 0 & q_4 & q_3 & 1 & . & . & . \\ 0 & 0 & 0 & 0 & q_5 & 1 & . & . \\ . & . & . & . & . & . & . & . \\ 0 & . & . & . & . & 0 & q_{N-1} & 1 \\ \alpha_N & \alpha_{N-1} & \alpha_{N-2} & . & . & . & . & \alpha_1 \end{bmatrix} \tag{4.56}$$

More generally, the *Q-operator canonical form*, which enables improved eigenvalue sensitivity with respect to M pairs of complex eigenvalues, has M diagonal blocks in $\boldsymbol{\mathcal{F}}$ of the form:

$$\begin{bmatrix} 0 & 1 \\ q_{2m} & q_{2m-1} \end{bmatrix} ; \; m = 1, 2 \dots, M$$

In all cases, the ijth component $f_{i,j}$ of $\mathcal{F}$ satisfy:

$$\begin{aligned} f_{i,i+1} &= 1 \; ; \; 1 \le i \le N-1 \\ f_{i,j} &= 0 \; ; \; |i-j| \ge 2 \\ f_{N,j} &= \alpha_{N+1-j} \; ; \; 1 \le j \le N \end{aligned} \tag{4.57}$$

In order to reduce sensitivity with respect to the complex eigenvalue pair $\rho_m e^{\pm j\theta_m}$ for $|\rho_m| \approx 1$, one should choose:

$$q_{2m} = -1 \; ; \; q_{2m-1} \approx 2\cos\theta_m \tag{4.58}$$

Example 4.3. The Q-canonical form for $\mathcal{F}$ corresponding to the pole parameterisation in Example 4.2 of $f(z) = (z^2 + 0.99)(z - 0.99)^2$ as in (4.46) is $\mathcal{F} = \mathcal{F}_1$ while the Q-canonical form corresponding to (4.49) (which is the delta operator structure) is $\mathcal{F} = \mathcal{F}_2$, where:

$$\mathcal{F}_1 = \begin{bmatrix} 0 & 1 & 0 & 0 \\ -1 & 0 & 1 & 0 \\ 0 & 0 & 1 & 1 \\ \alpha_{41} & \alpha_{31} & \alpha_{21} & \alpha_{11} \end{bmatrix} \; ; \; \mathcal{F}_2 = \begin{bmatrix} 1 & 1 & 0 & 0 \\ 0 & 1 & 1 & 0 \\ 0 & 0 & 1 & 1 \\ \alpha_{42} & \alpha_{32} & \alpha_{22} & 1+\alpha_{12} \end{bmatrix}$$

where $\{\alpha_{j1}\}$ and $\{\alpha_{j2}\}$ are defined in (4.47) and (4.50) respectively.

Using MATLAB®, the eigenvalues of $\mathcal{F}_1$ are calculated to be $\{0.00000000000000 \pm j0.99498743710662, 0.99000000000000, 0.99000000000000\}$, while the eigenvalues of $\mathcal{F}_2$ are calculated to be $\{0.00010152788793 \pm j0.99498796294987, 0.97568400703570, 1.00411293718844\}$. While the MATLAB® eigenvalue routine is not the most accurate, these calculations nevertheless reveal the inferior root sensitivity characteristics (*i.e.*, result in an unstable pole) of the delta operator structure compared with the more general Q-operator structure.

When all the coefficients $\{\alpha_{ij}\}$ are rounded to 16 bits, the eigenvalues of $\mathcal{F}_1$ are calculated to be $\{0.00000018818502 \pm j0.99498592771790, 0.98939286748194, 0.99060248368709\}$, while for 12 bits, the eigenvalues are calculated to be $\{-0.00001121562265 \pm j0.99499357632833, 0.98000289999530, 0.99999999999999\}$.

4.3.3 Transfer Function Accuracy

The corresponding FWL representation $\hat{H}_c(z)$ of $H_c(z)$ in (4.51), (4.52) is given by:

$$\hat{H}_c(z) = \hat{b}_c(z)/\hat{a}_c(z) \tag{4.59}$$

where:

$$\hat{a}_c(z) = \prod_{i=1}^{N}(z-q_i) - \sum_{n=1}^{N}\hat{\alpha}_n \prod_{i=1}^{N-n}(z-q_i) \; ; \; \hat{b}_c(z) = \sum_{n=0}^{N}\hat{\beta}_n \prod_{i=1}^{N-n}(z-q_i) \tag{4.60}$$

FWL Choice for Q-Parameters. In any FWL representation, we shall assume that all the Q-parameters $\{q_i\}$ are *exactly* represented in the available word-length. As a consequence, restrictions are then imposed on possible quantised values for q_i. Specifically, in an $[1, t_2]$ fixed-point representation, q_i is then restricted to fractional values of the form $q_i = \pm\ell_i 2^{-(t_2-1)}$ for some integer ℓ_i, while the special cases, $q_i = \pm 1$ can be realised as an addition/subtraction.

A special case of *FWL* Q-parameter selection is defined when all Q-parameters $\{\overline{q}_i\}$ are restricted to the *integer values* $\{0, \pm 1\}$. This case is useful since the "multiplication" $\overline{q}_i * x$ is then either ignored, or replaced by an addition/subtraction operation. (Examples 4.2 and 4.3 have highlighted the usefulness of using integer values.)

When it comes to *non-integral values*, it should be noted that the *optimal FWL* Q-parameters $\{\overline{q}_i\}$ that optimise sensitivity are *not necessarily* given by quantising the optimal Q-parameters $\{q_i^*\}$. The optimal solution is equivalent to solving a problem of *integer optimisation* for which there is no analytical solution. It also can be advantageous to restrict $\overline{q}_i$ to be of the form $\overline{q}_i = 2^{\overline{p}_i}$ for some integer $\overline{p}_i$ since then the "multiplication" $\overline{q}_i * x$ can be implemented simply by means of a shift operation on x.

Suboptimal Selection to Reduce Pole Sensitivity. With reference to the examples in (4.54)-(4.56), one suboptimal procedure is to choose the Q-parameters $\{q_i\}$ of the Nth-order controller $H_c(z)$ as follows:

1. Arrange the N poles in terms of their decreasing proximity to the unit circle $|z| = 1$.
2. Choose the coefficients $\{q_{2m}, q_{2m-1}\}$ as in (4.58) for complex poles $\{\rho_m e^{\pm\iota\theta_m}\}$ and $q_n = Q[\lambda_n]$ for real poles $\{\lambda_n\}$.

Example 4.4. Suppose the poles of $H_c(z)$ are arranged in decreasing proximity according to $\{0.99e^{\pm j46^o}, 0.97, 0.95^{\pm j62^o}, 0.7\}$. Then since $\{2\cos 46^o \approx \sqrt{2}, 2\cos 62^o \approx 1\}$, select:

$$\mathcal{F} = \begin{bmatrix} 0 & 1 & 0 & 0 & 0 & 0 \\ -1 & q_1 & 1 & 0 & 0 & 0 \\ 0 & 0 & 1 & 1 & 0 & 0 \\ 0 & 0 & 0 & 0 & 1 & 0 \\ 0 & 0 & 0 & -1 & 1 & 1 \\ \alpha_6 & \alpha_5 & \alpha_4 & \alpha_3 & \alpha_2 & \alpha_1 \end{bmatrix}$$

(One could also replace α_1 by $1+\tilde{\alpha}_1$.) If (say) 16-bit arithmetic is used, choose $q_1 = Q[\sqrt{2}] = 1.41421508789063$. Alternatively, one could choose $q_1 = 1.5$ and replace the "multiplication" $q_1 * x$ by $x + 2^{-1} * x$.

We have the following extension of Theorem 4.1.

Theorem 4.4. *Suppose the finite coefficient word-length controller $\hat{H}_c(z)$ as given by (4.59), (4.60) is used to approximate the ideal controller $H_c(z)$ in (4.51). Then the z-transform $\mathcal{Z}\{E_y\}$ of the resulting controller output error E_y due to controller coefficient error is given in terms of the error transfer function $G_{E,\boldsymbol{q}}(z)$ and the controller input z-transform $\mathcal{Z}\{u\}$ by:*

$$\mathcal{Z}\{E_y\} = \mathcal{G}_{E,\boldsymbol{q}}(z)\mathcal{Z}\{u\} \tag{4.61}$$

where:

$$\mathcal{G}_{E,\boldsymbol{q}}(z) = \frac{[\sum_{n=1}^{N}(\Delta\alpha_n)\prod_{i=1}^{N-n}(z-q_i)]\hat{H}_c(z) + \sum_{n=0}^{N}(\Delta\beta_n)\prod_{i=1}^{N-n}(z-q_i)}{a_c(z)} \tag{4.62}$$

Note that: (i) $\mathcal{G}_{E,\boldsymbol{q}}(z)$ reduces to $\mathcal{G}_E$ in (4.24) when $q_i = 0$ for all i, and (ii) the coefficient errors $\{\Delta\alpha_n, \Delta\beta_n\}$ are functions of $\{q_i\}$ and $\{|\Delta\alpha_n| \to 0;\ 1 \le n \le N-1\}$ as $\{q_i \to \lambda_i;\ 1 \le i \le N-1\}$ for all poles λ_i of $H_c(z)$.

Given the N poles (*i.e.*, roots of $a_c(z) = 0$) and the N zeros (*i.e.*, roots of $b_c(z) = 0$) as in (4.52), one possible optimal solution to minimise *controller transfer function sensitivity* is to choose the Q-parameters $\boldsymbol{q}^* = [q_i^*]$ such that:

$$\boldsymbol{q}^* \triangleq \arg\min_{q_i \in \mathcal{S}} \|\sum_{n=1}^{N}(\Delta\alpha_n)\prod_{i=1}^{N-n}(z-q_i)]\hat{H}_c(z) + \sum_{n=0}^{N}(\Delta\beta_n)\prod_{i=1}^{N-n}(z-q_i)\|$$

where $\mathcal{S}$ is the set of restricted values for the $\{q_i\}$. Possible examples are: $S = \mathcal{Z}$ (the set of integers) and $S = \{k \times 0.5;\ k \in \mathcal{Z}\}$.

4.4 Dynamically Scaled Controllers

In this section, a dynamically scaled recursive implementation of the digital controller (4.4) is first derived based on a time-varying state space representation. (This derivation is presented in [15].) We then go on to show how dynamic scaling can be used to eliminate overflow in a fixed-point implementation. It has already been shown [10] how dynamic scaling can both eliminate overflow and improve roundoff quantisation noise performance in an implementation of a direct form structure (4.8). Recent work [24] has extended these results and confirmed earlier conclusions that dynamically scaled fixed-point (or *block floating-point*) arithmetic is an excellent alternative for recursive filter or controller implementation.

To begin, define the invertible time-varying *state scaling matrix* $\boldsymbol{S}(k)$, the *incremental state scaling matrix* $\boldsymbol{\Gamma}(k)$ and the new state $\boldsymbol{w}(k)$ by:

$$\boldsymbol{w}(k) \triangleq \boldsymbol{S}(k)\boldsymbol{x}(k)\ ;\ \boldsymbol{S}(k+1) \triangleq \boldsymbol{\Gamma}(k+1)\boldsymbol{S}(k) \tag{4.63}$$

Then the controller state space representation (4.7) can be written in terms of an input/output scaling factor $s_{in}(k) \neq 0$ in the form:

$$\begin{aligned} y(k) &= s_{in}(k)[du(k) + \boldsymbol{h}^T \boldsymbol{S}^{-1}(k)\boldsymbol{w}(k)] \\ \boldsymbol{p}(k) &= \boldsymbol{S}(k)\boldsymbol{F}\boldsymbol{S}^{-1}(k)\boldsymbol{w}(k) + \boldsymbol{S}(k)\boldsymbol{g}s_{in}^{-1}(k)u(k) \\ \boldsymbol{w}(k+1) &= \boldsymbol{\Gamma}(k+1)\boldsymbol{p}(k) \\ \boldsymbol{S}(k+1) &= \boldsymbol{\Gamma}(k+1)\boldsymbol{S}(k) \end{aligned} \tag{4.64}$$

In particular, suppose $\boldsymbol{S}(k)$ and $\boldsymbol{\Gamma}(k)$ are both diagonal matrices whose diagonal elements are *integer powers of two*; that is, suppose:

$$\begin{aligned} \boldsymbol{S}(k) &= \text{diag}\{2^{\ell_1(k)}, 2^{\ell_2(k)}, \ \dots \ , 2^{\ell_N(k)}\} \\ \boldsymbol{\Gamma}(k) &= \text{diag}\{2^{m_1(k)}, 2^{m_2(k)}, \ \dots \ , 2^{m_N(k)}\} \\ s_{in}(k) &= 2^{\ell_N(k)} \end{aligned} \tag{4.65}$$

for some time-varying *integers* $\{\ell_i(k), m_i(k)\}$. Then from (4.64) and (4.65), the Q-operator canonical form (4.54) can be expressed as:

$$\begin{aligned} y(k) &= \beta_0 2^{\ell_N(k)} u(k) + \sum_{j=1}^{N} 2^{\ell_N(k)-\ell_j(k)} h_{N-j+1} w_j(k) \\ p_i(k) &= q_i w_i(k) + 2^{\ell_i(k)-\ell_{i+1}(k)} w_{i+1}(k) \ ; \ 1 \le i \le N-1 \\ p_N(k) &= \sum_{j=1}^{N} 2^{\ell_N(k)-\ell_j(k)} \alpha_{N-j+1} w_j(k) + u(k) \\ w_i(k+1) &= 2^{m_i(k+1)} p_i(k) \ ; \ 1 \le i \le N \\ \ell_i(k+1) &= m_i(k+1) + \ell_i(k) \ ; \ 1 \le i \le N \end{aligned} \tag{4.66}$$

A fixed-point implementation of (4.66) is given by:

$$\begin{aligned} \hat{y}(k) &= \hat{\beta}_0 2^{\ell_N(k)} u(k) + \sum_{j=1}^{N} 2^{\ell_N(k)-\ell_j(k)} \hat{h}_{N-j+1} Q[\hat{w}_j(k)] \\ \hat{p}_i(k) &= q_i Q[\hat{w}_i(k)] + 2^{\ell_i(k)-\ell_{i+1}(k)} Q[\hat{w}_{i+1}(k)] \ ; \ 1 \le i \le N-1 \\ \hat{p}_N(k) &= \sum_{j=1}^{N} 2^{\ell_N(k)-\ell_j(k)} \hat{\alpha}_{N-j+1} Q[\hat{w}_j(k)] + u(k) \\ \hat{w}_i(k+1) &= 2^{m_i(k+1)} \hat{p}_i(k) \ ; \ 1 \le i \le N \\ \ell_i(k+1) &= m_i(k+1) + \ell_i(k) \ ; \ 1 \le i \le N \\ Q[\hat{w}_j(k)] &= \hat{w}_j(k) - \epsilon_j(k) \end{aligned} \tag{4.67}$$

where:

$$Q[\hat{w}_j(k)] \approx [1, t_1] \ ; \ \hat{\alpha}_i \approx [1 + p_2, t_2]$$

Consequently, the magnitude of the roundoff quantisation error components $\{\epsilon_j(k)\}$ are bounded by $2^{-(t_1+1)}$. Note that once again, we assume that all the Q-parameters $\{q_i\}$ are selected so as to be exactly represented within the available coefficient precision $[1+p_2, t_2]$.

Given the initial scale parameters, $\{\ell_i(0); 1 \le i \le N\}$ in (4.67), the values $\{\ell_i(k); k \ge 1\}$ are uniquely determined by the incremental scaling integers $\{m_i(k); k \ge 1\}$. We now propose the following dynamic scaling algorithm for selecting the integers $\{m_i(k); k \ge 1\}$.

Dynamically Scaled Algorithm. For $k = 0, 1, 2, ...$:

1. Given the values $\{\ell_i(k), Q[\hat{w}_i(k)]\}$, compute the controller output $\hat{y}(k)$ by means of: (a) integer subtractions to determine $\{\ell_N(k) - \ell_i(k);\ 1 \le i \le N\}$, (b) multiplications $\hat{h}_{N-j+1}Q[\hat{w}_j(k)]$, and (c) shift operations on $\hat{h}_{N-j+1}Q[\hat{w}_j(k)]$ and additions to determine the inner product $\sum_{j=1}^{N} 2^{\ell_N(k)-\ell_i(k)}\hat{h}_{N-j+1}Q[\hat{w}_j(k)]$.
2. Compute $\hat{p}_i(k)$ as follows: (a) for $1 \le i \le N-1$, first compute $q_i Q[\hat{w}_i(k)]$, then use integer subtractions to determine $\{\ell_i(k) - \ell_{i+1}(k);\ 1 \le i \le N-1\}$ and then use a shift operation on $Q[\hat{w}_{i+1}(k)]$ to determine $2^{\ell_i(k)-\ell_{i+1}(k)}$ $Q[\hat{w}_{i+1}(k)]$, and (b) for $i = N$, complete multiplications $\hat{\alpha}_{N-j+1}Q[\hat{w}_j(k)]$, shift operations on $\hat{\alpha}_{N-j+1}Q[\hat{w}_j(k)]$ and additions to determine the inner product $\sum_{j=1}^{N} 2^{\ell_N(k)-\ell_i(k)}\hat{\alpha}_{N-j+1}Q[\hat{w}_j(k)]$.
3. For each $\hat{p}_i(k)$, find the integer $m_i(k+1)$ to guarantee $|\hat{w}_i(k+1)| < 1$ and compute $\hat{w}_i(k+1)$ as follows: for $q_i \ne 0$:

$$m_i(k+1) = \begin{cases} 2 & , \ |\hat{p}_i(k)| < 0.5 \\ 2^{m_i(k+1)}|\hat{p}_i(k)| < 1 < 2^{m_i(k+1)+1}|\hat{p}_i(k)| & , \ |\hat{p}_i(k)| \ge 0.5 \end{cases}$$

$$\hat{w}_i(k+1) = 2^{m_i(k+1)}\hat{p}_i(k)$$

   and for $q_i = 0$:

$$m_i(k+1) = 0 \ ; \hat{w}_i(k+1) = \hat{w}_{i+1}(k)$$

   (This step guarantees $m_i(k+1) \le 2$ for all i, k. Furthermore, when $q_i = 0$, then $\ell_i(k+1) = \ell_i(k)$ for all k.)
4. Given the integers $\{m_i(k+1), \ell_i(k)\}$, compute $\ell_i(k+1)$ by the integer addition $\ell_i(k+1) = m_i(k+1) + \ell_i(k)$.

In the case when $\{q_i = 0;\ 1 \le i \le N-1\}$, we have $\{\hat{\alpha}_i = \hat{a}_i;\ 1 \le i \le N-1\}$ and $h_{N+1-j} = b_j - b_0 a_i$ where $\{a_n, b_m\}$ are the original controller parameters of $H_c(z)$ in (4.5), (4.6). The FWL implementation (4.67) simplifies as follows:

$$\begin{aligned}
\hat{y}(k) &= b_0 u(k) + 2^{\ell_N(k)} \sum_{j=1}^{N} \hat{h}_{N-j+1} Q[\hat{w}_j(k)] \\
\hat{w}_i(k+1) &= \hat{w}_{i+1}(k) \ ; \ 1 \le i \le N-1 \\
\hat{p}_N(k) &= 2^{\ell_N(k)} \sum_{j=1}^{N} \hat{a}_{N-j+1} Q[\hat{w}_j(k)] + u(k) \\
\hat{w}_N(k+1) &= 2^{m_N(k+1)} \hat{p}_N(k) \\
\ell_N(k+1) &= m_N(k+1) + \ell_N(k)
\end{aligned} \tag{4.68}$$

where the scaling parameter $m_N(k+1)$ is determined by:

$$m_N(k+1) = \begin{cases} 2 & , \ |\hat{p}_N(k)| < 0.5 \\ 2^{m_N(k+1)} |\hat{p}_N(k)| < 1 < 2^{m_N(k+1)+1} |\hat{p}_N(k)| & , \ |\hat{p}_N(k)| \ge 0.5 \end{cases}$$

Overflow-free Behaviour. Without any loss of generality, we may assume that analogue scaling associated with the controller ADC guarantees $|u(k)| \le 1$ for all k. (This in fact was the reason for choosing $s_{in}(k)$ as in (4.65).) Also (for simplicity), suppose that the Q-parameters $\{q_i\}$ and the controller coefficients $\{\hat{\alpha}_i\}$ for all i are bounded by unity, that is, we assume:

$$|q_i| \le 1 \ ; \ |\hat{\alpha}_i| \le 1 \tag{4.69}$$

Then $|Q[\hat{w}_j(k)]| < 1$ implies $\{|\hat{q}_j Q[\hat{w}_j(k)]| < 1 \ ; \ |\hat{\alpha}_{N-j+1} Q[\hat{w}_j(k)]| < 1\}$, which in turn implies:

$$\begin{aligned}
|\hat{p}_i(k)| &< 1 + 2^{\ell_i(k)-\ell_{i+1}(k)} \ ; \ 1 \le i \le N-1 \\
|\hat{p}_N(k)| &< 1 + \sum_{j=1}^{N} 2^{\ell_N(k)-\ell_j(k)}
\end{aligned}$$

Note that *any* bounds on $\{|q_i|, |\alpha_i|\}$ will result in some bound on $\{|\hat{p}_i|\}$. However, it often turns out that many low sensitivity Q-parameterised structures have property (4.69).

Theorem 4.5. *Consider a stable algorithm (4.53) defined for $k \ge 0$. Suppose $|u(k)| < 1$, and define an integer L such that:*

$$2^{L-1} < \max_{i,k \ge 0} |x_i(k)| \le 2^L \tag{4.70}$$

Then in the dynamically scaled fixed-point algorithm (4.67), the $\{|p_i(k)|\}$ are bounded as follows:

$$|\hat{p}_i(k)| < \begin{cases} 2^{2L} + 1 \ , \ 1 \le i \le N-1 \\ N.2^{2L} + 1 \ , \ i = N \end{cases}$$

Consequently, overflow-free behaviour is guaranteed (i.e., $|Q[\hat{w}_i(k)| < 1$ *for all* i, k*) in the dynamically scaled algorithm provided:* $\ell_i \approx [1 + L, 0]$ *and:*

$$\hat{p}_i(k) \approx [1 + 2L + 1, t_p] \text{ and } \hat{p}_N(k) \approx [1 + 2L + 1 + M, t_p]$$

where the integer M *satisfies* $N \leq 2^M < N + 1$.

Note that this result says nothing about whether or not the output will overflow since this outcome will depend on the available word-length for $\hat{y}(k)$. However an "occasional" saturation of $\hat{y}(k)$ is unimportant compared with the elimination of overflow in the recursive component of the algorithm.

Floating-point Interpretation. As can be seen from (4.63) and (4.65), the consequence of the dynamic scaling algorithm is to select integers $\{\ell_j(k)\}$ to re-scale the state $x_j(k)$ in (4.53) at each time instant k so that the scaled state $w_j(k) = 2^{\ell_j(k)} x_j(k)$ satisfies the constraint $|w_j(k)| < 1$. The fraction $w_j(k)$ together with the integer $\ell_j(k)$ provide a "block floating" (or "dynamic fixed") point representation $2^{-\ell_j(k)} w_j(k)$ of $x_j(k)$.

The *scaled roundoff quantisation error* $\tilde{\epsilon}_j(k)$ where:

$$\tilde{\epsilon}_j(k) \triangleq 2^{-\ell_i(k)} \epsilon_i(k) \; ; \; \epsilon_i(k) = \hat{w}_j(k) - Q[\hat{w}_j(k)] \tag{4.71}$$

can also be likened to a floating-point quantisation $FL[\hat{x}_j(k)]$ of $\hat{x}_j(k)$; that is, as in (4.1):

$$\hat{x}_j(k) - FL[\hat{x}_j(k)] = Q_{RO}[\hat{w}_j(k)] \times 2^{-\ell_j(k)} = \tilde{\epsilon}_j(k)$$

A *roundoff noise model*, which can be used for predicting the roundoff noise performance of dynamically scaled fixed-point algorithms, has already been proposed and justified by extensive numerical simulation [10]. Analysis is based on the following model (where $\xi\{\cdot\}$ denotes mathematical expectation):

$$\begin{aligned} \xi\{\tilde{\epsilon}_j(k)\} &= 0 \\ \xi\{\tilde{\epsilon}_j(k)\tilde{\epsilon}_m(k)\} &= \begin{cases} 0 & ; \; j \neq m \\ \kappa_m \frac{1}{12} 2^{-2t_1} & ; \; j = m \end{cases} \end{aligned} \tag{4.72}$$

for some constant κ_m.

We also note that since $|x_i(k)|$ can be bounded according to (4.70), then a *constant scaled* fixed-point implementation can also be implemented with $x_i(k) \approx [1 + 2L + 1, t_p]$. The problem with constant scaling is that since the multiplier length is fixed, the fractional word-length t_p then needs to be decreased as L is increased, thereby leading to a loss in accuracy. The dynamic scaling algorithm has been shown [10] to improve the accuracy of a (constant scaled) fixed-point implementation but at the expense of extra computation and extra memory to store the scaling coefficients $\{\ell_i(k)\}$.

DSP56xxx Implementation. Since this DSP chip has a (dual) 24×24 multiplication and 56-bit accumulator, then we can take $Q[\hat{w}_i], \hat{q}_i, \alpha_i \approx [1, 24]$ and $\hat{w}_i \approx [1, 48]$. The following implementations are then possible:

(i) When $L = 4$ and $N = 12$, then $\{\hat{p}_i \approx [1+8, 47]\ ,\ 1 \leq i \leq 11\ ;\ \hat{p}_{12} \approx [1+8+4, 43]\}$, and (ii) when $L = 6$ and $N = 30$, then $\{\hat{p}_i \approx [1+12, 43]\ ,\ 1 \leq i \leq 29 \quad \hat{p}_{30} \approx [1+12+5, 38]\}$.

4.4.1 Numerical Accuracy with Signal Quantisation

As in (4.67), suppose there are now *no coefficient errors*; that is, $\{\hat{\alpha}_n = \alpha_n, \hat{\beta}_n = \beta_n\}$ for all n, and define the controller state numerical error $\tilde{E}_{w_i}$, the roundoff error $\tilde{\epsilon}_i$ and output numerical error $\tilde{E}_y$ by:

$$\begin{aligned} \tilde{E}_{w_i}(k) &\triangleq 2^{-\ell_i(k)}[w_i(k) - \hat{w}_i(k)]\ ;\ \tilde{\epsilon}_i(k) \triangleq 2^{-\ell_i(k)}\epsilon_i(k) \\ \tilde{E}_y(k) &\triangleq 2^{-\ell_N(k)}[y(k) - \hat{y}(k)] \end{aligned} \tag{4.73}$$

Then it follows from (4.66) and (4.67) that:

$$\begin{aligned} \tilde{E}_{w_i}(k+1) &= q_i\tilde{E}_{w_i}(k) + \tilde{E}_{w_{i+1}}(k) + q_i\tilde{\epsilon}_i(k) + \tilde{\epsilon}_{i+1}(k)\ ;\ 1 \leq i \leq N-1 \\ \tilde{E}_{w_N}(k+1) &= \sum_{j=1}^{N} \alpha_{N+1-j}\tilde{E}_{w_j}(k) + \sum_{j=1}^{N} \alpha_{N+1-j}\tilde{\epsilon}_j(k) \\ \tilde{E}_y(k) &= \sum_{j=1}^{N} h_{N+1-j}\tilde{E}_{w_j}(k) + \sum_{j=1}^{N} h_{N+1-j}\tilde{\epsilon}_j(k) \end{aligned} \tag{4.74}$$

These error equations can be written in the state space form:

$$\begin{aligned} \tilde{\boldsymbol{E}}(k+1) &= \boldsymbol{\mathcal{F}}\tilde{\boldsymbol{E}}(k) + \boldsymbol{\mathcal{F}}\tilde{\boldsymbol{\epsilon}}(k) \\ \tilde{E}_y(k) &= \boldsymbol{h}^T\tilde{\boldsymbol{E}}(k) + \boldsymbol{h}^T\tilde{\boldsymbol{\epsilon}}(k) \end{aligned} \tag{4.75}$$

where:

$$\tilde{\boldsymbol{E}}^T(k) = [\tilde{E}_{w_1}(k), \tilde{E}_{w_2}(k),\ \dots\ , \tilde{E}_{w_N}(k)]\ ;\ \tilde{\boldsymbol{\epsilon}}^T(k) = [\tilde{\epsilon}_1(k), \tilde{\epsilon}_2(k),\ \dots\ , \tilde{\epsilon}_N(k)]$$

with $\{\boldsymbol{\mathcal{F}}, \boldsymbol{h}\}$ given by (4.54).

Roundoff Noise Improvement. Modifications to the algorithm (4.67) can be made to further improve the roundoff noise performance. Specifically: (i) replace each term $q_iQ[\hat{w}_i(k)]$ by the term:

$$\delta_i\hat{w}_i(k) + \tilde{q}_iQ[w_i(k)] \tag{4.76}$$

where:

$$(\delta_i, \tilde{q}_i) = \begin{cases} (-1, q_i + 1)\ ; & -1.5 \leq q_i \leq -0.5 \\ (0, q_i)\ ; & -0.5 < q_i < 0.5 \\ (1, q_i - 1)\ ; & 0.5 \leq q_i < 1.5 \end{cases} \tag{4.77}$$

and (ii) replace each term $\alpha_{N+1-i}Q[\hat{w}_i(k)]$ by the term:

$$\gamma_{N+1-i}\hat{w}_i(k) + \tilde{\alpha}_{N+1-i}Q[\hat{w}_j(k)] \tag{4.78}$$

where:

$$(\gamma_j, \tilde{\alpha}_j) = \begin{cases} (-1, \alpha_j + 1) \ ; & -1.5 \le \alpha_j \le -0.5 \\ (0, \alpha_j) \ ; & -0.5 < \alpha_j < 0.5 \\ (1, \alpha_j - 1) \ ; & 0.5 \le \alpha_j < 1.5 \end{cases} \tag{4.79}$$

Note that: (i) each of the modifications (4.76), (4.78) replaces one multiplication by one multiplication plus one addition so has little effect on the overall complexity of the algorithm, and (ii) $\{|\tilde{q}_i| \le 0.5, |\tilde{\alpha}_j| \le 0.5\}$ with:

$$|\tilde{q}_i| \le |q_i| \ ; \ |\tilde{\alpha}_j| \le |\alpha_j| \tag{4.80}$$

Note also that (for example) when $0.5 < q_i < 1.5$ in (4.77) so that $\{\delta_i = 1, \tilde{q}_i = 1 - q_i\}$, then:

$$\delta_i\hat{w}_i(k) + \tilde{q}_iQ[\hat{w}_i(k)] = q_iQ[w_i(k)] + \epsilon_i(k)$$

where $\epsilon_i(k) = \hat{w}_i(k) - Q[\hat{w}_i(k)]$ is the roundoff quantisation error. These modifications (which are frequently used in signal processing to improve numerical accuracy) are known in the signal processing literature as either *error spectrum shaping* [5], *error feedback* or *residue feedback* [11].

After incorporating these modifications, it follows that the error equations (4.75) now become:

$$\begin{aligned} \tilde{\boldsymbol{E}}(k+1) &= \boldsymbol{\mathcal{F}}\tilde{\boldsymbol{E}}(k) + \tilde{\boldsymbol{\mathcal{F}}}\tilde{\boldsymbol{\epsilon}}(k) \\ \tilde{E}_y(k) &= \boldsymbol{h}^T\tilde{\boldsymbol{E}}(k) + \boldsymbol{h}^T\tilde{\boldsymbol{\epsilon}}(k) \end{aligned} \tag{4.81}$$

where:

$$\tilde{\boldsymbol{\mathcal{F}}} = \begin{bmatrix} \tilde{q}_1 & 1 & 0 & . & . & 0 \\ 0 & \tilde{q}_2 & 1 & . & . & 0 \\ 0 & 0 & \tilde{q}_3 & 1 & . & 0 \\ . & . & . & . & . & . \\ 0 & . & . & 0 & \tilde{q}_{N-1} & 1 \\ \tilde{\alpha}_N & \tilde{\alpha}_{N-1} & \tilde{\alpha}_{N-2} & . & . & \tilde{\alpha}_1 \end{bmatrix}$$

The roundoff noise improvement as evident in the change from (4.75) to (4.81) follows as a consequence of (4.80).

A final modification to both improve accuracy and simplify the implementation is to replace each term $2^{\ell_i(k)-\ell_{i+1}(k)}Q[\hat{w}_{i+1}(k)]$ in (4.67) by $2^{\ell_i(k)-\ell_{i+1}(k)}\ \hat{w}_{i+1}(k)$ when $\ell_i(k) \ge \ell_{i+1}(k)$ and by $Q[2^{\ell_i(k)-\ell_{i+1}(k)}\hat{w}_{i+1}(k)]$ when $\ell_i(k) < \ell_{i+1}(k)$ where this last quantisation rounds the shifted product $2^{\ell_i(k)-\ell_{i+1}(k)}\ \hat{w}_{i+1}(k)$ to t_1 fractional bits.

References

1. A. Fettweis, "Pseudopassivity, sensitivity, and stability of wave digital filters", *IEEE Trans. on Circuits Theory*, vol. CT-19, pp. 668-673, Nov. 1972.
2. R.E. Crochiere, "A new statistical approach to the coefficient wordlength problem for digital filters", *IEEE Trans. on Circuits & Syst.*, vol. CAS-22, pp. 190-196, Mar. 1975.
3. A.G. Agarwal and C.S. Burrus, "New recursive digital filter structures having very low sensitivity and roundoff noise", *IEEE Trans. on Circuits & Syst.*, vol. CAS-22, pp. 921-927, Dec. 1975.
4. C.T. Mullis and R.A. Roberts, "Synthesis of minimum roundoff noise fixed point digital filters", *IEEE Trans. on Circuits & Syst.*, vol. CAS-23, pp. 551-562, Sept. 1976.
5. D.C. Munson, Jr. and B. Liu, "Narrow-band recursive filters with error spectrum shaping", *IEEE Trans. on Circuits & Syst.*, vol. CAS-28, pp. 160-163, Feb. 1981.
6. G. Orlandi and G. Martinelli, "Low-sensitivity recursive digital filter structures having very low sensitivity and roundoff noise", *IEEE Trans. Circuits & Syst.*, vol. CAS-31, pp. 654-656, July 1984.
7. D. Williamson and S. Sridharan, "An approach to coefficient wordlength reduction in digital filters", *IEEE Trans. on Circuits & Syst.*, vol. CAS-32, no. 9, pp. 893-903, Sept. 1985.
8. D. Williamson, "Finite wordlength design of digital Kalman filters for state estimation", *IEEE Trans. on Auto. Control*, vol. AC-30, no. 10, pp. 930-939, Oct. 1985.
9. L. Thiele, "On the sensitivity of linear state space systems", *IEEE Trans. on Circuits & Syst.*, vol. CAS-33, pp. 502-510, May 1986.
10. S. Sridharan and D. Williamson, "Implementation of high order direct form digital filter structures", *IEEE Trans. on Circuits & Syst.*, vol. CAS-32, pp. 816-822, Aug. 1986.
11. D. Williamson, "Roundoff noise minimization and pole-zero sensitivity in fixed-point digital filters using residue feedback", *IEEE Trans. on Acoustics, Speech & Sig. Proc.*, vol. ASSP-34, pp. 1210-1220, Oct. 1986.
12. D. Williamson, "Delay replacement in direct form structures", *IEEE Trans. on Acoustics, Speech & Sig. Proc.*, vol. ASSP-36, pp. 453-460, April 1988.
13. D. Williamson and K. Kadiman, "Optimal finite wordlength linear quadratic regulation", *IEEE Trans. on Auto. Control*, vol. AC-34, no. 12, pp. 1218-1228, Dec 1989.
14. R.H. Middleton and G.C. Goodwin, *Digital Control and Estimation - A Unified Approach*, Prentice-Hall, Englewood Cliffs, NJ, 1990.
15. D. Williamson, *Digital Control and Implementation*, Prentice-Hall, Englewood Cliffs, NJ, 1991.
16. G.C. Goodwin, R.H. Middleton and H.V. Poor, "High speed digital signal processing and control", *Proc. IEEE*, vol. 80, pp. 240-259, Feb. 1992.
17. K. Liu, R. Skelton and K. Grigoriadis, "Optimal controllers for finite wordlength implementation," *IEEE Trans. on Auto. Control*, vol. AC-37, no. 9, pp. 1294-1304, 1992.
18. M. Gevers and G. Li, *Parametrizations in Control, Estimation and Filtering Problems: Accuracy Aspects*, Springer-Verlag, Berlin, 1993.

19. D. Williamson, "Sensitivity improvements using Q-operator representations", *Proc. 32nd IEEE Conf. Decision and Contr.*, San Antonio, TX. Dec. 1993.
20. M.A. Rotea and D. Williamson, "Optimal realizations of finite wordlength digital filters and controllers", *IEEE Trans. on Circuits & Syst.*, vol. CAS-42, pp. 61-72, Feb. 1995.
21. Bhaskar D. Rao, "Roundoff noise in floating point digital filters", *Control and Dynamic Systems*, vol. 75, pp. 79-103, 1996.
22. D. Williamson, *Discrete Time Signal Processing*, Springer, London, 1999.
23. A.D. Back, B.G. Horne and A. H. Tsoi, "Alternate discrete-time operators: an algorithm for optimal selection of parameters", *IEEE Trans. on Signal Processing*, vol. 47, pp. 2612-2615, Sept. 1999.
24. K.R. Ralev and P.H. Bauer, "Realization of block floating point digital filters and applications to block implementations", *IEEE Trans. on Signal Processing*, vol. 47, pp. 1076-1086, Sept. 1999.

CHAPTER 5

CONVEXITY AND DIAGONAL STABILITY: AN LMI APPROACH TO DIGITAL FILTER IMPLEMENTATION

Stéphane Dussy

Abstract. Implementation of a filter in a fixed-point digital hardware results in small errors during each step of the algorithm, induced by the inexact arithmetic structure of the real system (e.g., finite word-length effects). These errors can lead to unexpected behaviour since stability and performance are usually ensured for an idealised representation of the system, without taking into account the erratic effects of quantisation. This chapter addresses this finite-precision issue by proposing a convex formulation of the robust diagonal stability problem. This approach guarantees robust stability for a wide class of quantised systems, with direct application to the design of robust digital filters.

5.1 Introduction

5.1.1 Outline of the Approach

Most of the results in advanced control theory ignore the numerical issue raised by the implementation of controllers or filters in real hardware. Yet, during the last two decades, an extensive literature has studied the effects of finite word-length (FWL) or quantisation in digital control and signal processing. One of the most challenging issues resulting from this so-called finite-precision problem is the design of fixed-point arithmetic digital filters, with performance guarantee. In this chapter, the finite-precision problem is addressed in three steps. First, the concept of diagonal stability is defined and some linear matrix inequalities are derived in Section 5.2 for both the stability analysis and the closed-loop diagonal stabilisation problems; these convex formulations constitute the core results of the chapter. Then, in Section 5.3, the finite-precision problem is reviewed in more detail and is connected to the concept of diagonal stability. Finally, an illustrative example is provided with the design and the implementation of a robust digital filter.

5.1.2 Review of the Diagonal Stability Concept

Consider a parameter-dependent discrete time system of the form:

$$x(k+1) = A(p)x(k) \tag{5.1}$$

where p is the parameter vector. System (5.1) is said to be diagonally stable if there exists $P = P^T > 0$, with P diagonal, such that, for every admissible p:

$$A(p)^T PA(p) - P < 0 \tag{5.2}$$

In general, except for special structures of $A(p)$, the resolution of (5.2) is NP-hard [19]. NP-hard problems define a class that encompasses most of the untractable problems (in terms of required CPU time or memory size). This means that the computational complexity of the exact resolution of the problem is very high [2]. The purpose of the chapter is to propose a tractable solution to the robust diagonal stability problem for a large class of systems described by (5.1). This is achieved by deriving some equivalent conditions in the form of Linear Matrix Inequalities (LMI), *i.e.*, constraints that can be written as:

$$F(\xi) = F_0 + \sum_{i=1}^{m} \xi_i F_i > 0$$

where $\xi \in \mathbb{R}^m$ is the variable and $F_i = F_i^T \in \mathbb{R}^{m \times m}$, $i = 1, \dots m$ are given positive-definite matrices. Some LMI solvers can then provide a very efficient numerical solution (in acceptable computing time) via recently developed interior point techniques of optimisation [3].

As explained in Section 5.3, diagonal stability can imply some valuable properties of robust stability for a wide class of perturbed quantised systems [11,12,22]. In other words, this property can guarantee stability of the system even though the state of the system is quantised. More details of the finite-precision problem are also given in this section through the illustrative application of the digital filter implementation.

5.1.3 Notation

For a real matrix P, $P > 0$ (resp. $P \geq 0$) means P is symmetric and positive-definite (resp. positive semi-definite). We also use the notation $\mathbf{diag}(A, B)$ with $A \in \mathbb{R}^{p \times q}$ and $B \in \mathbb{R}^{m \times n}$ to denote the block-diagonal matrix written with A, B as its diagonal blocks. For $r = [r_1, \dots r_n]$, $r_i \in \mathbb{N}$, we define the sets:

$$\begin{aligned}
\mathcal{D}(r) &= \{\Delta = \mathbf{diag}(\delta_1 I_{r_1}, \dots, \delta_n I_{r_n}) \mid \delta_i \in \mathbb{R},\ \ i = 1, \dots n\} \\
\mathcal{B}(r) &= \{B = \mathbf{diag}(B_1, \dots, B_n) \mid B_i \in \mathbb{R}^{r_i \times r_i},\ \ i = 1, \dots n\} \\
\mathcal{S}(r) &= \{S \in \mathcal{B}(r) \mid S_i > 0,\ \ i = 1, \dots n\} \\
\mathcal{G}(r) &= \{G \in \mathcal{B}(r) \mid G_i + G_i^T = 0,\ \ i = 1, \dots n\} \\
\mathcal{P} &= \{p(t) \in \mathbb{R}^N \mid \forall t \geq 0, \forall j = 1, \dots N,\ |p_j(t)| \leq \overline{p}_j\}
\end{aligned}$$

5.2 Convex Approach to the Diagonal Stability Issue

5.2.1 LMI-based Analysis of Diagonal Stability

Let $a_i(p)$, $i = 1, \dots n$, be the coefficients of the characteristic polynomial of A. The main results of the chapter are based on a theorem that was originally stated in [20] for nominal systems (5.1) where A is in companion form. We first readily rewrite this theorem in a more general framework that relies on the characteristic polynomial of any matrix A.

Theorem 5.1. *[20] The Lyapunov equation for the nominal system (5.1) with $p = 0$ admits a diagonal solution P if and only if $\sum_{i=1}^{n} |a_i(0)| \leq 1$. Moreover, the diagonal elements P_i of P must satisfy $P_1 \geq P_2 \geq \dots \geq P_n > 0$.*

Assuming now that these coefficients $a_i(\cdot)$ are rational functions of the parameter $p \in \mathcal{P}$, they can be written via simple matrix operations as the following linear fractional representation (LFR) [7]:

$$\begin{aligned} a_i(p) &= \alpha_i + \beta_i \Delta_i(p)(I - \delta_i \Delta_i(p))^{-1} \gamma_i \\ \Delta_i(p) &= \mathbf{diag}(p_1 I_{r_{1,i}}, \dots p_N I_{r_{N,i}}) \in \mathcal{D}(r) \end{aligned} \tag{5.3}$$

where $\mathcal{N}_i = \sum_{k=1}^{N} r_{k,i}$, $\alpha_i \in \mathbb{R}$, $\beta_i \in \mathbb{R}^{1 \times \mathcal{N}_i}$, $\delta_i \in \mathbb{R}^{\mathcal{N}_i \times \mathcal{N}_i}$ and $\gamma_i \in \mathbb{R}^{\mathcal{N}_i \times 1}$.

The condition $p \in \mathcal{P}$ implies there exists an equivalent "normalised" representation of $a_i(p)$, where $||\Delta_i(p)|| \leq 1$. In the remainder, for simplicity, p will also denote the parameters corresponding to this new normalised representation. As a result of Theorem 5.1, we can state that the system (5.1) is diagonally stable if:

$$\forall\, p \in \mathcal{P}, \qquad \sum_{i=1}^{n} |a_i(p)| \leq 1 \tag{5.4}$$

The first idea would be to define $\underline{a}_i = \inf\{a_i(p) : p \in \mathcal{P}\}$ and $\overline{a}_i = \sup\{a_i(p) : p \in \mathcal{P}\}$, and then to check that $\sum_{i=1}^{n} \max(|\overline{a}_i|\ ,\ |\underline{a}_i|) \leq 1$. But this leads to a major problem: since $a_i(p)$ may not be convex in p, the computation of the global maximum $\overline{a}_i$ or minimum $\underline{a}_i$ of $a_i(p)$ is an NP-hard problem [2,19]. The following theorem proposes an LMI approach to solve the above problem:

Theorem 5.2. *[6] The parameter-dependent system (5.1), with $p \in \mathcal{P}$, is diagonally stable if we can find $z_i \in \mathbb{R}^+$, S_i, $T_i \in \mathcal{S}(r)$, G_i, $H_i \in \mathcal{G}(r)$, $i = 1, \dots n$, such that:*

$$\sum_{i=1}^{n} z_i \leq 1 \tag{5.5}$$

$$\begin{bmatrix} \alpha_i - z_i + \gamma_i^T S_i \gamma_i & \frac{1}{2}\beta_i + \gamma_i^T S_i \delta_i - \gamma_i^T G_i \\ \frac{1}{2}\beta_i^T + \delta_i^T S_i \gamma_i + G_i \gamma_i & \delta_i^T S_i \delta_i - S_i + G_i \delta_i - \delta_i^T G_i \end{bmatrix} < 0,\ i = 1, \dots n \tag{5.6}$$

$$\begin{bmatrix} -\alpha_i - z_i + \gamma_i^T T_i \gamma_i & -\frac{1}{2}\beta_i + \gamma_i^T T_i \delta_i - \gamma_i^T H_i \\ -\frac{1}{2}\beta_i^T + \delta_i^T T_i \gamma_i + H_i \gamma_i & \delta_i^T T_i \delta_i - T_i + H_i \delta_i - \delta_i^T H_i \end{bmatrix} < 0,\ i = 1, \dots n \tag{5.7}$$

where α_i, β_i, γ_i and δ_i are the coefficients of the LFR (5.3) of $a_i(p)$.

Proof. Let us introduce the variables $z_i \in \mathbb{R}^+$, $1 \leq i \leq n$, (5.4) is equivalent to:

$$\exists z_i \in \mathbb{R}^+,\ 1 \leq i \leq n,\ \text{ such that :}$$

$$\sum_{i=1}^{n} z_i \leq 1, \tag{5.8}$$

$$\forall p \in \mathcal{P},\ a_i(p) \leq z_i, \quad 1 \leq i \leq n \tag{5.9}$$

$$\forall p \in \mathcal{P},\ a_i(p) \geq -z_i, \quad 1 \leq i \leq n \tag{5.10}$$

We want to translate (5.9) and (5.10) into LMIs. To do so, the remainder of the proof relies on the following key idea: let us define the fictitious signals y_i and u_i and express the previous conditions on $a_i(p)$ through simple LMI-based conditions on these signals. The coefficients $a_i(\cdot)$ are rational functions of the parameter $p \in \mathcal{P}$, so that $y_i = (a_i(p) - z_i)u_i$ with $y_i \in \mathbb{R}^*$ and $u_i \in \mathbb{R}^*$, $i = 1, \dots n$, can be represented by the following "normalised" LFR:

$$\begin{aligned} y_i &= (\alpha_i - z_i)u_i + \beta_i \pi_i \\ \psi_i &= \gamma_i u_i + \delta_i \pi_i \\ \pi_i &= \Delta_i(p)\psi_i, \quad \Delta_i(p) \in \mathcal{D}(r), \quad ||\Delta_i(p)|| \leq 1 \end{aligned}$$

Then, (5.9) holds if and only if for every $p \in \mathcal{P}$, $y_i u_i \leq 0$, $1 \leq i \leq n$, with $y_i u_i = u_i(\alpha_i - z_i)u_i + \frac{1}{2}\pi_i^T \beta_i^T u_i + \frac{1}{2}u_i \beta_i \pi_i$, $i = 1, \dots n$.

Let us now introduce the symmetric scaling matrices $S_i \in \mathcal{S}(r)$, $i = 1, \dots n$ (to take the block-diagonal structure of Δ into account), and the skew-symmetric scaling matrices $G_i \in \mathcal{G}(r)$ (to take into account that p is real [7]), such that:

$$\begin{aligned} &\pi_i^T S_i \pi_i \leq \psi_i^T S_i \psi_i = (u_i \gamma_i^T + \pi_i^T \delta_i^T) S_i (\gamma_i u_i + \delta_i \pi_i), && i = 1, \dots n \\ &\pi_i^T G_i \psi_i - \psi_i^T G_i \pi_i = 0, && i = 1, \dots n \end{aligned}$$

Therefore, (5.9) holds if we can find $S_i \in \mathcal{S}(r)$, $G_i \in \mathcal{G}(r)$, $i = 1, \dots n$, such that:

$$\begin{bmatrix} u_i \\ \pi_i \end{bmatrix}^T \begin{bmatrix} \alpha_i - z_i & \frac{1}{2}\beta_i \\ \frac{1}{2}\beta_i^T & 0 \end{bmatrix} \begin{bmatrix} u_i \\ \pi_i \end{bmatrix} < 0 \text{ holds, } i = 1, \dots n$$

for all $u_i \neq 0$ and π_i, satisfying:

$$\begin{bmatrix} u_i \\ \pi_i \end{bmatrix}^T \begin{bmatrix} \gamma_i^T S_i \gamma_i & \gamma_i^T S_i \delta_i \\ \delta_i^T S_i \gamma_i & \delta_i^T S_i \delta_i - S_i \end{bmatrix} \begin{bmatrix} u_i \\ \pi_i \end{bmatrix} \geq 0$$

and:

$$\begin{bmatrix} u_i \\ \pi_i \end{bmatrix}^T \begin{bmatrix} 0 & -\gamma_i^T G_i \\ G_i \gamma_i & G_i \delta_i - \delta_i^T G_i \end{bmatrix} \begin{bmatrix} u_i \\ \pi_i \end{bmatrix} = 0$$

Using $\mathcal{S}$-procedure, (5.9) with $p \in \mathcal{P}$, holds if:

$$\begin{bmatrix} \alpha_i - z_i + \gamma_i^T S_i \gamma_i & \frac{1}{2}\beta_i + \gamma_i^T S_i \delta_i - \gamma_i^T G_i \\ \frac{1}{2}\beta_i^T + \delta_i^T S_i \gamma_i + G_i \gamma_i & \delta_i^T S_i \delta_i - S_i + G_i \delta_i - \delta_i^T G_i \end{bmatrix} < 0, \; i = 1, \dots n$$

Reiterating the same operation with (5.10) achieves the proof of Theorem 5.2. □

5.2.2 LMI Conditions For Output-feedback Diagonal Stabilisation

Let us now consider the following single-input multi-output, parameter-dependent, discrete time system:

$$\begin{aligned} x(k+1) &= A(p)x(k) + Bu(k) \\ y(k) &= Cx(k) \end{aligned} \tag{5.11}$$

with $A(p)$ in the companion form, $B = [1\; 0\; \dots\; 0]^T$, $C = (c_{ji})_{1 \leq i \leq n}^{1 \leq j \leq m}$ and $p \in \mathcal{P}$. We seek an output-feedback law $u(k) = Ky(k)$, with $K = [k_1\; k_2\; \dots\; k_m]$, such that the system described by (5.11) is diagonally stable. In the sequel, A is used to denote the nominal system $A(0)$ with a_i, $i = 1, \dots n$, the coefficients of its characteristic polynomial.

Theorem 5.3. *[5] The following propositions are equivalent:*

(i) There exists a static controller $K = [k_1\; k_2\; \dots\; k_m]$ such that the Lyapunov equation for the nominal closed-loop system (5.11):

$$(A + BKC)^T P (A + BKC) - P < 0$$

admits a diagonal solution P.

(ii) We can find $z_i \in \mathbb{R}^+$, $i = 1, \dots n$, and $k_j \in \mathbb{R}$, $j = 1, \dots m$, such that:

$$\sum_{i=1}^{n} z_i \leq 1 \tag{5.12}$$

$$\begin{bmatrix} z_i & -a_i + \sum_{j=1}^{m} c_{ji}k_j \\ -a_i + \sum_{j=1}^{m} c_{ji}k_j & z_i \end{bmatrix} \geq 0, \; i = 1, \dots n \tag{5.13}$$

Moreover, the diagonal elements P_i of P must satisfy $P_1 \geq P_2 \geq \dots \geq P_n > 0$.

Proof. Let us consider the nominal discrete time system in the companion form obtained by setting $p = 0$ in (5.11). The Lyapunov equation for the closed-loop system is:

$${A^{cl}}^T P A^{cl} - P < 0 \tag{5.14}$$

with $A^{cl} = A + BKC$. Then, A^{cl} is in companion form with a_i^{cl} denoting its first row coefficients such that:

$$a_i^{cl} = a_i - \sum_{j=1}^{m} k_j c_{ji}$$

From Theorem 5.1, it can be derived that $\exists K = [k_1 \; k_2 \; \dots \; k_m]$ such that (5.14) admits a diagonal solution if and only if:

$$\sum_{i=1}^{n} \left| -a_i + \sum_{j=1}^{m} c_{ji}k_j \right| \leq 1 \tag{5.15}$$

Moreover, the diagonal elements P_i of P must satisfy $P_1 \geq P_2 \geq \dots \geq P_n > 0$.

Let us introduce the variables $z_i \in \mathbb{R}^+$, $1 \leq i \leq n$. Then, (5.15) is equivalent to:

$\exists \, k_j \in \mathbb{R}$, $1 \leq j \leq m$, and $z_i \in \mathbb{R}^+$, $1 \leq i \leq n$, such that:

$$\sum_{i=1}^{n} z_i \leq 1 \tag{5.16}$$

$$\left| -a_i + \sum_{j=1}^{m} c_{ji}k_j \right| \leq z_i, \; i = 1, \dots n \tag{5.17}$$

It can be now readily stated that the conditions (5.17) and $z_i \geq 0$, $1 \leq i \leq n$, are equivalent to the conditions $(-a_i + \sum_{j=1}^{m} c_{ji}k_j)^2 - z_i^2 \leq 0$ and $z_i \geq 0$, $1 \leq i \leq n$. The Schur lemma completes the proof. □

Theorem 5.4. *[6] There exists a static controller $K = [k_1\ k_2\ \ldots\ k_m]$ diagonally stabilising (5.11), with $p \in \mathcal{P}$, if we can find $z_i \in \mathbb{R}^+$, S_i, $T_i \in \mathcal{S}(r)$, G_i, $H_i \in \mathcal{G}(r)$, $i = 1, \ldots n$, and $k_j \in \mathbb{R}$, $j = 1, \ldots m$, such that:*

$$\sum_{i=1}^{n} z_i \leq 1 \tag{5.18}$$

$$\begin{bmatrix} -\alpha_i - z_i + \sum_{j=1}^{m} c_{ji}k_j + \gamma_i^T S_i \gamma_i & -\frac{1}{2}\beta_i + \gamma_i^T S_i \delta_i - \gamma_i^T G_i \\ -\frac{1}{2}\beta_i^T + \delta_i^T S_i \gamma_i + G_i \gamma_i & \delta_i^T S_i \delta_i - S_i + G_i \delta_i - \delta_i^T G_i \end{bmatrix} < 0, \quad i = 1, \ldots n \tag{5.19}$$

$$\begin{bmatrix} \alpha_i - z_i - \sum_{j=1}^{m} c_{ji}k_j + \gamma_i^T T_i \gamma_i & \frac{1}{2}\beta_i + \gamma_i^T T_i \delta_i - \gamma_i^T H_i \\ \frac{1}{2}\beta_i^T + \delta_i^T T_i \gamma_i + H_i \gamma_i & \delta_i^T T_i \delta_i - T_i + H_i \delta_i - \delta_i^T H_i \end{bmatrix} < 0, \quad i = 1 \ldots n \tag{5.20}$$

where α_i, β_i, γ_i and δ_i are the coefficients of the LFR (5.3) of $a_i(p)$.

Proof. Let us consider the parameter-dependent discrete time system in the companion form described by (5.11), where the coefficients $a_i(\cdot)$ are rational functions of the parameter $p \in \mathcal{P}$. By analogy with the line of argument developed in the proof of Theorem 5.3 for the nominal system, it can be readily derived that there exists $K = [k_1\ k_2\ \ldots\ k_m]$ such that the Lyapunov equation $(A(p) + BKC)^T P (A(p) + BKC) - P < 0$, with $p \in \mathcal{P}$, admits a diagonal solution if, for every $p \in \mathcal{P}$, $\sum_{i=1}^{n} \left| -a_i(p) + \sum_{j=1}^{m} c_{ji}k_j \right| \leq 1$. From this point, the proof is equivalent to the one developed for Theorem 5.2 with $a_i(p)$ replaced by $a_i(p) - \sum_{j=1}^{m} c_{ji}k_j$. □

5.3 Application to Digital Filter Implementations

5.3.1 Overview of the Finite-precision Issue

From a functional point of view, digital and analogue equipment (for signal processing applications) are pretty much similar. However, from an operational point of view, their characteristics differ, which can lead to some major discrepancies between analogue and digital system behaviours. In a same connection, implementation of finite-precision equipment (e.g., DSP chips using fixed-point arithmetic) are potential sources of undesirable non-linear effects that could not be foreseen through an idealised mathematical representation of the system. The granular nature of digital signals (analogue-to-digital conversions) is responsible for erratic finite word-length effects. It is also well known that overflow, truncation resolution and size conversion of fixed- or floating-point numbers may result in computer word-length errors. Even though the severity of these effects can vary significantly (depending on system parameters such as the sampling rate or the Least Significant Bit), the impact of FWL phenomena on the system behaviour must be considered with high attention.

This notion of sensitivity to the finite-precision effects (or fragility [24]) is particularly relevant for the implementation of controllers or filters in fixed-point digital computers. In the "digital world", real numbers are represented with a finite word-length and computations are performed in finite precision, possibly subject to problems of adder overflow. These quantisation phenomena can deteriorate the performance of the system and be responsible for severe instabilities [8], limit cycles [22,25] or even chaos [9]. During the last decade, there has been an extensive literature on the quantisation effects in the field of the control [4,14,16,21], and more specifically in the field of digital filtering [8,17,23,25]. These effects can be represented through a quantisation operator defined as follows:

$$\mathcal{Q}_n = \left\{ \begin{array}{l} g(.,.) : \mathbb{R}^n \times \mathbb{N} \to \mathbb{R}^n, \\ \quad g(x(k),k) = [g_1(x_1(k),k) \; \dots \; g_n(x_n(k),k)]^T, \\ \quad g(0,k) = 0 \;\text{ such that } \forall i = 1,\dots n, \forall x_i \in \mathbb{R}, \forall k \in \mathbb{N}, \\ \quad |g_i(x_i(k),k)| \leq |x_i(k)| \end{array} \right\}$$

It is important to note that the operator g represents truncation and overflow (another description is given in [15] in terms of zeroing or saturation arithmetic) but that roundoff error is not considered here.

Then, a quantised system can be described by:

$$x(k+1) = g\left(A(p)x(k), k\right) \tag{5.21}$$

5.3.2 On the Key Role of Diagonal Stability for Finite-precision Problems

In [11,12,17,22], it is stated through the following theorem that the diagonal stability plays a core role in the robust stability of quantised systems. This theorem is the cornerstone of the proposed approach, since it allows to connect the notion of diagonal stability, for which some relevant results are presented in the previous sections, to the finite-precision issue.

Theorem 5.5. *If the discrete time system (5.1) is diagonally stable, then the quantised system (5.21) is asymptotically stable in spite of the finite-precision effects.*

The illustrative applications of this theorem are numerous: a stability condition for linear time-varying discrete *interval* systems is pointed out in [18] in terms of diagonal stability of a specific closed-loop matrix. Note that this result can be interpreted as a state feedback version for the diagonal stabilisation problem presented in Section 5.2.2, with the uncertainties in the form of interval elements. In [10,13], diagonal stability of an interconnection matrix leads to a simple proof of existence, uniqueness and *global asymptotic stability of a neural circuit.* The persistence of global asymptotic stability under perturbations is also stated. Last, [1] shows that simultaneous robust stability of a class of perturbed discrete time systems is implied by diagonal stability.

All these examples illustrate the significant role of diagonal stability, and thus of Theorems 5.2, 5.3 and 5.4, in the fields of control design and robust digital filtering. These problems can be rewritten using the characteristic polynomial framework. Once these illustrative examples are interpreted in terms of diagonal stability problems, the LMI approach proposed in the core theorems of the previous sections can provide an efficient numerical solution.

5.3.3 Application to Digital Filter Implementations

This approach appears to be well suited for the problem of direct-form digital filters implemented in fixed-point digital hardware, which can be represented by:

$$w(k+1) = g\left(\sum_{i=1}^{n} a_i(p)w(k+n-i), k\right) \tag{5.22}$$

By setting $w(k+j) = x_{n-j}(k)$, $j = 0, \ldots n-1$, and with:

$$x(k) = \left[x_1(k)\ x_2(k)\ \ldots\ x_n(k)\right]^T$$

(5.22) can be rewritten in the form of (5.21) and, in accordance with our knowledge of p in $a_i(p)$, the LMI formalism of Theorem 5.2 can be applied.

5.3.4 Numerical Example

To illustrate the methodology, let us define the coefficients $a_i(p)$, $i = 1, \ldots 3$, of the perturbed direct-form second order digital filter (5.22) as follows:

$$a_1(p) = \frac{-0.5 + \alpha p_1}{1 - \beta p_2^2}, \quad a_2(p) = \frac{0.08}{1 + \gamma p_2} \quad \text{and} \quad a_3(p) = 0.012 + \alpha p_1$$

with $|p_i| \leq 1, i = 1, 2$. The nominal system is stable: its poles are -0.6, 0.2 and -0.1. We seek the maximum values on the parameter α, β and γ (representing the bounds on the parameters p_1 and p_2 in $a_i(p)$, $i = 1, \ldots 3$), such that the

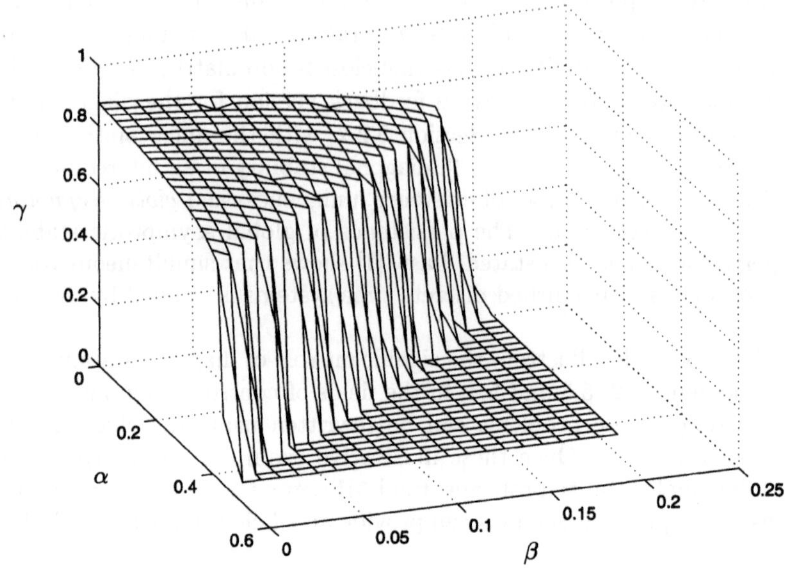

Fig. 5.1. Stability of the system: bounds on α, β and γ

quantised system (5.22) is still stable. We first derive the following LFR for the coefficients:

$$\begin{cases} a_1(p) = -0.5 + \begin{bmatrix}\alpha & -0.5\beta & 0\end{bmatrix} \Delta_1(p)\left(I_3 - \begin{bmatrix}0 & \beta & 0\\ 0 & 0 & 1\\ 0 & \beta & 0\end{bmatrix} \Delta_1(p)\right)^{-1} \begin{bmatrix}1\\0\\1\end{bmatrix} \\ a_2(p) = 0.08 - 0.08\gamma\Delta_2(p)(1+\gamma\Delta_2(p))^{-1} \\ a_3(p) = 0.012 + \alpha\Delta_3(p)(1+0\Delta_3(p))^{-1} \end{cases}$$

with $\Delta_1(p) = \mathbf{diag}(p_1, p_2, p_2)$, $\Delta_2(p) = p_2$ and $\Delta_3(p) = p_1$.

By performing a 3-dimensional gridding on $[\alpha \; ; \; \beta \; ; \; \gamma]$ in $\mathbb{R}^3_+$ and by testing for each grid-point the feasibility of the LMIs (5.5), (5.6) and (5.7), it is straightforward to find and plot in Figure 5.1 the maximum bounds on α, β and γ within which the system is diagonally stable. Thanks to the LMI framework of this approach, the required CPU time to perform the computation is not demanding.

5.4 Concluding Remarks

The finite-precision effects in digital filter implementation are often sources of performance degradation, or even severe instabilities. In this chapter, an original approach to this issue is proposed by considering the stability of uncertain quantised systems. This results in providing a unified framework for the robustness problem with respect to a wide class of nonlinear time-varying perturbations and to the effects induced by the FWL implementation. To do so, LMIs are derived, that ensure the diagonal stability and the output-feedback diagonal stabilisation for a parameter-dependent, discrete time system are given. It should be noted that this approach may still appear conservative when the quantisation perturbation is perfectly well-known and can be precisely described. However, this is a relevant approach for numerous illustrative applications, e.g., to design robust small-order digital filters that have to face some potential dramatic numerical issues due to the finite-precision effects.

References

1. A. Bhaya and E. Kaszkurewicz. Robust, diagonal and D-stability via QLF's: the discrete-time case. In *Proc. IEEE Conf. Decision & Contr.*, pages 2624–2629, Brighton, England, December 1991.
2. V. Blondel and J. Tsitsiklis. A survey of computational complexity results in systems and control. In *Automatica*, 36(9):1249–1274, September 2000.
3. S. Boyd, L. El Ghaoui, E. Feron, and V. Balakrishnan. *Linear Matrix Inequalities in System and Control Theory.* SIAM, Philadelphia, 1994.
4. D. Delchamps. Stabilizing a linear system with quantized state feedback. *IEEE Trans. Aut. Contr.*, 35(8):916–924, August 1990.
5. S. Dussy. Robust stabilization of discrete-time parameter-dependent systems: the finite precision problem. In *Proc. IEEE Conf. Decision & Contr.*, pages 3976–3981, Kobe, Japan, December 1996.
6. S. Dussy. On the robust control design of quantized systems. In *Proc. IEEE Conf. Decision & Contr.*, pages 1297–1298, San Diego, CA, December 1997.
7. S. Dussy. *An LMI Approach for Multiobjective Robust Control.* PhD thesis, University of Paris IX-Dauphine, Paris, March 1998.
8. P. Ebert, J. Mazo, and M. Taylor. Overflow oscillations in digital filters. *Bell Syst. Tech. Journ.*, 48:2999–3020, November 1969.
9. X. Feng and K. Loparo. A study of chaos in discrete time linear systems with quantized state feedback. In *Proc. IEEE Conf. Decision & Contr.*, pages 2107–2112, Tucson, AZ, December 1992.
10. M. Forti and A. Tesi. New conditions for global stability of neural networks with application to linear and quadratic programming problems. *IEEE Trans. Circ. & Syst. I: Fundamental Theory*, 42(7):354–356, July 1995.
11. E. Kaszkurewicz and A. Bhaya. Comments on "overflow oscillations in state-space digital filters". *IEEE Trans. Circ. & Syst. II: Analog & Digital Signal Processing*, 39(9):675–676, September 1992.
12. E. Kaszkurewicz and A. Bhaya. Robust stability and diagonal Liapunov functions. *SIAM Journ. Matrix Anal. Appl.*, 14(2):508–520, April 1993.
13. E. Kaszkurewicz and A. Bhaya. On a class of globally stable neural circuits. *IEEE Trans. Circ. & Syst. I: Fundamental Theory*, 41(2):171–174, February 1994.
14. G. Li and M. Gevers. Optimal finite precision implementation of a state-estimate feedback controller. *IEEE Trans. Aut. Contr.*, 37(12):1487–1498, December 1990.

15. D. Liu and A. Michel. Stability analysis of state-space realizations for two-dimensional filters with overflow nonlinearities. *IEEE Trans. Circ. & Syst.*, 41(2):127–135, February 1994.
16. K. Liu, R. Skelton, and K. Grigoriadis. Optimal controllers for finite wordlength implementation. *IEEE Trans. Aut. Contr.*, 37(9):1294–1304, September 1992.
17. W. Mills, C. Mullis, and R. Roberts. Digital filter realizations without overflow oscillations. *IEEE Trans. Acoustics, Speech, & Signal Proc.*, 26(4):334–338, August 1978.
18. F. Mota, E. Kaszkurewicz, and A. Bhaya. Robust stabilization of time-varying discrete interval systems. In *Proc. IEEE Conf. Decision & Contr.*, volume 1, pages 341–346, Tucson, AZ, December 1992.
19. A. Nemirovskii. Several NP-hard problems arising in robust stability analysis. *Math. of Contr., Signals & Syst.*, 6:99–105, 1993.
20. P. Regalia. On finite Lyapunov functions for companion matrices. *IEEE Trans. Aut. Contr.*, 37(10):1640–1644, October 1992.
21. M. Rotea and D. Williamson. Optimal realizations of finite wordlength digital filters and controllers. *IEEE Trans. Circ. & Syst.*, 42(2):61–72, February 1995.
22. P. Vaidyanathan and V. Liu. An improved sufficient condition for absence of limit cycles in digital filters. *IEEE Trans. Circ. & Syst.*, 34(3):319–332, March 1987.
23. M. Werter. Suppression of limit cycles in the first-order two-dimensional direct form digital filter with a controlled rounding arithmetic. *IEEE Trans. Sign. Proc.*, 40(6):1599–1601, June 1992.
24. J.F. Whidborne, R.S.H. Istepanian and J. Wu. Reduction of controller fragility by pole sensitivity minimization. *IEEE Trans. Aut. Contr.*, 46(2):320–325, February 2001
25. A. Willson. Limit cycles due to adder overflow in digital filters. *IEEE Trans. Circ. Theory*, 19(4):342–346, July 1972.

CHAPTER 6

THE DETERMINATION OF OPTIMAL FINITE-PRECISION CONTROLLER REALISATIONS USING A GLOBAL OPTIMISATION STRATEGY: A POLE-SENSITIVITY APPROACH

Sheng Chen and Jun Wu

Abstract. The pole-sensitivity approach is a general method for analysing the stability of discrete-time control systems with a finite word-length (FWL) implemented digital controller. It leads to a non-smooth and non-convex optimisation framework, where an optimal controller realisation can be designed by maximising some stability related measure. In this contribution, a new stability related measure is derived, which is more accurate in estimating the closed-loop stability robustness of an FWL implemented controller than the existing measures of pole-sensitivity analysis. This improved stability related measure provides a better criterion to find the optimal FWL realisations for a generic controller structure that includes output-feedback and observer-based controllers. An efficient global optimisation strategy called the adaptive simulated annealing (ASA) is adopted to solve for the resulting optimisation problem. A numerical example is included to verify the theoretical analysis and to illustrate the design procedure.

6.1 Introduction

The classical controller design methodology often assumes that the controller is implemented exactly, even though in reality a control law can only be realised in finite precision. The justification of this assumption is usually on the grounds that the plant uncertainty is the most significant source of uncertainty in the control system. However, researchers have realised that the controller uncertainty caused by finite-precision implementation has significant influence on the performance of the control system. A designed stable control system may achieve a lower than predicted performance or even become unstable when the control law is implemented with a finite-precision device due to the finite word-length (FWL) effects. This is highlighted in the so-called fragility puzzles [1–3]: certain high-performance robust optimal controllers are known to be fragile. Ironically, these controllers have been designed to tolerate uncertainty in the plant, and yet small perturbations on the controller parameters may cause the designed closed-loop system to go unstable.

The fragility issues are strongly related to and interconnected with the FWL controller implementation issues. Although the number of controller implementations using floating-point processors is increasing due to their reduced price, for reasons of cost, simplicity, speed, memory space and power consumption, the use of fixed-point processors is more desirable for many industrial and consumer applications. Furthermore, due to their reliability and well-understood properties, fixed-point processors predominate in safety-critical systems. With a fixed-point processor, however, the detrimental FWL effects are markedly increased due to a reduced precision. The problem can become serious when a high sampling rate and a high-order controller are used. It has been noted that a controller design can be implemented with different realisations and that the FWL effect on the closed-loop stability depends on the controller realisation structure. This property can be utilised to select a controller realisation in order to improve the robustness of closed-loop stability under controller perturbations. Currently, two approaches exist for determining the optimal controller realisations under different criteria, namely pole-sensitivity measures [4–8] and complex stability radius measures [9,10].

In the first approach, pole-sensitivity measures [5,6] are used to quantify the FWL effect, leading to a non-convex and non-smooth optimisation problem in finding an optimal FWL controller realisation. The need to solve for a non-convex and non-smooth optimisation problem had been seen as a disadvantage, as conventional optimisation algorithms [11,12], which are better known to the control community, may not guarantee to find a true optimal realisation. However, the efficient global optimisation techniques [13–18] to tackle this kind of difficult optimisation problems are now widely available. Moreover, the pole-sensitivity approach is very general and can be applied to output-feedback and observer-based controllers as well as the controllers that are parameterised either by the usual shift operator or the delta operator [8,19–21]. More recently, Fialho and Georgiou [10] used the complex stability radius measure to formulate an optimal FWL controller realisation problem that can be represented as a special $\mathcal{H}_\infty$-norm minimisation problem and solved for with the method of linear matrix inequality. In this second approach, the FWL perturbations are assumed to be complex-valued. Although this assumption is somewhat artificial and the approach can only be applied to shift operator-based output-feedback controllers, the method does not require to solve for a nonlinear optimisation problem and has certain attractive features. For a detailed treatment of this approach, see Chapter 7.

This contribution focuses on the pole-sensitivity analysis method and emphasises a unified approach for analysing the sensitivity of closed-loop stability with respect to FWL effects. A generic digital controller structure is considered that includes output-feedback and observer-based controllers, and a new stability related measure is proposed for the unified controller structure. An efficient global optimisation procedure based on the ASA algorithm

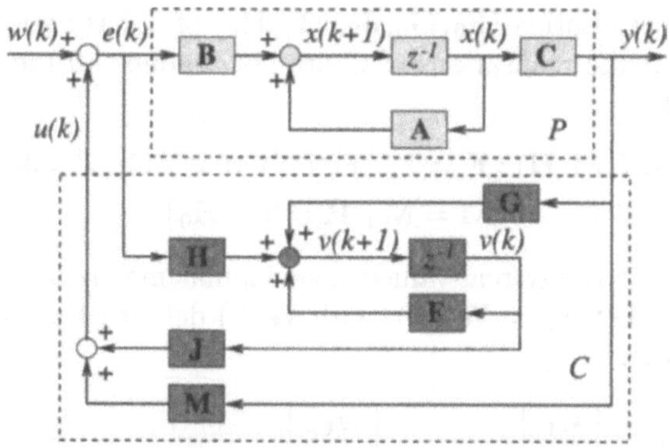

Fig. 6.1. Discrete time closed-loop system with a generic digital controller

[16–18] is developed to find the optimal controller realisation that maximises this new measure. Through theoretical analysis and numerical results, it is shown that this improved measure is less conservative in estimating the FWL closed-loop stability robustness of a controller realisation than the existing pole-sensitivity measures [5,6].

6.2 Problem Formulation

Consider the discrete time closed-loop control system shown in Figure 6.1, where the linear time-invariant plant P is described by:

$$\begin{cases} \mathbf{x}(k+1) = \mathbf{A}\mathbf{x}(k) + \mathbf{B}\mathbf{e}(k) \\ \mathbf{y}(k) = \mathbf{C}\mathbf{x}(k) \end{cases} \tag{6.1}$$

which is completely state controllable and observable with $\mathbf{A} \in \mathbb{R}^{n\times n}$, $\mathbf{B} \in \mathbb{R}^{n\times p}$ and $\mathbf{C} \in \mathbb{R}^{q\times n}$; and the generic digital stabilising controller C is described by:

$$\begin{cases} \mathbf{v}(k+1) = \mathbf{F}\mathbf{v}(k) + \mathbf{G}\mathbf{y}(k) + \mathbf{H}\mathbf{e}(k) \\ \mathbf{u}(k) = \mathbf{J}\mathbf{v}(k) + \mathbf{M}\mathbf{y}(k) \end{cases} \tag{6.2}$$

with $\mathbf{F} \in \mathbb{R}^{m\times m}$, $\mathbf{G} \in \mathbb{R}^{m\times q}$, $\mathbf{J} \in \mathbb{R}^{p\times m}$, $\mathbf{M} \in \mathbb{R}^{p\times q}$ and $\mathbf{H} \in \mathbb{R}^{m\times p}$. The output-feedback and observer-based controllers can be unified in this general structure: C is an output-feedback controller when $\mathbf{H} = \mathbf{0}$; a full-order observer-based controller when $\mathbf{F} = \mathbf{A} - \mathbf{G}\mathbf{C}$, $\mathbf{M} = \mathbf{0}$ and $\mathbf{H} = \mathbf{B}$; a reduced-order observer-based controller, otherwise [22,23]. Notice that, for notational simplicity, we have restricted the description to the controller structure with the shift operator z parameterisation. All the results, however, can readily be extended to the controller structure with the delta operator parameterisation [19–21].

Assume that a realisation $(\mathbf{F}_0, \mathbf{G}_0, \mathbf{J}_0, \mathbf{M}_0, \mathbf{H}_0)$ of C has been designed. It is well known that the realisations of C are not unique. All the realisations of C form the set:

$$\mathcal{S} = \{(\mathbf{F}, \mathbf{G}, \mathbf{J}, \mathbf{M}, \mathbf{H}) : \mathbf{F} = \mathbf{T}^{-1}\mathbf{F}_0\mathbf{T}, \mathbf{G} = \mathbf{T}^{-1}\mathbf{G}_0, \mathbf{J} = \mathbf{J}_0\mathbf{T}, \\ \mathbf{M} = \mathbf{M}_0, \mathbf{H} = \mathbf{T}^{-1}\mathbf{H}_0\} \tag{6.3}$$

where $\mathbf{T} \in \mathbb{R}^{m\times m}$ is any real-valued non-singular matrix, called a similarity transformation. Let $\mathbf{w}_\mathbf{F} = \mathrm{Vec}(\mathbf{F})$, with $\mathrm{Vec}(\cdot)$ defining the column stacking operator. Denote:

$$\mathbf{w} = \begin{bmatrix} w_1 \\ \vdots \\ w_N \end{bmatrix} \triangleq \begin{bmatrix} \mathbf{w}_\mathbf{F} \\ \mathbf{w}_\mathbf{G} \\ \mathbf{w}_\mathbf{J} \\ \mathbf{w}_\mathbf{M} \\ \mathbf{w}_\mathbf{H} \end{bmatrix}, \quad \mathbf{w}_0 \triangleq \begin{bmatrix} \mathbf{w}_{\mathbf{F}_0} \\ \mathbf{w}_{\mathbf{G}_0} \\ \mathbf{w}_{\mathbf{J}_0} \\ \mathbf{w}_{\mathbf{M}_0} \\ \mathbf{w}_{\mathbf{H}_0} \end{bmatrix} \tag{6.4}$$

where $N = (m+p)(m+q)+mp$. We also refer to $\mathbf{w}$ as a realisation of C. The stability of the closed-loop system in Figure 6.1 depends on the eigenvalues of the matrix:

$$\bar{\mathbf{A}}(\mathbf{w}) = \begin{bmatrix} \mathbf{A}+\mathbf{BMC} & \mathbf{BJ} \\ \mathbf{GC}+\mathbf{HMC} & \mathbf{F}+\mathbf{HJ} \end{bmatrix} = \begin{bmatrix} \mathbf{I} & \mathbf{0} \\ \mathbf{0} & \mathbf{T}^{-1} \end{bmatrix} \bar{\mathbf{A}}(\mathbf{w}_0) \begin{bmatrix} \mathbf{I} & \mathbf{0} \\ \mathbf{0} & \mathbf{T} \end{bmatrix} \tag{6.5}$$

All the different realisations $\mathbf{w}$ in $\mathcal{S}$ are completely equivalent and, in particular, achieve exactly the same set of closed-loop poles, if they are implemented with infinite precision. Since the closed-loop system will have been designed to be stable, the eigenvalues are:

$$|\lambda_i(\bar{\mathbf{A}}(\mathbf{w}))| = |\lambda_i(\bar{\mathbf{A}}(\mathbf{w}_0))| < 1, \ \forall i \in \{1, \ldots, m+n\} \tag{6.6}$$

When a $\mathbf{w}$ is implemented with a fixed-point processor, it is perturbed into $\mathbf{w} + \Delta\mathbf{w}$ due to the FWL effect. Each element of $\Delta\mathbf{w}$ is bounded by $\pm\epsilon/2$:

$$\|\Delta\mathbf{w}\|_\infty \triangleq \max_{i\in\{1,\cdots,N\}} |\Delta w_i| \le \epsilon/2 \tag{6.7}$$

For a fixed-point processor of B_s bits (the bit length), let $B_s = B_i + B_f$, where 2^{B_i} is a "normalisation" factor to make the absolute value of each element of $2^{-B_i}\mathbf{w}$ no larger than 1. Thus, B_i are bits required for the integer part of a number and B_f are bits used to implement the fractional part of a number. It can easily be seen that:

$$\epsilon = 2^{-B_f} \tag{6.8}$$

With the perturbation $\Delta\mathbf{w}$, $\lambda_i(\bar{\mathbf{A}}(\mathbf{w}))$ is moved to $\lambda_i(\bar{\mathbf{A}}(\mathbf{w} + \Delta\mathbf{w}))$. If an eigenvalue of $\bar{\mathbf{A}}(\mathbf{w} + \Delta\mathbf{w})$ is outside the open unit disk, the closed-loop system, designed to be stable, becomes unstable with B_s-bit implemented $\mathbf{w}$. It is therefore critical to know when the FWL error will cause closed-loop instability. This ultimately means that we would like to know the largest open "sphere" in the controller perturbation space, within which the closed-loop system remains stable. The size or radius of this "sphere" is defined by:

$$\mu_0(\mathbf{w}) \triangleq \inf\{\|\Delta\mathbf{w}\|_\infty : \bar{\mathbf{A}}(\mathbf{w}+\Delta\mathbf{w}) \text{ is unstable}\} \tag{6.9}$$

From the definition of $\mu_0(\mathbf{w})$, it is obvious that $\bar{\mathbf{A}}(\mathbf{w}+\Delta\mathbf{w})$ remains stable for any $\Delta\mathbf{w}$ with $\|\Delta\mathbf{w}\|_\infty < \mu_0(\mathbf{w})$. The larger $\mu_0(\mathbf{w})$ is, the larger FWL error the closed-loop stability can tolerate. Hence, $\mu_0(\mathbf{w})$ constitutes a FWL stability measure.

Let $B_s^{\min}$ be the smallest word-length, that when used to implement $\mathbf{w}$, can guarantee the closed-loop stability. An estimate of $B_s^{\min}$ can be obtained as:

$$\hat{B}_{s,0}^{\min} = B_i + \mathrm{Int}[-\log_2(\mu_0(\mathbf{w}))] - 1 \tag{6.10}$$

where the integer $\mathrm{Int}[x] \geq x$. It can easily be seen that the closed-loop system remains stable if $\mathbf{w}$ is implemented with a fixed-point processor of at least $\hat{B}_{s,0}^{\min}$. Moreover, $\mu_0(\mathbf{w})$ is a function of the controller realisation $\mathbf{w}$, we could search for an optimal realisation that maximises $\mu_0(\mathbf{w})$. However, it is not known yet how to compute the value of $\mu_0(\mathbf{w})$, given a realisation $\mathbf{w}$. A practical solution is to consider a lower bound of the stability measure $\mu_0(\mathbf{w})$ in some sense, which is computationally tractable. This in effect defines a smaller but known stable "sphere" or region in the $\Delta\mathbf{w}$ space. Obviously, the closer such a lower bound is to $\mu_0(\mathbf{w})$, the better. Two existing pole-sensitivity measures [5,6] can both be regarded as such lower bounds and, hence, termed stability related measures. It should be emphasised that the approach based on the complex stability radius measure [10] is also "conservative" in that the region defined by the complex stability radius measure is generally smaller than that defined by $\mu_0(\mathbf{w})$.

6.3 A New Pole-sensitivity Stability Related Measure

Roughly speaking, how easily the FWL error $\Delta\mathbf{w}$ can cause a stable control system to become unstable is determined by how close $|\lambda_i(\bar{\mathbf{A}}(\mathbf{w}))|$ are to 1 and how sensitive they are to the controller parameter perturbations. We propose the following FWL stability related measure[1]:

$$\mu_{1I}(\mathbf{w}) \triangleq \min_{i\in\{1,\cdots,m+n\}} \frac{1-|\lambda_i(\bar{\mathbf{A}}(\mathbf{w}))|}{\alpha_i(\mathbf{w})} \tag{6.11}$$

[1] This measure, as shown later, is an improved version of the existing measure μ_1 given in [6] and hence is denoted with μ_{1I}.

with:

$$\alpha_i(\mathbf{w}) \triangleq \sum_{\mathbf{X}=\mathbf{F},\mathbf{G},\mathbf{J},\mathbf{M},\mathbf{H}} \left\| \frac{\partial |\lambda_i(\bar{\mathbf{A}}(\mathbf{w}))|}{\partial \mathbf{X}} \right\|_1 \tag{6.12}$$

where, for a matrix $\mathbf{X} \in \mathbb{C}^{s\times v}$, the 1-norm $\|\mathbf{X}\|_1$ is defined as:

$$\|\mathbf{X}\|_1 \triangleq \sum_{i=1}^{s}\sum_{j=1}^{v} |x_{ij}| \tag{6.13}$$

It remains to be shown how $\mu_{1I}(\mathbf{w})$ can be regarded as an FWL stability related measure, under what conditions $\mu_{1I}(\mathbf{w})$ is a lower bound of $\mu_0(\mathbf{w})$, and that $\mu_{1I}(\mathbf{w})$ is computationally tractable.

In practice, those controller perturbations $\Delta\mathbf{w}$ that will not cause closed-loop instability are most important. These $\Delta\mathbf{w}$ lie in the bounded region:

$$\mathcal{Q}(\mathbf{w}) \triangleq \{\Delta\mathbf{w} : \|\Delta\mathbf{w}\|_\infty < \mu_0(\mathbf{w})\} \tag{6.14}$$

Define a perturbation subset to the controller realisation $\mathbf{w}$ to be:

$$\mathcal{P}(\mathbf{w}) \triangleq \{\Delta\mathbf{w} : \left|\lambda_i(\bar{A}(\mathbf{w}+\Delta\mathbf{w}))\right| - \left|\lambda_i(\bar{A}(\mathbf{w}))\right| \le \|\Delta\mathbf{w}\|_\infty \cdot \alpha_i(\mathbf{w}), \forall i\} \tag{6.15}$$

It is straightforward to prove the following proposition.

Proposition 6.1. $\bar{\mathbf{A}}(\mathbf{w}+\Delta\mathbf{w})$ *is stable if* $\Delta\mathbf{w} \in \mathcal{P}(\mathbf{w})$ *and* $\|\Delta\mathbf{w}\|_\infty < \mu_{1I}(\mathbf{w})$.

Thus, $\mu_{1I}(\mathbf{w})$ is a stability measure for $\Delta\mathbf{w} \in \mathcal{P}(\mathbf{w})$. The requirement for $\Delta\mathbf{w} \in \mathcal{P}(\mathbf{w})$ is not over-restrictive. Similar to the discussions in [8,19], it can be proved that $\mathcal{P}(\mathbf{w})$ exists and at least a large part of $\mathcal{Q}(\mathbf{w})$ is covered by $\mathcal{P}(\mathbf{w})$. Defining:

$$\rho\left(\mathcal{P}(\mathbf{w})\right) \triangleq \inf_{\Delta\mathbf{w}\notin\mathcal{P}(\mathbf{w})} \|\Delta\mathbf{w}\|_\infty \tag{6.16}$$

we have the following corollary, the proof of which is also straightforward.

Corollary 6.1. $\mu_{1I}(\mathbf{w}) \le \mu_0(\mathbf{w})$ *if* $\rho(\mathcal{P}(\mathbf{w})) > \mu_0(\mathbf{w})$.

It can be seen that $\mu_{1I}(\mathbf{w})$ is a lower bound of $\mu_0(\mathbf{w})$, provided that $\mu_0(\mathbf{w})$ is small enough. The assumption of small $\mu_0(\mathbf{w})$ is generally valid, especially for control systems with fast sampling. Given a controller realisation $\mathbf{w}$, the value of $\mu_{1I}(\mathbf{w})$ can readily be calculated. This is summarised in the following theorem, the proof of which is given in Appendix 6.A.

Theorem 6.1. *Let* $\mathbf{x}_i(\bar{\mathbf{A}}(\mathbf{w}))$ *and* $\mathbf{y}_i(\bar{\mathbf{A}}(\mathbf{w}))$ *be the right and reciprocal left eigenvectors related to the* $\lambda_i(\bar{\mathbf{A}}(\mathbf{w}))$*, respectively, and:*

$$\mathbf{L}_i(\mathbf{w}) = \frac{\mathrm{Re}\left[\lambda_i^*(\bar{\mathbf{A}}(\mathbf{w}))\mathbf{y}_i^*(\bar{\mathbf{A}}(\mathbf{w}))\mathbf{x}_i^T(\bar{\mathbf{A}}(\mathbf{w}))\right]}{|\lambda_i(\bar{\mathbf{A}}(\mathbf{w}))|} \tag{6.17}$$

where * *denotes the conjugate operator,* ^{T} *the transpose operator, and* $\mathrm{Re}[\cdot]$ *the real part. Then:*

$$\frac{\partial|\lambda_i(\bar{\mathbf{A}}(\mathbf{w}))|}{\partial \mathbf{F}} = [\mathbf{0}\ \mathbf{I}]\,\mathbf{L}_i(\mathbf{w})\begin{bmatrix}\mathbf{0}\\ \mathbf{I}\end{bmatrix} \tag{6.18}$$

$$\frac{\partial|\lambda_i(\bar{\mathbf{A}}(\mathbf{w}))|}{\partial \mathbf{G}} = [\mathbf{0}\ \mathbf{I}]\,\mathbf{L}_i(\mathbf{w})\begin{bmatrix}\mathbf{C}^T\\ \mathbf{0}\end{bmatrix} \tag{6.19}$$

$$\frac{\partial|\lambda_i(\bar{\mathbf{A}}(\mathbf{w}))|}{\partial \mathbf{J}} = [\mathbf{B}^T\ \mathbf{H}^T]\,\mathbf{L}_i(\mathbf{w})\begin{bmatrix}\mathbf{0}\\ \mathbf{I}\end{bmatrix} \tag{6.20}$$

$$\frac{\partial|\lambda_i(\bar{\mathbf{A}}(\mathbf{w}))|}{\partial \mathbf{M}} = [\mathbf{B}^T\ \mathbf{H}^T]\,\mathbf{L}_i(\mathbf{w})\begin{bmatrix}\mathbf{C}^T\\ \mathbf{0}\end{bmatrix} \tag{6.21}$$

$$\frac{\partial|\lambda_i(\bar{\mathbf{A}}(\mathbf{w}))|}{\partial \mathbf{H}} = [\mathbf{0}\ \mathbf{I}]\,\mathbf{L}_i(\mathbf{w})\begin{bmatrix}\mathbf{C}^T\mathbf{M}^T\\ \mathbf{J}^T\end{bmatrix} \tag{6.22}$$

Similar to (6.10), an estimate of $B_s^{\min}$ can be provided with $\mu_{1I}(\mathbf{w})$ by:

$$\hat{B}_{s,1I}^{\min} = B_i + \mathrm{Int}[-\log_2(\mu_{1I}(\mathbf{w}))] - 1 \tag{6.23}$$

Provided that the conditions of Proposition 6.1 and Corollary 6.1 are met, $\hat{B}_{s,1I}^{\min} \geq \hat{B}_{s,0}^{\min} \geq B_s^{\min}$. That is, $\hat{B}_{s,1I}^{\min}$ is a conservative estimate of the minimum bit length, compared with $\hat{B}_{s,0}^{\min}$. Unlike $\hat{B}_{s,0}^{\min}$, however, $\hat{B}_{s,1I}^{\min}$ can be computed easily. We now show that $\mu_{1I}(\mathbf{w})$ is a closer lower bound of $\mu_0(\mathbf{w})$ than the two existing pole-sensitivity measures [5,6], denoted as $\mu_2(\mathbf{w})$ and $\mu_1(\mathbf{w})$, respectively. Since it has been demonstrated [6] that $\mu_1(\mathbf{w})$ is a closer lower bound of $\mu_0(\mathbf{w})$ than $\mu_2(\mathbf{w})$, we only need to compare $\mu_{1I}(\mathbf{w})$ with $\mu_1(\mathbf{w})$. The stability related measure $\mu_1(\mathbf{w})$ is defined as [6]:

$$\mu_1(\mathbf{w}) \triangleq \min_{i\in\{1,\cdots,m+n\}} \frac{1-|\lambda_i(\bar{\mathbf{A}}(\mathbf{w}))|}{\beta_i(\mathbf{w})} \tag{6.24}$$

with:

$$\beta_i(\mathbf{w}) \triangleq \sum_{\mathbf{X}=\mathbf{F},\mathbf{G},\mathbf{J},\mathbf{M},\mathbf{H}} \left\|\frac{\partial\lambda_i(\bar{\mathbf{A}}(\mathbf{w}))}{\partial\mathbf{X}}\right\|_1 \tag{6.25}$$

An estimate of $B_s^{\min}$ is provided with $\mu_1(\mathbf{w})$ by:

$$\hat{B}_{s,1}^{\min} = B_i + \mathrm{Int}[-\log_2(\mu_1(\mathbf{w}))] - 1 \tag{6.26}$$

The key difference between $\mu_{1I}(\mathbf{w})$ and $\mu_1(\mathbf{w})$ is that the former considers the sensitivity of $\left|\lambda_i(\bar{\mathbf{A}}(\mathbf{w}))\right|$ while the latter considers the sensitivity of $\lambda_i(\bar{\mathbf{A}}(\mathbf{w}))$. It is well known that the stability of a linear discrete time system depends only on the moduli of its eigenvalues. As $\mu_1(\mathbf{w})$ includes the unnecessary eigenvalue arguments in consideration, it is expected that $\mu_1(\mathbf{w})$ is conservative in comparison with $\mu_{1I}(\mathbf{w})$. This can strictly be verified with the following. Noting:

$$\frac{\partial\left|\lambda_i(\bar{\mathbf{A}}(\mathbf{w}))\right|}{\partial w_j} = \mathrm{Re}\left[\lambda_i^*(\bar{\mathbf{A}}(\mathbf{w}))\frac{\partial\lambda_i(\bar{\mathbf{A}}(\mathbf{w}))}{\partial w_j}\right] / \left|\lambda_i(\bar{\mathbf{A}}(\mathbf{w}))\right| \tag{6.27}$$

gives rise to:

$$\left|\frac{\partial\left|\lambda_i(\bar{\mathbf{A}}(\mathbf{w}))\right|}{\partial w_j}\right| \le \frac{\left|\lambda_i^*(\bar{\mathbf{A}}(\mathbf{w}))\frac{\partial\lambda_i(\bar{\mathbf{A}}(\mathbf{w}))}{\partial w_j}\right|}{\left|\lambda_i(\bar{\mathbf{A}}(\mathbf{w}))\right|} = \left|\frac{\partial\lambda_i(\bar{\mathbf{A}}(\mathbf{w}))}{\partial w_j}\right| \tag{6.28}$$

which means that $\alpha_i(\mathbf{w}) \le \beta_i(\mathbf{w})$. This leads to:

Theorem 6.2. $\mu_1(\mathbf{w}) \le \mu_{1I}(\mathbf{w})$ *and* $\hat{B}^{\min}_{s,1} \ge \hat{B}^{\min}_{s,1I}$.

6.4 Optimisation Procedure

When a controller is designed, it will have satisfied certain performance criteria and, in particular, ensures closed-loop stability. The design, however, is usually done under the infinite or at least high-precision assumption. As actual implementation can only be finite precision, the real controller performance may be quite different from the designed one and, if the bit length is too small, the closed-loop stability may even be lost. Given a designed controller realisation, denoted as $\mathbf{w}_0$, there are infinite realisations $\mathbf{w}$ related to $\mathbf{w}_0$ by (6.3). All these realisations are completely equivalent under infinite-precision implementation, but they may perform differently under FWL implementation. The problem naturally arisen is to find an "optimal" realisation, denoted as $\mathbf{w}_{\mathrm{opt}}$, such that $\mu_{1I}(\mathbf{w})$ is maximised. This is of practical importance, since the controller implemented with $\mathbf{w}_{\mathrm{opt}}$ can tolerate a maximum FWL error. This optimal realisation problem is formally defined as:

$$\mathbf{w}_{\mathrm{opt}} = \arg\max_{\mathbf{w}\in\mathcal{S}} \mu_{1I}(\mathbf{w}) \tag{6.29}$$

Given the design $\mathbf{w}_0$, $\forall i \in \{1, \cdots, m+n\}$, partition $\mathbf{x}_i(\bar{\mathbf{A}}(\mathbf{w}_0))$ and $\mathbf{y}_i(\bar{\mathbf{A}}(\mathbf{w}_0))$:

$$\mathbf{x}_i(\bar{\mathbf{A}}(\mathbf{w}_0)) = \begin{bmatrix}\mathbf{x}_{i,1}(\bar{\mathbf{A}}(\mathbf{w}_0))\\ \mathbf{x}_{i,2}(\bar{\mathbf{A}}(\mathbf{w}_0))\end{bmatrix} \tag{6.30}$$

$$\mathbf{y}_i(\bar{\mathbf{A}}(\mathbf{w}_0)) = \begin{bmatrix} \mathbf{y}_{i,1}(\bar{\mathbf{A}}(\mathbf{w}_0)) \\ \mathbf{y}_{i,2}(\bar{\mathbf{A}}(\mathbf{w}_0)) \end{bmatrix} \tag{6.31}$$

where $\mathbf{x}_{i,1}(\bar{\mathbf{A}}(\mathbf{w}_0)), \mathbf{y}_{i,1}(\bar{\mathbf{A}}(\mathbf{w}_0)) \in \mathbb{C}^n$ and $\mathbf{x}_{i,2}(\bar{\mathbf{A}}(\mathbf{w}_0)), \mathbf{y}_{i,2}(\bar{\mathbf{A}}(\mathbf{w}_0)) \in \mathbb{C}^m$. It is easily seen from (6.5) that:

$$\mathbf{x}_i(\bar{\mathbf{A}}(\mathbf{w})) = \begin{bmatrix} \mathbf{x}_{i,1}(\bar{\mathbf{A}}(\mathbf{w}_0)) \\ \mathbf{T}^{-1}\mathbf{x}_{i,2}(\bar{\mathbf{A}}(\mathbf{w}_0)) \end{bmatrix} \tag{6.32}$$

$$\mathbf{y}_i(\bar{A}(\mathbf{w})) = \begin{bmatrix} \mathbf{y}_{i,1}(\bar{\mathbf{A}}(\mathbf{w}_0)) \\ \mathbf{T}^T\mathbf{y}_{i,2}(\bar{\mathbf{A}}(\mathbf{w}_0)) \end{bmatrix} \tag{6.33}$$

From (6.18)–(6.22), we have:

$$\frac{\partial \left|\lambda_i(\bar{\mathbf{A}}(\mathbf{w}))\right|}{\partial \mathbf{F}} = \mathbf{T}^T \mathbf{L}_{i,2,2}(\mathbf{w}_0)\mathbf{T}^{-T} \tag{6.34}$$

$$\frac{\partial \left|\lambda_i(\bar{\mathbf{A}}(\mathbf{w}))\right|}{\partial \mathbf{G}} = \mathbf{T}^T \mathbf{L}_{i,2,1}(\mathbf{w}_0)\mathbf{C}^T \tag{6.35}$$

$$\frac{\partial \left|\lambda_i(\bar{\mathbf{A}}(\mathbf{w}))\right|}{\partial \mathbf{J}} = \left(\mathbf{B}^T \mathbf{L}_{i,1,2}(\mathbf{w}_0) + \mathbf{H}_0^T \mathbf{L}_{i,2,2}(\mathbf{w}_0)\right)\mathbf{T}^{-T} \tag{6.36}$$

$$\frac{\partial \left|\lambda_i(\bar{\mathbf{A}}(\mathbf{w}))\right|}{\partial \mathbf{M}} = \left(\mathbf{B}^T \mathbf{L}_{i,1,1}(\mathbf{w}_0) + \mathbf{H}_0^T \mathbf{L}_{i,2,1}(\mathbf{w}_0)\right)\mathbf{C}^T \tag{6.37}$$

$$\frac{\partial \left|\lambda_i(\bar{\mathbf{A}}(\mathbf{w}))\right|}{\partial \mathbf{H}} = \mathbf{T}^T\left(\mathbf{L}_{i,2,1}(\mathbf{w}_0)\mathbf{C}^T\mathbf{M}_0^T + \mathbf{L}_{i,2,2}(\mathbf{w}_0)\mathbf{J}_0^T\right) \tag{6.38}$$

where:

$$\mathbf{L}_{i,j,l}(\mathbf{w}_0) = \frac{\mathrm{Re}\left[\lambda_i^*(\bar{\mathbf{A}}(\mathbf{w}_0))\mathbf{y}_{i,j}^*(\bar{\mathbf{A}}(\mathbf{w}_0))\mathbf{x}_{i,l}^T(\bar{\mathbf{A}}(\mathbf{w}_0))\right]}{\left|\lambda_i(\bar{\mathbf{A}}(\mathbf{w}_0))\right|}, \quad j,l = 1,2 \tag{6.39}$$

Define the following cost function:

$$f(\mathbf{T}) \triangleq \min_{i\in\{1,\cdots,m+n\}} \frac{1-|\lambda_i(\bar{\mathbf{A}}(\mathbf{w}_0))|}{\alpha_i(\mathbf{w})} = \mu_{1I}(\mathbf{w}) \tag{6.40}$$

The optimal realisation problem (6.29) can then be posed as the following optimisation problem of finding an optimal similarity transformation matrix:

$$\mathbf{T}_{\mathrm{opt}} = \arg \max_{\substack{\mathbf{T}\in\mathbb{R}^{m\times m}\\ \det(\mathbf{T})\neq 0}} f(\mathbf{T}) \tag{6.41}$$

Although $f(\mathbf{T})$ is non-smooth and non-convex, efficient global optimisation methods exist for solving for this kind of optimisation problem. The ASA [17,18] is such an algorithm and is adopted to search for a true global optimum $\mathbf{T}_{\mathrm{opt}}$ of the problem (6.41). The detailed ASA algorithm is provided in Appendix B. With $\mathbf{T}_{\mathrm{opt}}$, the optimal controller realisation $\mathbf{w}_{\mathrm{opt}}$ can readily be obtained using the relationship (6.3).

6.5 A Numerical Example

A numerical example was used to illustrate the FWL optimal design procedure based on the pole-sensitivity approach. The plant model used was a modification of the plant studied in [5], which was a single-input single-output system. One more output, the first state in the original plant model, was added. The state space model of this modified plant was given by:

$$\mathbf{A} = \begin{bmatrix} 3.2439e-1 & -4.5451e+0 & -4.0535e+0 & -2.7003e-3 & 0 \\ 1.4518e-1 & 4.9477e-1 & -4.6945e-1 & -3.1274e-4 & 0 \\ 1.6814e-2 & 1.6491e-1 & 9.6681e-1 & -2.2114e-5 & 0 \\ 1.1889e-3 & 1.8209e-2 & 1.9829e-1 & 1.0000e+0 & 0 \\ 6.1301e-5 & 1.2609e-3 & 1.9930e-2 & 2.0000e-1 & 1 \end{bmatrix}$$

$$\mathbf{B} = \begin{bmatrix} 1.4518e-1 \\ 1.6814e-2 \\ 1.1889e-3 \\ 6.1301e-5 \\ 2.4979e-6 \end{bmatrix}$$

$$\mathbf{C} = \begin{bmatrix} 0 & 0 & 1.6188e+0 & -1.5750e-1 & -4.3943e+1 \\ 1 & 0 & 0 & 0 & 0 \end{bmatrix}$$

The closed-loop poles as given in [5] were used in design, and the designed reduced-order observer-based controller obtained using a standard design procedure [23] had the form:

$$\mathbf{F}_0 = \begin{bmatrix} 0 & 1 \\ -9.3303e-1 & 1.9319e+0 \end{bmatrix}$$

$$\mathbf{G}_0 = \begin{bmatrix} 4.1814e-2 & 2.7132e+2 \\ 3.9090e-2 & 1.0167e+3 \end{bmatrix}$$

$$\mathbf{J}_0 = \begin{bmatrix} 3.0000e-4 & 5.0000e-4 \end{bmatrix}$$

$$\mathbf{M}_0 = \begin{bmatrix} 0 & 6.1250e-1 \end{bmatrix}, \quad \mathbf{H}_0 = \begin{bmatrix} 7.8047e+1 \\ 7.3849e+1 \end{bmatrix}$$

With this initial controller realisation $\mathbf{w}_0$ and the plant model, the optimisation problem (6.41) was formed and solved for, giving rise to the following optimal similarity transformation matrix:

$$\mathbf{T}_{\text{opt}} = \begin{bmatrix} 1.4714e+1 & 3.2071e+1 \\ 1.3588e+1 & 3.0531e+1 \end{bmatrix}$$

From $\mathbf{T}_{\text{opt}}$, the corresponding optimal controller realisation $\mathbf{w}_{\text{opt}}$ was determined:

$$\mathbf{F}_{\text{opt}} = \begin{bmatrix} 9.8677e-1 & 1.4943e-2 \\ -2.9047e-2 & 9.4511e-1 \end{bmatrix}$$

$$\mathbf{G}_{\text{opt}} = \begin{bmatrix} 1.7066e-3 & -1.8080e+3 \\ 5.2084e-4 & 8.3794e+2 \end{bmatrix}$$

$$\mathbf{J}_{\text{opt}} = \begin{bmatrix} 1.1208e-2 & 2.4887e-2 \end{bmatrix}$$

$$\mathbf{M}_{\text{opt}} = \begin{bmatrix} 0 & 6.1250e-1 \end{bmatrix}, \quad \mathbf{H}_{\text{opt}} = \begin{bmatrix} 1.0691e+0 \\ 1.9430e+0 \end{bmatrix}$$

For the initial and optimal controller realisations, the true minimal bit lengths $B_s^{\min}$ that can guarantee the closed-loop stability were also determined using a computer simulation method. Table 6.1 compares the values of the two stability related measures, corresponding estimated minimum bit lengths and true minimum bit lengths for the initial and optimal controller realisations. The results clearly show that the new measure μ_{1I} is much less conservative than the existing measure μ_1 in estimating the true minimum bit length.

Table 6.1. Comparison of the two stability related measures, corresponding estimated minimum bit lengths and true minimum bit lengths for the initial and optimal controller realisations.

realisation	μ_{1I}	$\hat{B}_{s,1I}^{\min}$	μ_1	$\hat{B}_{s,1}^{\min}$	$B_s^{\min}$
$\mathbf{w}_0$	2.556877e-6	28	4.050854e-7	31	22
$\mathbf{w}_{\text{opt}}$	8.696940e-5	24	3.012354e-6	29	21

The unit impulse response of the closed-loop control system when the controllers were the infinite-precision implemented $\mathbf{w}_0$ and various FWL implemented realisations were also computed. Notice that any realisation $\mathbf{w} \in \mathcal{S}$, implemented in infinite precision, will achieve the exact performance of the infinite-precision implemented $\mathbf{w}_0$, which is the *designed* controller performance. For this reason, the infinite-precision implemented $\mathbf{w}_0$ is referred to as the *ideal* controller realisation $\mathbf{w}_{\text{ideal}}$. Figure 6.2 compares the unit impulse response of the first plant output $y_1(k)$ for the ideal controller implementation $\mathbf{w}_{\text{ideal}}$ with those of 21-bit implemented realisations $\mathbf{w}_0$ and $\mathbf{w}_{\text{opt}}$. It can be seen that the closed-loop became unstable with a 21-bit implemented controller realisation $\mathbf{w}_0$. However, the closed-loop system remained stable with the 21-bit implemented $\mathbf{w}_{\text{opt}}$.

6.6 Conclusions

The pole-sensitivity approach has been adopted to address the stability issue of the closed-loop discrete time control system where a digital controller is implemented with a fixed-point processor. A new FWL closed-loop stability related measure has been derived, which is a less conservative lower bound

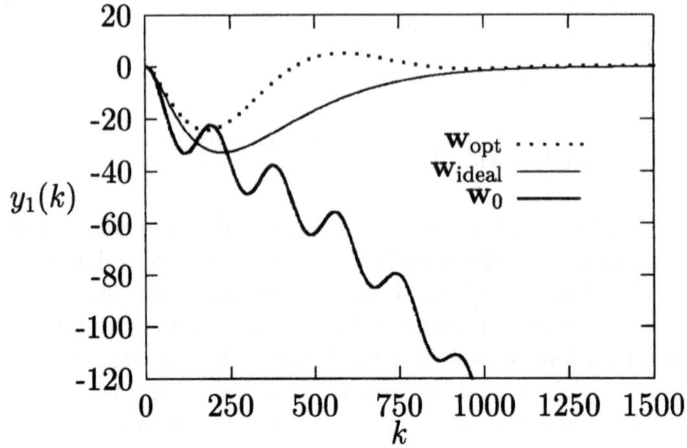

Fig. 6.2. Comparison of unit impulse response for the infinite-precision controller implementation $\mathbf{w}_{\text{ideal}}$ with those for the 21-bit implemented controller realisations $\mathbf{w}_0$ and $\mathbf{w}_{\text{opt}}$

of the computationally intractable true stability measure than other existing measures of the pole-sensitivity approach. As this new stability related measure is a function of the controller realisation, it can be used as a cost function for obtaining an optimal controller realisation that maximises the proposed measure. An efficient optimisation strategy has been developed for optimising a unified controller structure that includes output-feedback and observer-based controllers.

Acknowledgements

The authors wish to thank the support of the UK Royal Society under a KC Wong fellowship (RL/ART/CN/XFI/KCW/11949).

References

1. L.H. Keel and S.P. Bhattacharryya, "Robust, fragile, or optimal?" *IEEE Trans. Automatic Control*, Vol.42, No.8, pp.1098–1105, 1997.
2. P.M. Mäkilä, "Comments on 'Robust, fragile, or optimal?'," *IEEE Trans. Automatic Control*, Vol.43, No.9, pp.1265–1267, 1998.
3. P.M. Mäkilä, "Fragility and robustness puzzles," in *Proc. American Control Conf.* (San Diego, CA, USA), June 2-4, 1999, pp.2914–2919.
4. M. Gevers and G. Li, *Parametrizations in Control, Estimation and Filtering Problems: Accuracy Aspects*. London: Springer Verlag, 1993.
5. G. Li, "On the structure of digital controllers with finitez word length consideration," *IEEE Trans. Automatic Control*, Vol.43, pp.689–693, 1998.

6. R.H. Istepanian, G. Li, J. Wu and J. Chu, "Analysis of sensitivity measures of finite-precision digital controller structures with closed-loop stability bounds," *IEE Proc. Control Theory and Applications*, Vol.145, No.5, pp.472–478, 1998.
7. S. Chen, J. Wu, R.H. Istepanian and J. Chu, "Optimizing stability bounds of finite-precision PID controller structures," *IEEE Trans. Automatic Control*, Vol.44, No.11, pp.2149–2153, 1999.
8. J. Wu, S. Chen, G. Li and J. Chu, "Optimal finite-precision state-estimate feedback controller realization of discrete-time systems," *IEEE Trans. Automatic Control*, Vol.45, No.8, pp.1550–1554, 2000.
9. I.J. Fialho and T.T. Georgiou, "On stability and performance of sampled data systems subject to word length constraint," *IEEE Trans. Automatic Control*, Vol.39, No.12, pp.2476–2481, 1994.
10. I.J. Fialho and T.T. Georgiou, "Optimal finite wordlength digital controller realization," in *Proc. American Control Conf.* (San Diego, CA, USA), June 2-4, 1999, pp.4326-4327.
11. G.S.G. Beveridge and R.S. Schechter, *Optimization: Theory and Practice.* McGraw-Hill, 1970.
12. L.C.W. Dixon, *Nonlinear Optimisation.* London: English Universities Press, 1972.
13. D.E. Goldberg, *Genetic Algorithms in Search, Optimisation and Machine Learning.* Addison Wesley, 1989.
14. K.F. Man, K.S. Tang and S. Kwong, *Genetic Algorithms: Concepts and Design.* London: Springer-Verlag, 1998.
15. C.M. Fonseca and P.J. Fleming, "Multiobjective optimization and multiple constraint handling with evolutionary algorithms – Part I: A unified formulation," *IEEE Trans. Systems, Man, and Cybernetics Part A: Systems and Humans*, Vol.28, No.1, pp.26–37, 1998.
16. L. Ingber, "Simulated annealing: practice versus theory," *Mathematical and Computer Modelling*, Vol.18, No.11, pp.29–57, 1993.
17. L. Ingber, "Adaptive simulated annealing (ASA): lessons learned," *J. Control and Cybernetics*, Vol.25, No.1, pp.33-54, 1996.
18. S. Chen and B.L. Luk, "Adaptive simulated annealing for optimization in signal processing applications," *Signal Processing*, Vol.79, No.11, pp.117-128, 1999.
19. S. Chen, J. Wu, R.H. Istepanian, J. Chu and J.F. Whidborne, "Optimizing stability bounds of finite-precision controller structures for sampled-data systems in the delta operator domain," *IEE Proc. Control Theory and Applications*, Vol.146, No.6, pp.517–526, 1999.
20. S. Chen, R.H. Istepanian, J. Wu and J. Chu, "Comparative study on optimizing closed-loop stability bounds of finite-precision controller structures with shift and delta operators," *Systems and Control Letters*, Vol.40, No.3, pp.153–163, 2000.
21. J. Wu, S. Chen, G. Li, R.H. Istepanian and J. Chu, "Shift and delta operator realizations for digital controllers with finite-word-length considerations," *IEE Proc. Control Theory and Applications*, Vol.147, No.6, pp.664–672, 2000.
22. T. Kailath, *Linear Systems.* Prentice-Hall, 1980.
23. J. O'Reilly, *Observers for Linear Systems.* London: Academic Press, 1983.

6.A Appendix - Theorem Proof

Proof. Let the real-valued square matrix $\bar{\mathbf{A}} = \mathbf{V}_0 + \mathbf{V}_1\mathbf{X}\mathbf{V}_2$ be diagonalisable, where all the matrices concerned are real-valued with proper dimensions, and $\mathbf{V}_0$, $\mathbf{V}_1$ and $\mathbf{V}_2$ are independent of $\mathbf{X}$. From Lemma 1 in [5],

$$\frac{\partial \lambda_i(\bar{\mathbf{A}})}{\partial \mathbf{X}} = \mathbf{V}_1^T \mathbf{y}_i^*(\bar{\mathbf{A}})\mathbf{x}_i^T(\bar{\mathbf{A}})\mathbf{V}_2^T \tag{6.42}$$

where $\lambda_i(\bar{\mathbf{A}})$ denotes the i-th eigenvalue of $\bar{\mathbf{A}}$, $\mathbf{x}_i(\bar{\mathbf{A}})$ and $\mathbf{y}_i(\bar{\mathbf{A}})$ the related right and reciprocal left eigenvectors, respectively. Noting:

$$|\lambda_i(\bar{\mathbf{A}})| = \sqrt{\lambda_i^*(\bar{\mathbf{A}})\lambda_i(\bar{\mathbf{A}})} \tag{6.43}$$

leads to:

$$\begin{aligned}
\frac{\partial |\lambda_i(\bar{\mathbf{A}})|}{\partial \mathbf{X}} &= \frac{1}{2\sqrt{\lambda_i^*(\bar{\mathbf{A}})\lambda_i(\bar{\mathbf{A}})}} \left(\frac{\partial \lambda_i^*(\bar{\mathbf{A}})}{\partial \mathbf{X}} \lambda_i(\bar{\mathbf{A}}) + \lambda_i^*(\bar{\mathbf{A}}) \frac{\partial \lambda_i(\bar{\mathbf{A}})}{\partial \mathbf{X}} \right) \\
&= \frac{1}{2|\lambda_i(\bar{\mathbf{A}})|} \left(\left(\frac{\partial \lambda_i(\bar{\mathbf{A}})}{\partial \mathbf{X}} \right)^* \lambda_i(\bar{\mathbf{A}}) + \lambda_i^*(\bar{\mathbf{A}}) \frac{\partial \lambda_i(\bar{\mathbf{A}})}{\partial \mathbf{X}} \right) \\
&= \frac{1}{|\lambda_i(\bar{\mathbf{A}})|} \mathrm{Re} \left[\lambda_i^*(\bar{\mathbf{A}}) \frac{\partial \lambda_i(\bar{\mathbf{A}})}{\partial \mathbf{X}} \right] \\
&= \frac{1}{|\lambda_i(\bar{\mathbf{A}})|} \mathbf{V}_1^T \mathrm{Re} \left[\lambda_i^*(\bar{\mathbf{A}}) \mathbf{y}_i^*(\bar{\mathbf{A}}) \mathbf{x}_i^T(\bar{\mathbf{A}}) \right] \mathbf{V}_2^T
\end{aligned} \tag{6.44}$$

The closed-loop system matrix (6.5) has the following equivalent forms:

$$\bar{\mathbf{A}}(\mathbf{w}) = \begin{bmatrix} \mathbf{A} + \mathbf{BMC} & \mathbf{BJ} \\ \mathbf{GC} + \mathbf{HMC} & \mathbf{HJ} \end{bmatrix} + \begin{bmatrix} \mathbf{0} \\ \mathbf{I} \end{bmatrix} \mathbf{F} \begin{bmatrix} \mathbf{0} & \mathbf{I} \end{bmatrix} \tag{6.45}$$

$$\bar{\mathbf{A}}(\mathbf{w}) = \begin{bmatrix} \mathbf{A} + \mathbf{BMC} & \mathbf{BJ} \\ \mathbf{HMC} & \mathbf{F} + \mathbf{HJ} \end{bmatrix} + \begin{bmatrix} \mathbf{0} \\ \mathbf{I} \end{bmatrix} \mathbf{G} \begin{bmatrix} \mathbf{C} & \mathbf{0} \end{bmatrix} \tag{6.46}$$

$$\bar{\mathbf{A}}(\mathbf{w}) = \begin{bmatrix} \mathbf{A} + \mathbf{BMC} & \mathbf{0} \\ \mathbf{GC} + \mathbf{HMC} & \mathbf{F} \end{bmatrix} + \begin{bmatrix} \mathbf{B} \\ \mathbf{H} \end{bmatrix} \mathbf{J} \begin{bmatrix} \mathbf{0} & \mathbf{I} \end{bmatrix} \tag{6.47}$$

$$\bar{\mathbf{A}}(\mathbf{w}) = \begin{bmatrix} \mathbf{A} & \mathbf{BJ} \\ \mathbf{GC} & \mathbf{F} + \mathbf{HJ} \end{bmatrix} + \begin{bmatrix} \mathbf{B} \\ \mathbf{H} \end{bmatrix} \mathbf{M} \begin{bmatrix} \mathbf{C} & \mathbf{0} \end{bmatrix} \tag{6.48}$$

$$\bar{\mathbf{A}}(\mathbf{w}) = \begin{bmatrix} \mathbf{A} + \mathbf{BMC} & \mathbf{BJ} \\ \mathbf{GC} & \mathbf{F} \end{bmatrix} + \begin{bmatrix} \mathbf{0} \\ \mathbf{I} \end{bmatrix} \mathbf{H} \begin{bmatrix} \mathbf{MC} & \mathbf{J} \end{bmatrix} \tag{6.49}$$

Using (6.44) in (6.45)–(6.49) leads to (6.18)–(6.22). □

6.B Appendix - Adaptive Simulated Annealing

The ASA is a global optimisation scheme for solving for the following general optimisation problem:

$$\min_{\mathbf{x}\in\mathcal{X}} J(\mathbf{x}) \tag{6.50}$$

It evolves a single point $\mathbf{x} = [x_1 \cdots x_D]^T$ in the parameter or state space $\mathcal{X}$. The seemingly random search is guided by certain underlying probability distributions. Specifically, the general algorithm is described by three functions.

1. Generating probability density function

$$G(x_i^{\text{old}}, x_i^{\text{new}}, T_i; 1 \le i \le D) \tag{6.51}$$

This determines how a new state $\mathbf{x}^{\text{new}}$ is created, and from what neighbourhood and probability distributions it is generated, given the current state $\mathbf{x}^{\text{old}}$. The generating "temperatures" T_i describe the widths or scales of the generating distribution along each dimension x_i of the state space.

Often a cost function has different sensitivities along different dimensions of the state space. Ideally, the generating distribution used to search a steeper and more sensitive dimension should have a narrower width than that of the distribution used in searching a dimension less sensitive to change. The ASA adopts a so-called reannealing scheme to periodically re-scale T_i, so that they optimally adapt to the current status of the cost function. This is an important mechanism, which not only speeds up the search process but also makes the optimisation process robust to different problems.

2. Acceptance function

$$P_{\text{accept}}(J(\mathbf{x}^{\text{old}}), J(\mathbf{x}^{\text{new}}), T_c) \tag{6.52}$$

This gives the probability of $\mathbf{x}^{\text{new}}$ being accepted. The acceptance temperature T_c determines the frequency of accepting new states of poorer quality.

Probability of acceptance is very high at very high temperature T_c, and it becomes smaller as T_c is reduced. At every acceptance temperature, there is a finite probability of accepting the new state. This produces occasionally uphill move, enables the algorithm to escape from local minima, and allows a more effective search of the state space to find a global minimum. The ASA also periodically adapts T_c to best suit the status of the cost function. This helps to improve convergence speed and robustness.

3. Reduce temperatures or annealing schedule

$$\left.\begin{array}{l} T_c(k_c) \longrightarrow T_c(k_c+1) \\ T_i(k_i) \longrightarrow T_i(k_i+1),\ 1 \le i \le D \end{array}\right\} \tag{6.53}$$

where k_c and k_i are some annealing time indexes. The reduction of temperatures should be sufficiently gradual in order to ensure that the algorithm finds a global minimum.

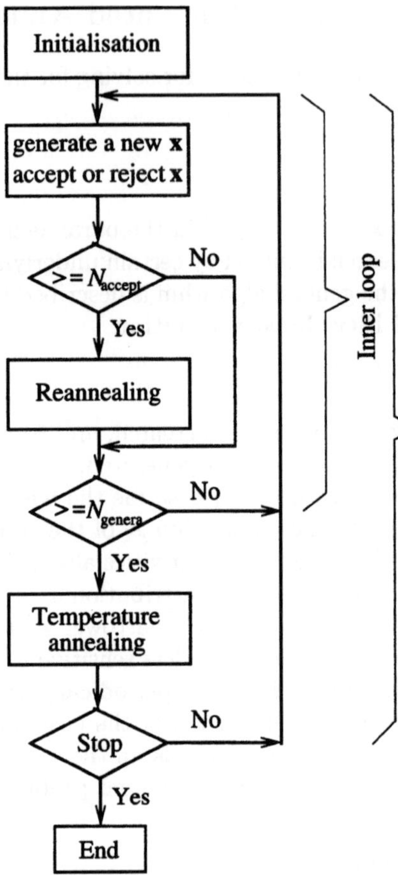

Fig. 6.3. Flow chart of the adaptive simulated annealing

This mechanism is based on the observations of the physical annealing process. When the metal is cooled from a high temperature, if the cooling is sufficiently slow, the atoms line themselves up and form a crystal, which is the state of minimum energy in the system. The annealing process usually must converge very slowly to ensure a global optimum. The ASA, however, can employ a very fast annealing schedule, as it has self adaptation ability to re-scale temperatures.

An implementation of the ASA algorithm, shown in Figure 6.3, is detailed as follows.

(i) Initialisation. An initial $\mathbf{x}$ is randomly generated, the initial temperature of the acceptance probability function, $T_c(0)$, is set to the initial value of the cost function $J(\mathbf{x})$, and the initial temperatures of the parameter generating probability functions, $T_i(0)$, $1 \le i \le D$, are set to 1.0.

A control parameter c in annealing process is given, and the annealing times, k_i for $1 \le i \le D$ and k_c, are all set to 0.

(ii) Generating. The algorithm generates a new point in the parameter space with:

$$x_i^{\text{new}} = x_i^{\text{old}} + g_i\,(U_i - V_i) \text{ and } \\ x_i^{\text{new}} \in [U_i,\ V_i],\ 1 \le i \le D. \tag{6.54}$$

Here U_i and V_i are the lower and upper bounds for x_i, g_i is calculated as:

$$g_i = \operatorname{sgn}\left(u_i - \frac{1}{2}\right) T_i(k_i)\left(\left(1 + \frac{1}{T_i(k_i)}\right)^{|2u_i - 1|} - 1\right) \tag{6.55}$$

and u_i a uniformly distributed random variable in [0, 1]. The value of the cost function $J(\mathbf{x}^{\text{new}})$ is then evaluated and the acceptance probability function of $\mathbf{x}^{\text{new}}$ is given by:

$$P_{\text{accept}} = \frac{1}{1 + \exp\left(\left(J(\mathbf{x}^{\text{new}}) - J(\mathbf{x}^{\text{old}})\right)/T_c(k_c)\right)} \tag{6.56}$$

A uniform random variable P_{unif} is generated in [0, 1]. If $P_{\text{unif}} \le P_{\text{accept}}$, $\mathbf{x}^{\text{new}}$ is accepted; otherwise it is rejected.

(iii) Reannealing. After every N_{accept} acceptance points, calculating the sensitivities:

$$s_i = \left|\frac{J(\mathbf{x}^{\text{best}} + \mathbf{1}_i\,\delta) - J(\mathbf{x}^{\text{best}})}{\delta}\right|, \quad 1 \le i \le D \tag{6.57}$$

where $\mathbf{x}^{\text{best}}$ is the best point found so far, δ is a small step size, the D-dimensional vector $\mathbf{1}_i$ has unit ith element and the rest of elements of $\mathbf{1}_i$ are all zeros. Let $s_{\max} = \max\{s_i,\ 1 \le i \le D\}$. Each T_i is scaled by a factor $s_{\max}/s_i$ and the annealing time k_i is reset:

$$T_i(k_i) = \frac{s_{\max}}{s_i} T_i(k_i),\ k_i = \left(-\frac{1}{c}\log\left(\frac{T_i(k_i)}{T_i(0)}\right)\right)^D. \tag{6.58}$$

Similarly, $T_c(0)$ is reset to the value of the last accepted cost function, $T_c(k_c)$ is reset to $J(\mathbf{x}^{\text{best}})$ and the annealing time k_c is rescaled accordingly:

$$k_c = \left(-\frac{1}{c}\log\left(\frac{T_c(k_c)}{T_c(0)}\right)\right)^D \tag{6.59}$$

(iv) Annealing. After every N_{genera} generated points, annealing takes place with:

$$\left.\begin{aligned} k_i &= k_i + 1 \\ T_i(k_i) &= T_i(0)\exp\left(-c k_i^{\frac{1}{D}}\right) \end{aligned}\right\} \quad 1 \le i \le D \tag{6.60}$$

and:

$$\left.\begin{aligned} k_c &= k_c + 1 \\ T_c(k_c) &= T_c(0) \exp\left(-c k_c^{\frac{1}{D}}\right) \end{aligned}\right\} \tag{6.61}$$

Otherwise, go to step (ii).

(v) **Termination.** The algorithm is terminated if the parameters have remained unchanged for a few successive reannealings or a preset maximum number of cost function evaluations has been reached; otherwise, go to step (ii).

CHAPTER 7

COMPUTATIONAL ALGORITHMS FOR SPARSE OPTIMAL DIGITAL CONTROLLER REALISATIONS

Ian J. Fialho and Tryphon T. Georgiou

Abstract. This chapter deals with the computation of optimal finite-precision digital controller realisations. The statistical word-length is used as a measure of optimality. Minimisation of the statistical word-length involves maximisation of robustness to structured perturbations in the digital controller, and maximisation of controller sparseness. The robustness issue is reformulated as a Linear Matrix Inequality problem for which efficient numerical solution methods exist. Sparseness maximisation is formulated as an appropriately constrained evolution that converges towards maximal sparseness. Two sparsing algorithms are presented, one of which is shown to rely purely on linear optimisation methods. Numerical issues associated with the algorithms are discussed and illustrated by means of an example.

7.1 Introduction

It is well known that finite word-length (FWL) digital controller implementation errors such as coefficient and signal quantisation can have a detrimental effect on the performance of a digital control systems [5,7]. The effect of coefficient quantisation on closed-loop performance can be addressed using metrics that capture the robustness of the closed-loop system to structured controller perturbations. To this end, several metrics such as the pole-sensitivity measure in [9], the linear-quadratic (LQ) measure in [10], and the $\mathcal{H}_\infty$/statistical measures in [5], have been proposed in the literature. When finite-precision effects are considered, all controller realisations are no longer equivalent as would be the case if infinite precision were assumed. It is well known that certain controller realisations exhibit superior finite-precision performance compared to others. All the above metrics depend on the controller realisation, and thus capture this fact. Optimisation algorithms that minimise the above metrics by computing optimal controller coordinate transformations can be found in the literature [3,7,9,12].

In this chapter we present computational algorithms for optimal finite-precision digital controller realisations, using the statistical word-length [5] as a measure of optimality. In Section 7.2 we formulate the problem at hand and introduce the statistical word-length for closed-loop stability. Minimisation of the statistical word-length involves maximisation of robustness to structured perturbations in the digital controller, and maximisation of controller sparseness. This is discussed in Section 7.3. The robustness issue is reformulated as a Linear Matrix Inequality problem, for which efficient numerical solution methods exist. Sparseness maximisation is formulated as an appropriately constrained evolution that converges towards maximal sparseness. Numerical issues associated with the algorithms are discussed in Section 7.4. Finally conclusions and scope for further work may be found in Section 7.5.

The notation in the chapter is standard. For a matrix $A \in \mathbb{R}^{n\times m}$, $\bar{\sigma}(A)$ denotes its largest singular value. $A^{-1/2}$ denotes the Hermitian square root of a symmetric positive-semidefinite matrix $A \in \mathbb{R}^{n\times n}$. The notation $A > 0/A < 0$ is used to denote positive/negative definiteness of the symmetric matrix $A \in \mathbb{R}^{n\times n}$. $\|T(z)\|_\infty := \sup_{z\in\partial\mathcal{D}} \bar{\sigma}(T(z))$, where $\partial\mathcal{D}$ is the boundary of the unit disc $\{z : |z| < 1\}$, denotes the $\mathcal{H}_\infty$-norm of the stable transfer function matrix $T(z)$. $\mathcal{N} \cap \mathcal{S}$ denotes the set intersection of $\mathcal{N}$ and $\mathcal{S}$. Finally, by a sparse matrix we mean a matrix that has "trivial" elements, *i.e.*, elements that are either $0, \pm 1$ or other powers of two.

7.2 Digital Controller Coefficient Quantisation

We consider the typical sampled-data feedback configuration shown in Figure 7.1, where **P** denotes a continuous time finite-dimensional linear plant, $\mathbf{K}_d$ a discrete time finite-dimensional linear controller, $\mathbf{H}_h$ and $\mathbf{S}_h$ synchronised hold and sampler at a sampling period h, and **W** an anti-aliasing (stable and strictly proper) filter.

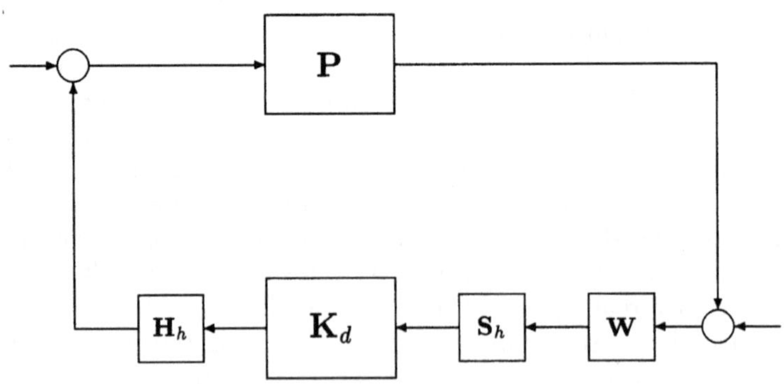

Fig. 7.1. Standard sampled-data feedback system

Let $[A_c, B_c, C_c, 0] := C_c(sI - A_c)^{-1}B_c$ denote a canonical realisation of the continuous time component **WP** and $[A_d, B_d, C_d, D_d] := D_d + C_d(\lambda I -$

$A_d)^{-1}B_d$ a canonical realisation of the discrete time controller $\mathbf{K}_d$ with n states, q inputs, and r outputs. Recall from [4] that under generic assumptions, namely **WP** having no unstable pole/zero cancellation and the sampling rate being non-pathological, the feedback system $[\mathbf{P}, \mathbf{H}_h\mathbf{K}_d\mathbf{S}_h\mathbf{W}]$ is $\mathcal{L}_2$-input/output stable if and only if the discrete time feedback system $[\mathbf{S}_h\mathbf{WPH}_h, \mathbf{K}_d]$ is stable. The latter feedback system is stable if the corresponding state matrix is stable, *i.e.*, has all eigenvalues in the interior of the unit disc $\mathcal{D} := \{\lambda : |\lambda| < 1\}$. For the natural realisation $[A_{cl}, B_{cl}, C_{cl}, D_{cl}]$ of this discrete time closed-loop system we have:

$$A_{cl} = A + BMC$$

where:

$$A := \begin{bmatrix} e^{A_c h} & 0 \\ 0 & 0 \end{bmatrix}, \quad B := \begin{bmatrix} \int_0^h e^{A_c \tau} B_c d\tau & 0 \\ 0 & I \end{bmatrix}$$

$$C := \begin{bmatrix} C_c & 0 \\ 0 & I \end{bmatrix}, \quad M := \begin{bmatrix} D_d & C_d \\ B_d & A_d \end{bmatrix}$$

Due to finite-precision fixed-point implementation of the digital controller we expect the controller system matrix M to assume the value $M + \Delta$. The matrix Δ contains all the controller coefficient quantisation errors. Thereby, A_{cl} will be perturbed to $A_{cl} + B\Delta C$. Due to this perturbation, the stability and performance properties of the as-implemented closed-loop system will differ from that of the ideal (infinite-precision) closed-loop. The size of the perturbation Δ depends on the word-length used for controller implementation, being larger for smaller word-lengths and vice versa. For a given digital controller $\mathbf{K}_d$, the following are some of the questions that arise in the process of fixed-point implementation:

1. For a given realisation of $\mathbf{K}_d$, compute the implementation word-length required to ensure closed-loop stability, and performance degradation within a specified tolerance ϵ.
2. Compute realisations of $\mathbf{K}_d$ that minimise the word-length needed for stable and ϵ-performance degradation controller implementation.

For a given realisation and word-length the perturbation matrix Δ can be explicitly computed, and hence (1) is easily addressed. In order to address (2) an analytical and computationally tractable estimate of the word-length necessary for stable and ϵ-performance degradation controller implementation is needed. One such estimate, the statistical word-length (SWL) for stability and performance was developed in [5] using robust stability theory and statistical arguments. In the development to follow we shall restrict our attention to the stability issue. The SWL for stable closed-loop controller implementation, as presented in [5], is given by the expression:

$$W_s = \left\lceil \log_2 \frac{2\sqrt{\frac{N}{12} + \sqrt{\frac{N}{45}}}}{\eta_1} \right\rceil \text{ in [bits]} \tag{7.1}$$

where, $\eta_1 := \{\|C(sI - A_{cl})^{-1}B\|_\infty\}^{-1}$ is the complex stability radius [8] for the triple $[A_{cl}, B, C]$, and N is the number of non-trivial elements in the controller realisation $[A_d, B_d, C_d, D_d]$. Under the assumptions in [5], if the controller $[A_d, B_d, C_d, D_d]$ is implemented with at least W_s [bits], the probability that the as-implemented closed-loop system will be stable is ≥ 0.9777. Observe that in order for a controller realisation to have a low SWL, η_1 must be large and N must be small, *i.e.*, the realisation must have a relatively large number of trivial elements. The stability radius η_1 is a natural measure of robustness of the closed-loop system to structured perturbations in the digital controller, while a large number of trivial elements is a desirable feature in any controller realisation. Since both the above features are desirable in a controller realisation we refer to realisations that minimise W_s as being ***stability-optimal*** and develop numerical algorithms for computing such realisations. We remark that several alternative metrics that capture optimality of the controller realisation can be found in the literature and refer the reader to [3,7,10–12] and the references therein.

7.3 Stability-optimal Controller Realisations

As discussed above, the SWL, W_s, depends on the $\mathcal{H}_\infty$-norm $\|C(zI - A_{cl})^{-1}B\|_\infty$, and on the number N of non-trivial elements in the realisation. Observe that under a controller coordinate transformation T the closed-loop A-matrix is given by:

$$A_{cl,T} = \begin{bmatrix} I & 0 \\ 0 & T^{-1} \end{bmatrix} A_{cl} \begin{bmatrix} I & 0 \\ 0 & T \end{bmatrix}$$

and hence the above $\mathcal{H}_\infty$-norm is strongly dependent on the digital controller realisation. We follow a two step algorithm in order to compute digital controller realisations $[A_d, B_d, C_d, D_d]$ that minimise W_s:

Step 1. Compute a controller realisation that minimises the above $\mathcal{H}_\infty$-norm. Note that the complete set of controller realisations is:

$$\{[T^{-1}A_dT, T^{-1}B_d, C_dT, D_d] : T \text{ invertible}\}$$

where $[A_d, B_d, C_d, D_d]$ is any realisation of the controller in question.

Step 2. Sparse (maximise the number of trivial elements) the realisation computed in Step 1 without altering the value of the $\mathcal{H}_\infty$-norm that was achieved.

7.3.1 Minimising the $\mathcal{H}_\infty$-norm

We begin with Step 1. The optimisation problem at hand is:

$$\min_{T \text{ invertible}} \|C(zI - A_{cl,T})^{-1}B\|_\infty \tag{7.2}$$

We denote the optimal value in (7.2) as γ_{opt} and the corresponding optimal controller coordinate transformation as T_{opt}.

Proposition 7.1. *The optimal value γ_{opt} is the smallest γ for which there exists a symmetric positive-definite matrix P of the form:*

$$P = \begin{bmatrix} P_1 & 0 & 0 \\ 0 & I & 0 \\ 0 & 0 & P_2 \end{bmatrix}, \; P > 0, P_1, P_2 \textit{ full symmetric}$$

such that:

$$[M_\gamma]^T P [M_\gamma] < P \tag{7.3}$$

where:

$$M_\gamma := \begin{bmatrix} A_{cl} & B/\gamma \\ C & 0 \end{bmatrix}$$

The optimal coordinate transformation matrix is given by $T_{opt} = P_{2,opt}^{-1/2}$ and the optimal controller realisation is given as:

$$[A_{d,opt}, B_{d,opt}, C_{d,opt}, D_{d,opt}] = [T_{opt}^{-1} A_d T_{opt}, T_{opt}^{-1} B_d, C_d T_{opt}, D_d]$$

A bisection search over γ can be performed to compute the smallest γ for which (7.3) holds. For a given γ value the condition (7.3) is a convex feasibility problem, specifically a Linear Matrix Inequality (LMI). Efficient interior point methods exist to numerically solve such LMIs [2,6] and these can be used to solve the optimisation problem as stated above.

Proof of Proposition 7.1. Using a constant matrix characterisation of the $\mathcal{H}_\infty$-norm [13] we observe that $\|C(zI - A_{cl,T})^{-1}B\|_\infty < \gamma$ iff $\exists$ an invertible matrix J such that:

$$\bar{\sigma}\left(\begin{bmatrix} J^{-1} & 0 \\ 0 & I \end{bmatrix} \begin{bmatrix} A_{cl,T} & B/\gamma \\ C & 0 \end{bmatrix} \begin{bmatrix} J & 0 \\ 0 & I \end{bmatrix} \right) < 1 \tag{7.4}$$

Extracting T from (7.4) we can rewrite it as:

$$\bar{\sigma}\left(\begin{bmatrix} J^{-1} & 0 \\ 0 & I \end{bmatrix}\begin{bmatrix} \begin{bmatrix} I & 0 \\ 0 & T^{-1} \end{bmatrix} & 0 \\ 0 & \begin{bmatrix} I & 0 \\ 0 & T^{-1} \end{bmatrix} \end{bmatrix}[M_\gamma] \right.$$

$$\left. \times \begin{bmatrix} \begin{bmatrix} I & 0 \\ 0 & T \end{bmatrix} & 0 \\ 0 & \begin{bmatrix} I & 0 \\ 0 & T \end{bmatrix} \end{bmatrix}\begin{bmatrix} J & 0 \\ 0 & I \end{bmatrix}\right) < 1 \tag{7.5}$$

Appealing to the polar decomposition theorem we can, without loss of generality, simplify (7.5) to:

$$\bar{\sigma}\left(\begin{bmatrix} \tilde{J}^{-1} & 0 \\ 0 & \begin{bmatrix} I & 0 \\ 0 & T^{-1} \end{bmatrix} \end{bmatrix}[M_\gamma]\begin{bmatrix} \tilde{J} & 0 \\ 0 & \begin{bmatrix} I & 0 \\ 0 & T \end{bmatrix} \end{bmatrix}\right) < 1 \tag{7.6}$$

where $\tilde{J}$ and T are symmetric positive-definite matrices. Using the fact that $\bar{\sigma}(F) < 1$ iff $F^T F - I < 0$ we can rewrite (7.6) as:

$$\begin{bmatrix} \tilde{J} & 0 \\ 0 & \begin{bmatrix} I & 0 \\ 0 & T \end{bmatrix} \end{bmatrix}^T [M_\gamma]^T \begin{bmatrix} \tilde{J}^{-1} & 0 \\ 0 & \begin{bmatrix} I & 0 \\ 0 & T^{-1} \end{bmatrix} \end{bmatrix}^T$$

$$\times \begin{bmatrix} \tilde{J}^{-1} & 0 \\ 0 & \begin{bmatrix} I & 0 \\ 0 & T^{-1} \end{bmatrix} \end{bmatrix}[M_\gamma]\begin{bmatrix} \tilde{J} & 0 \\ 0 & \begin{bmatrix} I & 0 \\ 0 & T \end{bmatrix} \end{bmatrix} < I \tag{7.7}$$

Equation (7.7) further simplifies to:

$$[M_\gamma]^T \begin{bmatrix} \tilde{J}^{-1} & 0 \\ 0 & \begin{bmatrix} I & 0 \\ 0 & T^{-1} \end{bmatrix} \end{bmatrix}^T \begin{bmatrix} \tilde{J}^{-1} & 0 \\ 0 & \begin{bmatrix} I & 0 \\ 0 & T^{-1} \end{bmatrix} \end{bmatrix}[M_\gamma]$$

$$< \begin{bmatrix} \tilde{J}^{-1} & 0 \\ 0 & \begin{bmatrix} I & 0 \\ 0 & T^{-1} \end{bmatrix} \end{bmatrix}^T \begin{bmatrix} \tilde{J}^{-1} & 0 \\ 0 & \begin{bmatrix} I & 0 \\ 0 & T^{-1} \end{bmatrix} \end{bmatrix} \tag{7.8}$$

Defining:

$$P := \begin{bmatrix} \tilde{J}^{-1} & 0 \\ 0 & \begin{bmatrix} I & 0 \\ 0 & T^{-1} \end{bmatrix} \end{bmatrix}^T \begin{bmatrix} \tilde{J}^{-1} & 0 \\ 0 & \begin{bmatrix} I & 0 \\ 0 & T^{-1} \end{bmatrix} \end{bmatrix}$$

the proposition follows. □

Example 7.1. We consider the closed-loop system $[\mathbf{S}_h \mathbf{WPH}_h, \mathbf{K}_d]$, where the discretised plant is described by:

$$A_{cd} = \begin{bmatrix} 9.987801244528510e-01 & -4.877144682830336e+00 & 1.219875547148251e-03 \\ 4.995170419650730e-04 & 9.971028117567653e-01 & -4.995170419650730e-04 \\ 1.677312696085738e-03 & 6.706009245190812e+00 & 9.983226873039137e-01 \end{bmatrix}$$

$$B_{cd} = \begin{bmatrix} 1.244643592795073e-01 \\ 3.111371466738237e-05 \\ 6.963031894512889e-05 \end{bmatrix}$$

$$C_{cd} = \begin{bmatrix} 1 \; 0 \; 0 \end{bmatrix}$$

$$D_{cd} = \begin{bmatrix} 0 \end{bmatrix}$$

and the stabilising digital controller $\mathbf{K}_d$ has the initial realisation:

$$A_d = \begin{bmatrix} 6.065306597126336e-01 & 0 \\ 0 & 1 \end{bmatrix}$$

$$B_d = \begin{bmatrix} -2.040736993091496e-02 \\ 1.888120758849921e-03 \end{bmatrix}$$

$$C_d = \begin{bmatrix} 5.186520991955975e+01 & -3.776241517699841e+00 \end{bmatrix}$$

$$D_d = \begin{bmatrix} 2.255000000000000e+00 \end{bmatrix}$$

Observe that this initial realisation has $N = 6$ non-trivial elements. The stability radius is computed to be $\eta_1 = 6.880607695271646e-04$, and hence the SWL, as defined in equation (7.1), is computed to be $W_s = 12$ [bits].

Solving the optimal realisation LMI, as given in Proposition 7.1, the following optimal realisation of $\mathbf{K}_d$ is achieved:

$$A_d = \begin{bmatrix} 6.065304992398700e-01 & 2.659671734539570e-04 \\ -2.374019972436824e-04 & 1.000000160472764e+00 \end{bmatrix}$$

$$B_d = \begin{bmatrix} -1.040186915718599e+00 \\ 8.528057666046039e-02 \end{bmatrix}$$

$$C_d = \begin{bmatrix} 1.017534391176377e+00 & -8.368336211035821e-02 \end{bmatrix}$$

$$D_d = \begin{bmatrix} 2.255000000000000e+00 \end{bmatrix}$$

The optimal realisation is fully parameterised, *i.e.*, $N = 9$. The optimal stability radius achieved was $\eta_1 = 1.174536058256988e-02$, which is approximately 17 times larger than that computed for the initial realisation above. The corresponding SWL is computed to be $W_s = 8$ [bits], a significant reduction from the 12 bits as computed above. This example will be continued in Section 7.4.

7.3.2 Sparse Controller Realisations

As seen in the previous example, the controller realisation that results from the $\mathcal{H}_\infty$ optimisation of Step 1 may be fully parameterised, *i.e.*, there may be no trivial elements. Hence, in order to further minimise the SWL the realisation must be sparsed without affecting the optimal value γ_{opt} achieved in Step 1. The sparsing algorithms presented herein are based on the evolution approach discussed in [1]. We present two such algorithms and discuss their relative numerical merits.

We begin with the observation that the optimal state coordinate transformation T_{opt} is not unique. Indeed, any transformation $T = T_{opt}U$, where U is a unitary matrix, would achieve the optimal value γ_{opt}. This freedom can be used to search for sparser realisations within the set of realisations that minimise the $\mathcal{H}_\infty$-norm in (7.2). The algorithms are based on an evolution $U(t)$ of unitary matrices and a corresponding evolution of controller realisations:

$$\begin{aligned}&[A_d(t), B_d(t), C_d(t), D_d(t)] := \\ &\qquad [U(t)^{-1}A_{d,opt}U(t), U(t)^{-1}B_{d,opt}, C_{d,opt}U(t), D_{d,opt}]\end{aligned}$$

The evolution $U(t)$ is chosen to drive the controller realisation evolution $[A_d(t), B_d(t), C_d(t), D_d(t)]$ in the direction of increasing sparseness. At each step of the evolution, $U(t)$ will be chosen so as to drive one element of the realisation towards a trivial value, while at the same time leaving already trivial elements unaltered.

We assume that the evolution $U(t)$ is governed by the differential equation:

$$\dot{U}(t) = U(t)F(t) \tag{7.9}$$

where $F(t)$ is chosen at each step to drive the evolution as described above. With this definition it can be shown that the differential equations that describe the evolution of controller realisations are:

$$\begin{aligned}\dot{A}_d(t) &= A_d(t)F(t) - F(t)A_d(t) \\ \dot{B}_d(t) &= -F(t)B_d(t) \\ \dot{C}_d(t) &= C_d(t)F(t)\end{aligned} \tag{7.10}$$

Remark 7.1. We point out that during implementation of the algorithm described below, the matrix equations (7.10) will have to be rewritten in vector form:

$$\begin{bmatrix} \dot{a}_{d,11} \\ \dot{a}_{d,21} \\ \cdot \\ \dot{a}_{d,n1} \\ \cdot \\ \dot{a}_{d,1n} \\ \dot{a}_{d,2n} \\ \cdot \\ \dot{a}_{d,nn} \\ \dot{b}_{d,11} \\ \dot{b}_{d,21} \\ \cdot \\ \dot{b}_{d,n1} \\ \cdot \\ \dot{b}_{d,1q} \\ \dot{b}_{d,2q} \\ \cdot \\ \dot{b}_{d,nq} \\ \dot{c}_{d,11} \\ \dot{c}_{d,21} \\ \cdot \\ \dot{c}_{d,r1} \\ \cdot \\ \dot{c}_{d,1n} \\ \dot{c}_{d,2n} \\ \cdot \\ \dot{c}_{d,rn} \end{bmatrix} = [\Gamma] \begin{bmatrix} f_{11} \\ f_{21} \\ \cdot \\ f_{n1} \\ \cdot \\ f_{1n} \\ f_{2n} \\ \cdot \\ f_{nn} \end{bmatrix}$$

where $\Gamma \in \mathbb{R}^{((n\times n)+(n\times q)+(r\times n))\times(n\times n)}$. For notational simplicity however, we will use the more compact format of (7.10) in the presentation to follow.

In order to facilitate numerical computations we replace the unitary evolution in (7.9) with its first-order Euler approximation:

$$U_{i+1} = U_i + \Delta_i U_i F_i \tag{7.11}$$

where Δ_i is the step size at the ith step.

Assume that at the ith step of the discrete evolution we have a set of elements of the realisation $\{\theta_j : j = 1, \ldots m\}$ that are already trivial elements. Hence, the first constraint on the choice of F_i is that these elements remain trivial, *i.e.*,

$$\dot{\theta}_j = 0, \text{ for } j = 1, \ldots m \tag{7.12}$$

This means that F_i must be of the form:

$$F_i = \sum_{l=1}^{p} \lambda_l F^l \tag{7.13}$$

where $\{F^l\}$ is a basis for the solution space $\mathcal{N}$ of equation (7.12).

The second constraint on the choice of F_i is that it must be chosen so that U_{i+1} remains unitary. It is easy to see that this is equivalent to the condition $F_i^T + F_i + \Delta_i F_i^T F_i = 0$. Using the form of F_i from (7.13) we see that this reduces to:

$$\sum_{l=1}^{p} \lambda_l ({F^l}^T + F^l) + \Delta_i (\sum_{l=1}^{p} \lambda_l {F^l}^T)(\sum_{l=1}^{p} \lambda_l F^l) = 0$$

which is a nonlinear constraint on the choice of the λ_ls.

Now, let β denote the element that we are trying to drive towards a trivial value. Hence, depending on the current value of β and the trivial value towards which we seek to drive it, the λ_ls must be chosen to minimise (or maximise) the rate of change of β towards the trivial value. This is stated as a nonlinear minimisation (or maximisation) problem:

$$\begin{array}{l} \min_{\{\lambda_l\}} \dot{\beta} \\ \text{subject to} \\ \sum_{l=1}^{p} \lambda_l ({F^l}^T + F^l) + \Delta_i (\sum_{l=1}^{p} \lambda_l {F^l}^T)(\sum_{l=1}^{p} \lambda_l F^l) = 0 \end{array} \tag{7.14}$$

Observe that $\lambda_l = 0$ is an initial feasible point. Numerical algorithms such as the MATLAB® function `fmincon` can be used to efficiently find local minima (or maxima) for (7.14). We proceed to summarise the algorithm.

Sparsing Algorithm 7.1. Set $U_0 = I$. The ith step of the algorithm is as follows:

- Let $\{\theta_j : j = 1, \dots m\}$ be the set of elements that have achieved trivial value. Compute a basis $\{F^l : l = 1, \dots p\}$ for the solution space of $\dot{\theta}_j = 0,\ j = 1, \dots m$.
- Let β denote the element that we are trying to drive towards a trivial value. Compute a numerical solution to the nonlinear minimisation (or maximisation) problem:

$$\begin{array}{l} \min_{\{\lambda_l\}} \dot{\beta} \\ \text{subject to} \\ \sum_{l=1}^{p} \lambda_l ({F^l}^T + F^l) + \Delta_i (\sum_{l=1}^{p} \lambda_l {F^l}^T)(\sum_{l=1}^{p} \lambda_l F^l) = 0 \end{array}$$

- Set $F_i = \sum_{l=1}^{p} \lambda_l F^l$. Update U_i to:

$$U_{i+1} = U_i + \Delta_i U_i F_i$$

and:

$$\begin{aligned} &[A_{d,i+1}, B_{d,i+1}, C_{d,i+1}, D_{d,i+1}] := \\ &\qquad [U_{i+1}^{-1} A_{d,i} U_{i+1}, U_{i+1}^{-1} B_{d,i}, C_{d,i} U_{i+1}, D_{d,i}] \end{aligned}$$

We proceed to discuss a second version of the sparsing algorithm that is based on an alternate formulation of the unitarity constraint on the matrix evolution $U(t)$. Assume that at "time" t the matrix $U(t)$ satisfies the unitary constraint $U^T(t)U(t) = I$. The constraint that ensures that this continues to be the case is $\frac{d}{dt}[U^T(t)U(t)] = 0$. This condition can be rewritten as:

$$[\frac{d}{dt}U^T(t)]U(t) + U^T(t)[\frac{d}{dt}U(t)] = 0$$

Using the form of $\dot{U}(t)$ from (7.9) and the fact that $U^T(t)U(t) = I$, the above equation reduces to:

$$F^T(t) = -F(t)$$

i.e., at every t the matrix $F(t)$ must be chosen to be skew-symmetric in order to maintain unitarity of $U(t)$. Combining this with the constraint imposed on the choice of F by equation (7.12) we see that at every step of the discretised algorithm the matrix F_i must be chosen from the intersection of the solution space $\mathcal{N}$ of equation (7.12) and the space of skew-symmetric matrices $\mathcal{S}$. This means that F_i must be of the form:

$$F_i = \sum_{l=1}^{p} \lambda_l {F^l}_{\mathcal{N}\cap\mathcal{S}} \tag{7.15}$$

where $\{{F^l}_{\mathcal{N}\cap\mathcal{S}}\}$ is a basis for $\mathcal{N}\cap\mathcal{S}$. We refer the reader to the next section for a numerical algorithm to compute a basis set for the intersection of two linear subspaces. We proceed to summarise the algorithm.

Sparsing Algorithm 7.2. Set $U_0 = I$. The ith step of the algorithm is as follows:

- Let $\{\theta_j : j = 1, \ldots m\}$ be the set of elements that have achieved trivial value. Compute a basis $\{{F^l}_{\mathcal{N}\cap\mathcal{S}} : l = 1, \ldots p\}$ for $\mathcal{N}\cap\mathcal{S}$.
- Let β denote the element that we are trying to drive towards a trivial value. Compute a numerical solution to the linear minimisation (or maximisation) problem:

  $$\min_{\{\lambda_l\}} \dot{\beta}$$

- Set $F_i = \sum_{l=1}^{p} \lambda_l F^l$. Update U_i to:

  $$U_{i+1} = U_i + \Delta_i U_i F_i$$

  and:

  $$[A_{d,i+1}, B_{d,i+1}, C_{d,i+1}, D_{d,i+1}] := [U_{i+1}^{-1} A_{d,i} U_{i+1}, U_{i+1}^{-1} B_{d,i}, C_{d,i} U_{i+1}, D_{d,i}]$$

7.4 Numerical Issues

We proceed to discuss some of the numerical issues associated with the implementation of the two sparsing algorithms presented above. In both algorithms, the constraint that trivial elements remain trivial, *i.e.*, equation (7.12), is imposed directly in the "continuous" domain. The unitarity constraint on $U(t)$, which is key to ensuring that the $\mathcal{H}_\infty$-norm remains constant, is imposed in the "discrete" domain in Algorithm 7.1. The same constraint is imposed directly in the "continuous" domain in Algorithm 7.2. Due to this fact, Algorithm 7.1 can be run with a significantly larger time step Δ_i than Algorithm 7.2, which requires tight control on the time step to ensure that unitarity is maintained. Both algorithms however, do require that the time step decreases as the element converges to its trivial value.

The iterations in Algorithm 7.1 can continue as long as the solution space of (7.12) is non-trivial. This would imply that the algorithm is capable of trivialising at least $(n \times n)$ elements in the controller realisation. As a practical matter though, as the dimension of the solution space of (7.12) reduces, the gradient minima that can be achieved in the optimisation problem (7.14) reduces as well. The convergence of the algorithm slows considerably, and hence the theoretical expectation of $(n \times n)$ trivial elements is not always realisable.

Similarly, the iterations in Algorithm 7.2 can continue as long as the subspace $\mathcal{N} \cap \mathcal{S}$ is non-trivial. This, in general, is a more conservative constraint than that imposed on the iterations in Algorithm 7.1, especially in the case of low-order (small n) controllers. For example, if $n = 2$ the dimension of the space of skew-symmetric matrices $\mathcal{S}$ is 1. As more elements are trivialised the dimension of the subspace $\mathcal{N}$ reduces and hence the likelihood that $\mathcal{N} \cap \mathcal{S}$ is non-trivial decreases. Hence, it is more appropriate to use Algorithm 7.2 while sparsing higher order controllers. The advantage of Algorithm 7.2 is the fact that the constrained nonlinear optimisation problem (7.14) is replaced by an unconstrained linear one whose global optima can be efficiently computed. Given the fact that Algorithm 7.2 is more suited to sparsing higher order controllers this feature takes on added significance. The additional step in Algorithm 7.2 is the computation of a basis for $\mathcal{N} \cap \mathcal{S}$, an algorithm for which is discussed next.

Let $\{F^1, F^2, \ldots, F^p\}$ denote a basis for $\mathcal{N}$ and $\{S^1, S^2, \ldots, S^k\}$ denote a basis for $\mathcal{S}$. Observe that there exists a non-zero $x \in \mathcal{N} \cap \mathcal{S}$ iff there exists a non-zero $\lambda \in \mathbb{R}^p$ and a non-zero $\alpha \in \mathbb{R}^k$ such that:

$$x = [F^1\ F^2\ \ldots\ F^p]\lambda = [S^1\ S^2\ \ldots\ S^k]\alpha \tag{7.16}$$

Defining:

$$\phi := \begin{bmatrix} \lambda \\ \alpha \end{bmatrix}$$

we can rewrite equation (7.16) as:

$$[F^1 \ F^2 \ \dots \ F^p \ -S^1 \ -S^2 \ \dots \ -S^k]\phi = 0 \tag{7.17}$$

Let:

$$\left\{ \begin{matrix} P_\lambda^1 \ P_\lambda^2 \dots P_\lambda^s \\ P_\alpha^1 \ P_\alpha^2 \dots P_\alpha^s \end{matrix} \right\}$$

denote a basis for the solution space of equation (7.17). Then x is of the form:

$$x = [F^1 \ F^2 \ \dots \ F^p][P_\lambda^1 \ P_\lambda^2 \ \dots \ P_\lambda^s]\zeta$$

for some $\zeta \in \mathbb{R}^s$. This implies that the columns of the matrix:

$$[F^1 \ F^2 \ \dots \ F^p][P_\lambda^1 \ P_\lambda^2 \ \dots \ P_\lambda^s]$$

span the subspace $\mathcal{N} \cap \mathcal{S}$. We proceed to illustrate the above algorithms by sparsing the optimal realisation computed in the example in Section 7.3.1.

Example 7.2. Using Sparsing Algorithm 7.1, three elements of the optimal realisation were driven to trivial values. First, $A_d(2,2)$ was driven to 1. Then $A_d(2,1)$ was driven to 0, and finally $C_d(1,1)$ was driven to 1. The convergence of the algorithm was found to be extremely slow while attempting to sparse a fourth element, and hence was abandoned. The final sparse optimal realisation is as follows:

$$A_d = \begin{bmatrix} 6.065306597126329e-01 & 1.167645293413626e-03 \\ 0 & 1 \end{bmatrix}$$

$$B_d = \begin{bmatrix} -1.058177588357172e+00 \\ 8.590785232910426e-02 \end{bmatrix}$$

$$C_d = \begin{bmatrix} 1 & -8.596346917804266e-02 \end{bmatrix}$$

$$D_d = \begin{bmatrix} 2.255000000000000e+00 \end{bmatrix}$$

The sparse optimal realisation has $N = 6$ non-trivial elements. The stability radius was computed to be $\eta_1 = 1.174536058256988e-02$, which is the same as that for the optimal realisation above. The corresponding SWL is computed to be $W_s = 8$ [bits], which is the same as that computed for the optimal realisation.

Figures 7.2 to 7.4 illustrate the convergence of the sparsing algorithm for the three elements $A_d(2,2)$, $A_d(2,1)$, and $C_d(1,1)$. These plots show the iteration step size and distance to the trivial value versus iteration number. Notice the relatively large step size used in all three sparsing passes, as well as the reduction in step size as the element nears its trivial value. Convergence was restricted to be one-sided (either from above or below) and the step size was reduced whenever undershoot or overshoot of the trivial value occured.

The properties of the three controller realisations described above are summarised in Table 7.1. In addition to the SWL W_s, the table also lists the actual word-length $W_{s,act}$ necessary for stable closed-loop implementation. This illustrates that the optimal and sparse optimal realisations are significantly superior from the point of view of actual implementation.

Table 7.1. Summary of realisation properties

Realisation	W_s [bits]	$W_{s,act}$ [bits]	N
Initial	12	10	6
Optimal	8	5	9
Sparse Optimal	8	5	6

Finally, in order to illustrate Sparsing Algorithm 7.2, the optimal realisation described above was sparsed using this approach. As discussed above, due to the conservativeness of imposing the unitarity constraint in the "con-

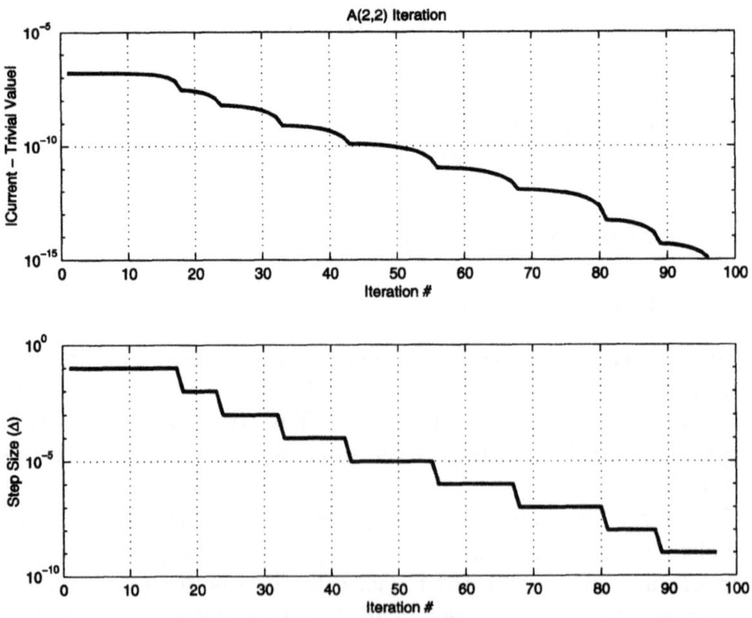

Fig. 7.2. $A_d(2,2)$ convergence (Algorithm 7.1)

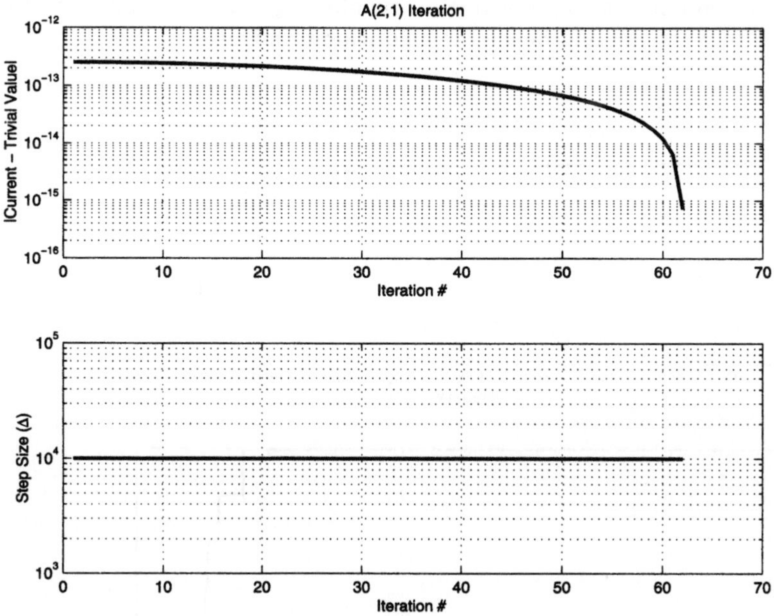

Fig. 7.3. $A_d(2,1)$ convergence (Algorithm 7.1)

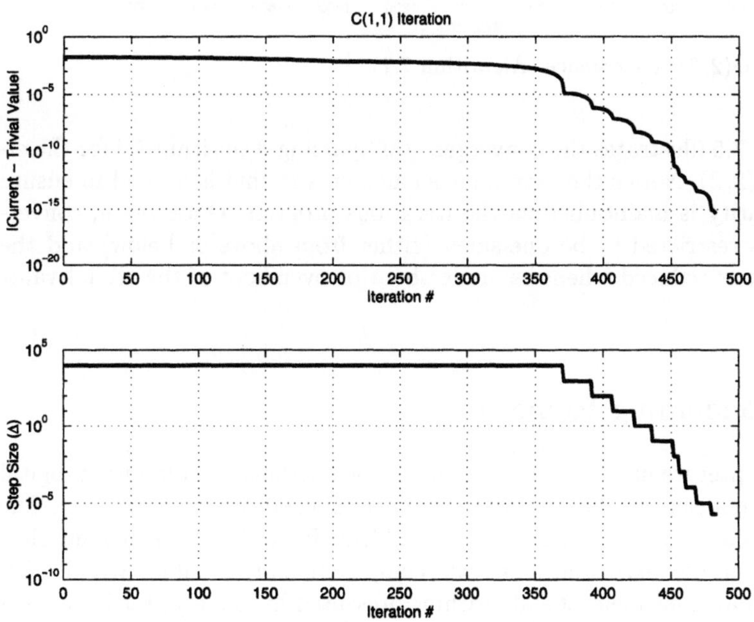

Fig. 7.4. $C_d(1,1)$ convergence (Algorithm 7.1)

tinuous" domain, it was only possible to trivialise $A_d(2,2)$. Thereafter, $\mathcal{N} \cap \mathcal{S}$ was found to be trivial and hence the algorithm terminates.

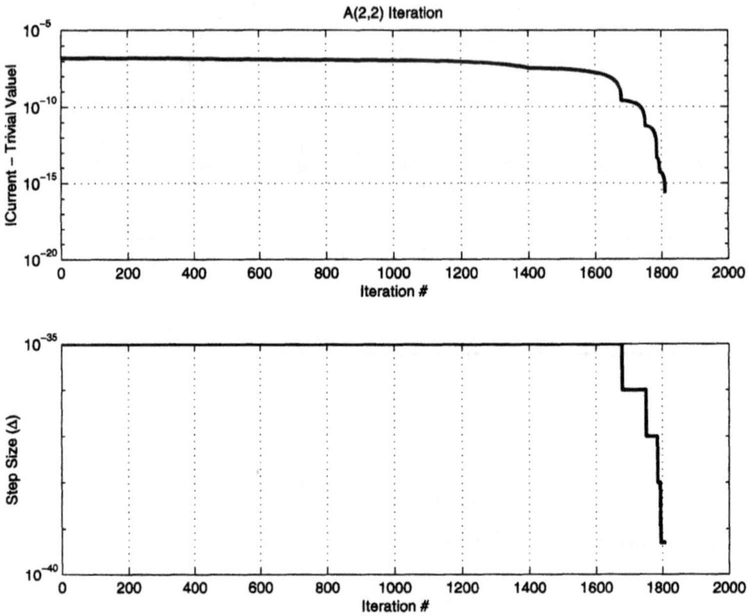

Fig. 7.5. $A_d(2,2)$ convergence (Algorithm 7.2)

Figure 7.5 illustrates the convergence of Sparsing Algorithm 7.2 for the element $A_d(2,2)$. Notice the extremely small step size that is needed to ensure that unitarity is maintained as the iterations progress. Once again, convergence was restricted to be one-sided (either from above or below) and the step size was reduced whenever undershoot or overshoot of the trivial value occurred.

7.5 Concluding Remarks

In this chapter we have presented numerical algorithms for computing digital controller realisations that minimise the effect of coefficient quantisation. The algorithms are based on optimisation problems for which efficient numerical solution methods exist and hence are computationally feasible. MATLAB® routines that implement the algorithms presented in this chapter have been developed and are available from the first author upon request. Performance issues were not addressed in this chapter, our focus being the stability of the closed-loop system. Performance under coefficient quantisation, along with other finite-precision effects such as roundoff noise and signal quantisation effects need to be addressed in future work. Ultimately, the above FWL constraints should be incorporated into the control design process itself, in order to capture errors that are inevitable in the post-design implementation phase.

References

1. B.W. Bomar and J.C. Hung, "Minimum Roundoff Noise Digital Filters With Some Power-Of-Two Coefficients", *IEEE Transactions on Circuits and Systems*, vol. CAS-31, pp. 833-840, 1984.
2. S. Boyd, L. El Ghaoui, E. Feron and V. Balakrishnan, *Linear Matrix Inequalities in System and Control Theory*, Studies in Applied Mathematics, vol. 15, SIAM, 1994.
3. S. Chen, J. Wu, R.H. Istepanian and J. Chu, "Optimizing Stability Bounds of Finite-Precision PID Controller Structures", *IEEE Transactions on Automatic Control*, vol. 44, no. 11, pp. 2149-2153, 1999.
4. T. Chen and B.A. Francis, "Input-Output Stability Of Sampled Data Systems", *IEEE Transactions on Automatic Control*, vol. 36, no. 1, pp 50-58, 1991.
5. I.J. Fialho and T.T. Georgiou, "On Stability and Performance of Sampled-Data Systems Subject to Wordlength Constraint", *IEEE Transactions on Automatic Control*, vol. 39, no.12, pp. 2476-2481, 1994.
6. P. Gahinet, A. Nemirovski, A.J. Laub and M. Chilali, *LMI Control Toolbox*, The MathWorks Inc., 1995.
7. M. Gevers and G. Li, *Parametrizations in Control, Estimation and Filtering Problems: Accuracy Aspects*, London: Springer-Verlag, 1993.
8. D. Hinrichsen and A.J. Pritchard, "Stability Radius for Structural Perturbations and the Algebraic Riccati Equation", *System and Control Letters*, vol. 8, pp. 105-113, 1986.
9. G. Li, "On the Structure of Digital Controllers with Finite Wordlength Consideration", *IEEE Transactions on Automatic Control*, vol. 43, pp. 689-693 , 1998.
10. P. Moroney, A.S. Willsky and P.K. Houpt, "The Digital Implementation of Control Compensators: the Coefficient Wordlength Issue", *IEEE Transactions on Automatic Control*, vol. 25, no. 4, pp. 621-630, 1980.
11. D. Williamson and K. Kadiman, "Optimal Finite Wordlength Linear Quadratic Regulation", *IEEE Transactions on Automatic Control*, vol. 34, no. 12, pp. 1218-1228, 1989.
12. J. Wu, S. Chen, G. Li and J. Chu, "Optimal Finite-Precision State-Estimate Feedback Controller Realizations of Discrete-Time Systems", *IEEE Transactions on Automatic Control*, vol. 45, no. 8, pp. 1550-1554, 2000.
13. K. Zhou, J.C. Doyle and K. Glover, *Robust and Optimal Control*, Prentice Hall, 1996.

CHAPTER 8

ON THE STRUCTURE OF DIGITAL CONTROLLERS IN SAMPLED-DATA SYSTEMS WITH STABILITY CONSIDERATION

Gang Li

Abstract. In this chapter, we address the digital controller structure problem with stability robustness consideration for a sampled-data control system, where the digital controller is implemented in a state space realisation with finite word-length. A pole-sensitivity based stability measure is derived. Noting that this measure is realisation dependent, the optimal controller realisation problem is to find those realisations that maximise this measure. An efficient numerical algorithm is proposed to solve this problem. For real time applications, it would be better to implement the controller using such a realisation that not only yields a good stability robustness but also possesses as many trivial parameters (such as $0, \pm 1$) as possible. An algorithm is developed for finding sparse realisations of nice stability behaviour.

8.1 Introduction

A digitally controlled system consists of the continuous time plant $P(s)$, and a sampled-data controller composed of a sampler (A/D converter) S, the digital controller $C_d(z)$ to be designed and a hold (D/A converter) H. An anti-aliasing filter $F(s)$ must be introduced before the sampler S. See Figure 8.1.

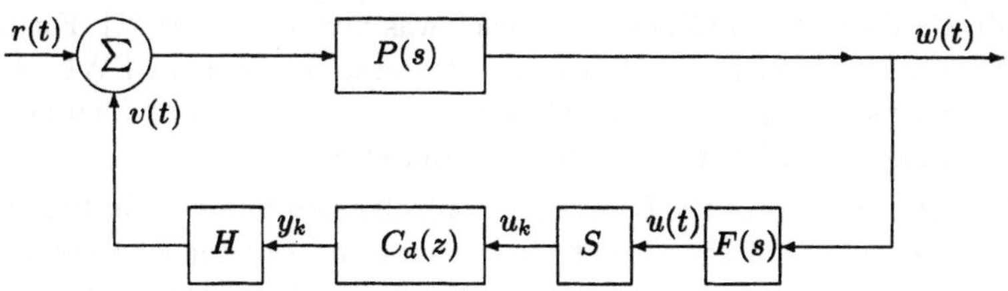

Fig. 8.1. Block-diagram of a sampled-data feedback system

There are basically three ways to design the controller. The first is to design the controller in the continuous time domain and then perform a digital implementation of the controller. The second way is to design the digital controller based on a discretised model of the plant. The third way is to do the digital design directly in the continuous time domain. The designed digital controller, no matter which way it is used, has to be implemented with a digital device such as a digital control processor (DCP). Due to the finite word-length (FWL) effects, the actually implemented controller is different from the designed one and consequently, the performance of the sampled-data system degrades.

Traditionally, the FWL effects are classified into two types, which are studied separately. The first type of FWL error is the roundoff error that occurs in arithmetic operations and is usually measured with the so-called roundoff noise gain. The second one is perturbation of system parameters implemented with FWL. These effects of FWL errors are classically studied with a transfer function sensitivity measure. It is well known that a linear system/controller can be implemented with many different structures such as cascade/parallel forms and state space realisations. In the infinite-precision situation, all the structures are totally equivalent. However, different structures have different numerical properties. This implies that they are no longer equivalent in finite-precision situations. It is interesting to find those structures that minimise the FWL effects in a certain sense. Therefore, the optimal structure problem consists of two aspects: (i) to define an appropriate measure, (ii) to optimise this measure over a class of controller structures.

It should be pointed out that the effects of FWL effects have been well studied in digital signal processing, particularly in digital filter implementation [1,2]. It was not until the late 1980s that the FWL effects on control systems were seriously addressed. In [3], the "optimal" controller realisation was computed with the loop opened, that is, the controller was considered as an isolated digital filter. This realisation is obviously not optimal in the sense that it does not minimise the roundoff noise in the closed-loop system. A roundoff noise gain was derived for a control system with state-estimate feedback controller and the corresponding optimal realisation problem was solved in [4]. The effect of FWL errors of the regulator parameters on linear quadratic Gaussian (LQG) performance was investigated in [5]. For the roundoff error effect on the same control strategy, the optimal FWL-LQG design problem was studied and a sub optimal solution was provided in [6], while the optimal solution was obtained by Liu *et al.* [7].

In [2,4], the analysis of FWL errors in controllers on the transfer function was performed based on the discrete time counterpart of the sampled-data system. The corresponding sensitivity measure does not take the inter-sample behaviour into account. To overcome this, Madievski *et al.* [8] derived a sensitivity measure based on a hybrid operator (transfer function) of the sampled-data system.

The stability of the sampled-data system may be lost due to the FWL errors, which are not considered when the digital controller is designed. In this chapter, we adopt a stability robustness related approach and consider the situation where the digital controller is implemented with a state space realisation. One of the main objectives is to derive a stability robustness related measure for the sampled-data system and to find those controller realisations that maximise this measure. For real time applications, it would be better to implement the controller using such a structure that not only yields a good stability robustness against the FWL errors but also possesses as many trivial parameters (such as $0, \pm 1$) as possible. The second topic in this chapter is to find such a structure.

8.2 Digital Controller State Space Implementation

It is well known that the digital controller $C_d(z)$ can be implemented with its state space equations:

$$\begin{aligned} x_{k+1} &= Ax_k + Bu_k \\ y_k &= Cx_k + du_k \end{aligned} \tag{8.1}$$

where $u_k = u(kT_s)$, $A \in \mathbb{R}^{n_c \times n_c}$, $B \in \mathbb{R}^{n_c \times 1}$, $C \in \mathbb{R}^{1 \times n_c}$ and $d \in \mathbb{R}$. $R \triangleq (A, B, C, d)$ is called a realisation of $C_d(z)$, satisfying:

$$C_d(z) = d + C(zI - A)^{-1}B \tag{8.2}$$

Denote S_{C_d} as the set of all realisations (A, B, C, d). It should be pointed out that S_{C_d} is an infinite set. In fact, if $(A_0, B_0, C_0, d) \in S_{C_d}$, $S_{C_d} = \{(A, B, C, d)\}$ can be characterised with:

$$A = T^{-1}A_0T, \qquad B = T^{-1}B_0, \qquad C = C_0T \tag{8.3}$$

where $T \in \mathbb{R}^{n_c \times n_c}$ is any non-singular matrix. Usually, such a T is called a similarity transformation.

It should be pointed out that (8.1) is the digital controller implemented with infinite precision. Though there exist different state space realisations, they yield exactly the same performance — the desired one. In real time applications, however, the well designed digital controller has to be implemented with finite precision. In fact, the practical problems of numerical computation begin with the finite representation of real numbers. There are two basic ways of representing real numbers. One is fixed-point representation, the other is floating-point representation. The latter has a large dynamic range and hence is more accurate. For real time applications, fixed-point implementation is usually adopted since it is faster and cheaper. The FWL problem for fixed-point representation, however, is more serious. In the sequel, a fixed-point implementation of B_c bits for coefficients is assumed.

The stability may be lost when the digital controller $C_d(z)$ is implemented in (8.1), where (A, B, C, d) is replaced with its FWL version. This problem is more serious when fast sampling is used [9]. Chen and Francis [10] showed that under generic assumptions, namely, $P(s)F(s)$ having no unstable pole/zero cancellation and the sampling rate being non-pathological [11], the sampled-data system depicted in Figure 8.1 is L_2 input/output stable *if and only if* the discrete time feedback system shown in Figure 8.2 is exponentially stable. This is a very important result, which allows us to study the stability behaviour of a sampled-data system via the corresponding equivalent discrete time feedback system.

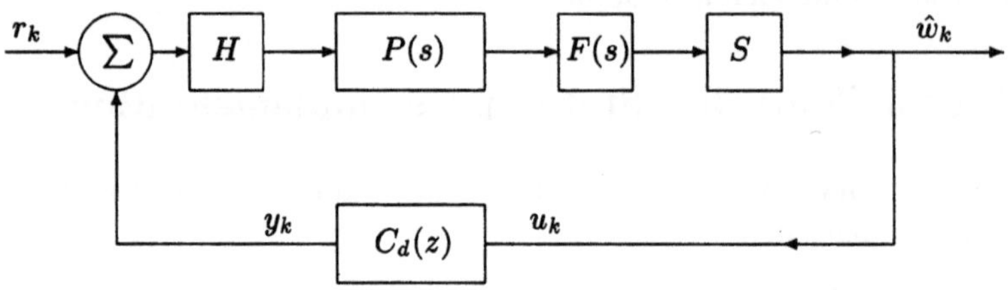

Fig. 8.2. Block-diagram of the equivalent discrete time feedback system

Suppose $P(s)F(s)$ is strictly proper. Let (A_s, B_s, C_s) be one of its realisations. Then (A_z, B_z, C_z) is the corresponding realisation of the discrete time system $HP(s)F(s)S$, where:

$$A_z = e^{A_s T_s}, \qquad B_z = \int_0^{T_s} e^{A_s \tau} B_s d\tau, \qquad C_z = C_s \tag{8.4}$$

where $T_s = 1/F_s$ is the sampling period.

This discrete time system can also be represented by its state space equations:

$$\begin{aligned} x_{k+1}^p &= A_z x_k^p + B_z(r_k + y_k) \\ \hat{w}_k &= C_z x_k^p \end{aligned} \tag{8.5}$$

Noting $u_k = \hat{w}_k$, it is easy to show that the corresponding state space equations of the equivalent discrete time closed-loop system, in which the digital controller is implemented with (8.1), are:

$$\begin{aligned} \hat{x}_{k+1} &= \bar{A}\hat{x}_k + \bar{B} r_k \\ \hat{w}_k &= \bar{C}\hat{x}_k \end{aligned} \tag{8.6}$$

where $\hat{x}_k \triangleq \begin{pmatrix} x_k^p \\ x_k \end{pmatrix}$ and:

$$\bar{A} \triangleq \begin{pmatrix} A_z + dB_z C_z & B_z C \\ BC_z & A \end{pmatrix}, \qquad \bar{B} \triangleq \begin{pmatrix} B_z \\ 0 \end{pmatrix}, \qquad \bar{C} \triangleq (C_z \ 0) \tag{8.7}$$

It can be shown that:

$$\bar{A} = \begin{pmatrix} A_z & \mathbf{0} \\ \mathbf{0} & \mathbf{0} \end{pmatrix} + \begin{pmatrix} B_z & \mathbf{0} \\ \mathbf{0} & \mathbf{I} \end{pmatrix} \begin{pmatrix} d & C \\ B & A \end{pmatrix} \begin{pmatrix} C_z & \mathbf{0} \\ \mathbf{0} & \mathbf{I} \end{pmatrix}$$
$$\triangleq M_0 + M_1 X M_2 \triangleq \bar{A}(X) \tag{8.8}$$

where X is the ***system matrix*** of the digital controller $C_d(z)$, [1] and all the **0**s/**I**s are zero/identity matrices of proper dimension.

Therefore, the sampled-data control system is stable *if and only if* all the eigenvalues of $\bar{A}$ are in the interior of the unit circle.

The system matrix of the actually implemented digital controller is $X + \Delta X$ instead of $X \triangleq \{x_{ij}\}$, where each element of ΔX is bounded by $\epsilon/2$:

$$||\Delta X||_0 \triangleq \max_{i,j} |\Delta x_{ij}| \leq \epsilon/2 \tag{8.9}$$

where:

$$\epsilon \triangleq 2^{-(B_c - B_X)} \tag{8.10}$$

with 2^{B_X} a normalisation factor such that $2^{-B_X}|x_{ij}| \leq 1, \forall(i,j)$.

With X replaced by $X + \Delta X$ in (8.8), one has:

$$\bar{A}(X + \Delta X) = \bar{A}(X) + M_1 \Delta X M_2. \tag{8.11}$$

A question to be asked is: what is the "smallest" perturbation ΔX such that the closed-loop system becomes unstable? This is a stability robustness problem which can be formulated as:

$$\rho_p(X) \triangleq \inf \{||\Delta X||_p : \ \bar{A}(X) + M_1 \Delta X M_2 \text{ is unstable}\} \tag{8.12}$$

where $||\Delta X||_p$ is a certain norm/measure of perturbation ΔX.

Let $||\Delta X||_2$ denote the maximal singular value of ΔX, (8.12) with $p = 2$ becomes the well known ***real stability radius*** problem:

$$\rho_2(X) \triangleq \inf \{||\Delta X||_2 : \ \bar{A}(X) + M_1 \Delta X M_2 \text{ is unstable}\} \tag{8.13}$$

The problem of computing ρ_2 for a given structure X was solved by Qiu *et al.* in [12].

There are two main aspects involved in controller implementations: the first one is concerned with estimating the smallest word-length B_c^{min} that ensures stability. Although the closed-loop remains stable for perturbations ΔX satisfying $||\Delta X||_2 < \rho_2(x)$, it should be noted that since $||\Delta X||_0 < ||\Delta X||_2$ (with the difference potentially quite large depending on ΔX), $\rho_2(X)$ and $||\Delta X||_0$ cannot be compared directly to assess robustness of stability [13].

---

[1] Clearly, X is an alternative way to represent a realisation. In the sequel, X is also called a realisation.

The most relevant measure to digital controller implementation, where the parameter perturbation obeys (8.9), is:

$$\rho_0(X) \triangleq \inf\{\|\Delta X\|_0 : \bar{A}(X) + M_1 \Delta X M_2) \text{ is unstable}\} \tag{8.14}$$

with which $\|\Delta X\|_0$ can be compared directly to assess robustness of stability.

It follows from $\epsilon/2 \leq \rho_0(X)$ with ϵ given by (8.10) that B_c^{min} should not be smaller than $\log_2 \rho_0^{-1}(X) - 1 + B_X$. Computing explicitly the value for $\rho_0(X)$, however, seems very hard and is still an open problem. With a statistical approach considering the elements of ΔX to be normally distributed in the interval $[-\frac{\epsilon}{2}, \frac{\epsilon}{2}]$, a lower bound of $\rho_0(X)$ was derived in terms of $\rho_2(X)$ by Fialho and Georgiou in [13], which leads to a ***statistical word-length*** as the estimate of B_c^{min}.

The other aspect is concerned with the optimal controller structure problem. We note that the controller realisations are not unique. In fact, if (A_0, B_0, C_0, d) and (A, B, C, d) are related with T, as indicated in (8.3), the controller system matrix X can be characterised by:

$$X = \begin{pmatrix} \mathbf{I} & 0 \\ 0 & T \end{pmatrix}^{-1} \begin{pmatrix} d & C_0 \\ B_0 & A_0 \end{pmatrix} \begin{pmatrix} \mathbf{I} & 0 \\ 0 & T \end{pmatrix} \triangleq \begin{pmatrix} \mathbf{I} & 0 \\ 0 & T \end{pmatrix}^{-1} X_0 \begin{pmatrix} \mathbf{I} & 0 \\ 0 & T \end{pmatrix} \tag{8.15}$$

Once an initial realisation is given, different ***structures*** (***realisations***)[2] X correspond to different similarity transformations T.

Noting that $\rho_0(X)$ is a function of the controller structure X, the interesting problem is to find out those structures such that $\rho_0(X)$ is maximised:

$$\max_{X \in S_{C_d}} \rho_0(X) \tag{8.16}$$

Determining B_c^{min} for a given structure X is relatively easy even though $\rho_0(X)$ is not available, since one can easily find the ***integer*** B_c^{min} by computer simulation. The ***optimal structure*** problem defined by (8.16) is, however, of great importance. In fact, the digital controller implemented with an X that is a solution to (8.16) implies that the system has the best stability robustness against the FWL errors and can be implemented with a low cost. Since $\rho_0(X)$ is not a tractable function of X (that is the transformation matrix T), the problem defined in (8.16) cannot be solved so far.

8.3 A Stability Robustness Related Measure

Let X be a controller structure. For a stable pole $\lambda_k^0 = \lambda_k(\bar{A}(X))$, the corresponding ***stability margin*** is $m_k \triangleq 1 - |\lambda_k^0|$. Let ΔX be a perturbation of structure X such that $\|\Delta X\|_0 = \rho_0(X)$. Therefore, one has $m_k \leq$

[2] The state space equations provide a class of realisations (Φ, L, K, D) for the controller. These parameter matrices can take different structures. Throughout this paper, by saying a "***structure***" we mean a state space "***realisation***", and vice versa.

$|\lambda_k(\bar{A}(X+\Delta X)) - \lambda_k(\bar{A}(X))|$ for some k. Now, based on a ***first-order approximation***, *i.e.*, sensitivity approach, we propose the following measure (see [14]) [3]:

$$\mu_2(X) \triangleq \min_k \frac{1-|\lambda_k^0|}{\sqrt{N\Psi_k}} \tag{8.17}$$

where N is the number of non-zero elements in ΔX, and $\delta(x_{ij}) = 0$ for $x_{ij} = 0, \pm 1$, [4] otherwise $\delta(x_{ij}) = 1$, and Ψ_k is called the ***partial pole-sensitivity measure***:

$$\Psi_k \triangleq \sum_{i,j=1} \delta(x_{ij}) \left| \frac{\partial \lambda_k}{\partial x_{ij}} \right|^2 \tag{8.18}$$

Denote:

$$P_{\Delta X} \triangleq \left\{ \Delta X : \ |\lambda_k(\bar{A}(X+\Delta X)) - \lambda_k(\bar{A}(X))| < \|\Delta X\|_0 \sqrt{N\Psi_k}, \ \forall k \right\} \tag{8.19}$$

$\mu_2(X)$ has the following property over $P_{\Delta X}$:

Proposition 8.1. *$\bar{A}(X+\Delta X)$ is stable if $\|\Delta X\|_0 < \mu_2(X)$ for any $\Delta X \in P_{\Delta X}$.*

Proof. First of all, one has:

$$|\lambda_k(\bar{A}(X+\Delta X))| \leq |\lambda_k(\bar{A}(X))| + |\lambda_k(\bar{A}(X+\Delta X)) - \lambda_k(\bar{A}(X))|$$

For any $\Delta X \in P_{\Delta X}$, the above inequality yields:

$$\begin{aligned} |\lambda_k(\bar{A}(X+\Delta X))| &\leq |\lambda_k(\bar{A}(X))| + \|\Delta X\|_0 \sqrt{N\Psi_k} \\ &< |\lambda_k(\bar{A}(X))| + \mu_2(X)\sqrt{N\Psi_k} \\ &\leq |\lambda_k(\bar{A}(X))| + \frac{m_k}{\sqrt{N\Psi_k}} \sqrt{N\Psi_k} = 1, \ \forall k \end{aligned}$$

□

The newly defined measure $\mu_2(X)$ can be computed for a given structure X with (8.17) as long as the pole-sensitivities $\{\partial \lambda_k / \partial x_{ij}\}$ are known. As will be shown later, $\mu_2(X)$ is a *tractable* function of X.

Remark 8.1. It should be pointed out that $\mu_2(X)$ has no rigorous connection with $\rho_0(X)$ since its derivation is based on the first-order perturbation theory, but it is a reasonable measure for studying the controller structure

---

[3] Recently, some related measures have been proposed, e.g., [15] and [16].

[4] For those parameters, there is no FWL error. Here, it is assumed that 0 and ±1 are the only trivial parameters.

problem given that there is no *tractable* rigorous measure available so far. It is argued in [17] that the estimates obtained from first-order perturbation theory are often more realistic than rigorous bounds obtained by other means. Therefore, this measure is in fact a trade-off between rigour and computational tractability. The structures that maximise this measure provide a better stability behaviour if not the best in the sense given by (8.16).

In order to evaluate $\mu_2(X)$, the pole-sensitivities are required. We now compute the pole-sensitivities with respect to a structure X. To do so, we need the following technical lemma.

Lemma 8.1. *Let $f(M) \in \mathbb{C}$ be a differentiable function of a matrix $M \triangleq \{m_{ij}\}$ and M_k, $k = 0, 1, 2$ be constant matrices. Let $M = g(X)$, where X is a matrix of appropriate dimension. Denote $\partial f(M)/\partial M$ as the sensitivity matrix of $f(M)$ with respect to M, whose (i,j)th element is $\{\partial f(M)/\partial m_{ij}\}$, and $F(X) = f(g(X))$.*

- *If $M = M_0 + M_1 X M_2$, then:*

$$\frac{\partial F(X)}{\partial X} = M_1^{\mathcal{T}} \frac{\partial f(M)}{\partial M} M_2^{\mathcal{T}} \tag{8.20}$$

where $\mathcal{T}$ denotes the transpose operation.
- *If $M = M_0 + M_1 X^{-1} M_2$, then:*

$$\frac{\partial F(X)}{\partial X} = -(M_1 X^{-1})^{\mathcal{T}} \frac{\partial f(M)}{\partial M} (X^{-1} M_2)^{\mathcal{T}} \tag{8.21}$$

Proof. Denote e_k as the kth elementary vector (whose elements are all zero except the kth one which is 1). Since $M = M_0 + M_1 X M_2$, $m_{ij} = e_i^{\mathcal{T}} M_0 e_j + e_i^{\mathcal{T}} M_1 X M_2 e_j$, it follows from:

$$\frac{\partial F(X)}{\partial x_{qp}} = \sum_{i,j} \frac{\partial f(M)}{\partial m_{ij}} \frac{\partial m_{ij}}{\partial x_{qp}} = \sum_{i,j} \frac{\partial f(M)}{\partial m_{ij}} e_i^{\mathcal{T}} M_1 e_q e_p^{\mathcal{T}} M_2 e_j$$

that $\frac{\partial F(X)}{\partial x_{qp}} = M_2(p,:)(\frac{\partial f(M)}{\partial M})^{\mathcal{T}} M_1(:,q)$, where $M_1(:,q)$ and $M_2(p,:)$ denote the qth column and the pth row vector of M_1 and M_2, respectively, and hence (8.20) follows.

Similarly, $m_{ij} = e_i^{\mathcal{T}} M_0 e_j + e_i^{\mathcal{T}} M_1 X^{-1} M_2 e_j$ with $M = M_0 + M_1 X^{-1} M_2$. It then follows from $\frac{\partial X^{-1}}{\partial x_{qp}} = -X^{-1} e_q e_p^{\mathcal{T}} X^{-1}$ that:

$$\frac{\partial F(X)}{\partial x_{qp}} = -\sum_{i,j} \frac{\partial f(M)}{\partial m_{ij}} e_i^{\mathcal{T}} M_1 X^{-1} e_q e_p^{\mathcal{T}} X^{-1} M_2 e_j$$

that $\frac{\partial F(X)}{\partial x_{qp}} = -(X^{-1}M_2)(p,:)(\frac{\partial f(M)}{\partial M})^{\mathcal{T}} (M_1 X^{-1})(:,q)$, which leads to (8.21). □

Applying the above technical lemma to λ_k, which is a function of $\bar{A} = M_0 + M_1 X M_2$ defined by (8.8), we can now compute the kth pole-sensitivity of the closed-loop system with respect to the controller system matrix X:

$$\frac{\partial \lambda_k}{\partial X} = M_1^{\mathcal{T}} \frac{\partial \lambda_k}{\partial \bar{A}} M_2^{\mathcal{T}} \tag{8.22}$$

where $\partial \lambda_k / \partial \bar{A}$ can be computed using the following theorem, which can be found from many textbooks [2,17–19].

Theorem 8.1. *Let $M \in \mathbb{R}^{m \times m}$ be diagonalisable and have $\{\lambda_k\} = \lambda(M)$ as its eigenvalues, let x_k be a right eigenvector of M corresponding to the eigenvalue λ_k. Denote $M_x \triangleq \begin{pmatrix} x_1\ x_2 \ldots x_m \end{pmatrix}$ and $M_y = \begin{pmatrix} y_1\ y_2 \ldots y_m \end{pmatrix} \triangleq M_x^{-\mathcal{H}}$. Then:*

$$\left(\frac{\partial \lambda_k}{\partial M}\right)^{\mathcal{T}} = x_k y_k^{\mathcal{H}}, \quad k = 1, \ldots, m \tag{8.23}$$

where y_k is called the reciprocal left eigenvector corresponding to x_k and "$\mathcal{H}$" denotes the transpose and conjugate operation.

Remark 8.2. In Theorem 8.1, M is assumed to be non-defective. If M is defective, the sensitivity of the defective eigenvalue is infinite [19,20]. That is why such a situation should be avoided in controller design, especially when a FWL implementation has to be carried out.

Let $x_k = \begin{pmatrix} x_k^{\mathcal{T}}(1)\ x_k^{\mathcal{T}}(2) \end{pmatrix}^{\mathcal{T}}$ be a right eigenvector of $\bar{A}$ corresponding to pole λ_k^0 and $y_k = \begin{pmatrix} y_k^{\mathcal{T}}(1)\ y_k^{\mathcal{T}}(2) \end{pmatrix}^{\mathcal{T}}$ be the corresponding left eigenvector with the partitions corresponding to the block partitioned structure of $\bar{A}$ in (8.8). It follows from (8.22) and (8.23) that $(\frac{\partial \lambda_k}{\partial X})^{\mathcal{T}} = M_2 (\frac{\partial \lambda_k}{\partial \bar{A}})^{\mathcal{T}} M_1 = M_2 x_k y_k^{\mathcal{H}} M_1$ with M_1 and M_2 defined in (8.8). It can be shown that:

$$\begin{aligned} \left(\frac{\partial \lambda_k}{\partial A}\right)^{\mathcal{T}} &= x_k(2) y_k^{\mathcal{H}}(2), & \left(\frac{\partial \lambda_k}{\partial C}\right)^{\mathcal{T}} &= x_k(2) y_k^{\mathcal{H}}(1) B_z \\ \left(\frac{\partial \lambda_k}{\partial B}\right)^{\mathcal{T}} &= C_z x_k(1) y_k^{\mathcal{H}}(2), & \left(\frac{\partial \lambda_k}{\partial d}\right)^{\mathcal{T}} &= C_z x_k(1) y_k^{\mathcal{H}}(1) B_z \end{aligned} \tag{8.24}$$

Therefore, for a given structure (realisation) of the controller, one can compute the corresponding pole-sensitivity functions and hence $\mu_2(X)$.

8.4 Optimal Controller Structures

In this section, we assume that the controller structure is fully parameterised, which is the most general case. This assumption allows us to search for interesting structures within a wide range in the structure set S_{C_d}. Therefore,

$\delta(x_{ij}) = 1, \forall(i,j)$ and $N = n_c^2 + 2n_c + 1 \triangleq N_c$. It then follows from (8.18) that:

$$\Psi_k = \left\|\frac{\partial\lambda_k}{\partial A}\right\|_F^2 + \left\|\frac{\partial\lambda_k}{\partial B}\right\|_F^2 + \left\|\frac{\partial\lambda_k}{\partial C}\right\|_F^2 + \left\|\frac{\partial\lambda_k}{\partial d}\right\|_F^2$$

where $||\cdot||_F$ denotes the *Frobenius* norm. Noting (8.24), one has:

$$\Psi_k = ||x_k(2)||_2^2||y_k(2)||_2^2 + \alpha_k^2||y_k(2)||_2^2 + \beta_k^2||x_k(2)||_2^2 + \alpha_k^2\beta_k^2 \tag{8.25}$$

where:

$$\alpha_k \triangleq ||C_z x_k(1)||_2, \qquad \beta_k \triangleq ||y_k^{\mathcal{H}}(1)B_z||_2 \tag{8.26}$$

With (8.15), the closed-loop transition matrix $\bar{A}$ given by (8.8) can be expressed as:

$$\bar{A} = \bar{T}^{-1}\bar{A}_0\bar{T}, \qquad \bar{T} \triangleq \begin{pmatrix} I & 0 \\ 0 & T \end{pmatrix} \tag{8.27}$$

where $\bar{A}_0$ is the closed-loop transition matrix corresponding to (A_0, B_0, C_0, d).

Now, let $x_k^0 = \left(x_k^{0\mathcal{T}}(1)\; x_k^{0\mathcal{T}}(2)\right)^{\mathcal{T}}$ be a right eigenvector of $\bar{A}_0$ corresponding to the eigenvalue λ_k^0, and let $y_k^0 = \left(y_k^{0\mathcal{T}}(1)\; y_k^{0\mathcal{T}}(2)\right)^{\mathcal{T}}$ be the corresponding left eigenvector. It is easy to see that $x_k = \bar{T}^{-1}x_k^0$ is a right eigenvector and $y_k = \bar{T}^{\mathcal{T}}y_k^0$ is the corresponding left eigenvector of $\bar{A}$ for the same eigenvalue λ_k^0. Clearly:

$$\begin{aligned} x_k(1) &= x_k^0(1), & y_k(1) &= y_k^0(1) \\ x_k(2) &= T^{-1}x_k^0(2), & y_k(2) &= T^{\mathcal{T}}y_k^0(2) \end{aligned} \tag{8.28}$$

It is interesting to note that the two positive scalar sets $\{\alpha_k\}$ and $\{\beta_k\}$ given in (8.26) are controller structure independent.

Now, looking at $\eta_k \triangleq \sqrt{\frac{m_k^2}{N_c\Psi_k}}$, one can see that $\mu_2(X) = \min_k \eta_k$ is a function of T. An interesting problem is to identify those structures that maximise this measure:

$$\max_{X\in S_{C_d}} \mu_2(X) \Longleftrightarrow \max_{T:detT\neq 0} \mu_2(X) \tag{8.29}$$

where the solution T, if any, is the similarity transformation from the arbitrary initial structure $X_0 \in S_{C_d}$ to the corresponding optimal structure.

Remark 8.3. The "*optimal structure*", denoted as X_{opt}^f with "f" indicating *fully parameterised*, in this section is generally different from the *rigorous optimal structure* defined in (8.16). In the sequel, the term "*optimal structure*" means a solution to (8.29).

8.4.1 A Lower Bound for Partial Pole-sensitivity Measure Ψ_k

The following theorem gives a lower bound for Ψ_k.

Theorem 8.2. *Let Ψ_k be given by (8.25). Then $\Psi_k \geq (|1 - x_k^{\mathcal{H}}(1)y_k(1)| + \alpha_k\beta_k)^2 \triangleq \Psi_k^{min}$. This lower bound is achieved if and only if:*

$$y_k(2) = c_k \frac{\beta_k}{\alpha_k} x_k(2) \tag{8.30}$$

where c_k is some constant satisfying $|c_k| = 1$.

Proof. First of all, we note that according to the definition of M_y in Theorem 8.1, one has $x_k^{\mathcal{H}} y_k = 1$, that is $x_k^{\mathcal{H}}(1)y_k(1) + x_k^{\mathcal{H}}(2)y_k(2) = 1$, which leads to $x_k^{\mathcal{H}}(2)y_k(2) = 1 - x_k^{\mathcal{H}}(1)y_k(1)$. Consider expression (8.25). Using the Cauchy Schwartz inequality, the first term on the right hand side is bounded from below by $|x_k^{\mathcal{H}}(2)y_k(2)|^2$, that is by $|1 - x_k^{\mathcal{H}}(1)y_k(1))|^2$ with equality if and only if $y_k(2) = r_k x_k(2)$ for some constant r_k.

Now consider the second and third terms of (8.25). We have:

$$||\alpha_k y_k(2)||_2^2 + ||\beta_k x_k(2)||_2^2 \geq 2\alpha_k\beta_k ||y_k(2)||_2 ||x_k(2)||_2 \tag{8.31}$$

and the equality holds if and only if:

$$\alpha_k ||y_k(2)||_2 = \beta_k ||x_k(2)||_2 \tag{8.32}$$

Now by the Cauchy Schwartz inequality as before, the expression on the right hand side of (8.31) can be bounded from below by:

$$2\alpha_k\beta_k ||y_k(2)||_2 ||x_k(2)||_2 \geq 2\alpha_k\beta_k |1 - x_k^{\mathcal{H}}(1)y_k(1)| \tag{8.33}$$

To achieve both bounds with equality one needs to satisfy both $y_k(2) = r_k x_k(2)$ and (8.32). This implies $|r_k| = \frac{\beta_k}{\alpha_k}$. This completes the proof. □

In order to achieve the lower bound, it is assumed in (8.30) that β_k should not be equal to zero if $\alpha_k \neq 0$. The following lemma shows that this is in fact the case under a mild assumption.

Lemma 8.2. *The above-defined scalars $\{\alpha_k\}$ and $\{\beta_k\}$, $k = 1, \ldots, m$, are positive if the closed-loop system has no common poles with the discrete time plant as well as with the discrete time controller.*

Proof. Let λ_k be a pole of the closed-loop system, and $x_k = \begin{pmatrix} x_k^{\mathcal{T}}(1) & x_k^{\mathcal{T}}(2) \end{pmatrix}^{\mathcal{T}}$ and $y_k = \begin{pmatrix} y_k^{\mathcal{T}}(1) & y_k^{\mathcal{T}}(2) \end{pmatrix}^{\mathcal{T}}$ be a corresponding right and left eigenvector, respectively, that is $\bar{A}x_k = \lambda_k x_k$ and $y_k^{\mathcal{H}}\bar{A} = \lambda_k y_k^{\mathcal{H}}$ with $\bar{A}$ given by (8.8). Then $\alpha_k = 0$ means $C_z x_k(1) = 0$, which leads to $\Phi x_k(2) = \lambda_k x_k(2)$ if $x_k(2) \neq 0$ or to $A_z x_k(1) = \lambda_k x_k(1)$ if $x_k(2) = 0$. Both cases imply that the controller and the plant have the same pole λ_k, respectively, which contradicts the assumption. In the same way, we can show that $\beta_k > 0$. □

We mention that the assumption in the above lemma is a sufficient condition only.

8.4.2 Minimal Sensitivity of a Single Closed-loop Pole

It follows from (8.30) there exist digital controller realisations that achieve the lower bound if and only if the following equation:

$$TT^{\mathcal{T}} y_k^0(2) = c_k \frac{\beta_k}{\alpha_k} x_k^0(2) \tag{8.34}$$

has solutions for $T \in \mathbb{R}^{n_c \times n_c}$ satisfying $\det T \neq 0$. It is easy to construct an example for which (8.34) has no solutions no matter how the constant c is chosen. In fact, (8.30) or (8.34) is a *necessary* and *sufficient* (NS) condition for achieving the lower bound and should not be considered as an NS condition for achieving the *minimal sensitivity of a single closed-loop eigenvalue.*

With $x_k(2) = T^{-1} x_k^0(2)$ and $y_k(2) = T^{\mathcal{T}} y_k^0(2)$, (8.25) can be rewritten as:

$$\Psi_k = \text{tr}(P^{-1}Q_k)\text{tr}(W_k P) + \alpha_k^2 \,\text{tr}(W_k P) + \beta_k^2 \,\text{tr}(P^{-1}Q_k) + \alpha_k^2 \beta_k^2 \tag{8.35}$$

where $\text{tr}(\cdot)$ denotes the trace operation, and:

$$P \triangleq TT^{\mathcal{T}}, \qquad Q_k \triangleq R_{x_k} R_{x_k}^{\mathcal{T}} + I_{x_k} I_{x_k}^{\mathcal{T}}, \qquad W_k \triangleq R_{y_k} R_{y_k}^{\mathcal{T}} + I_{y_k} I_{y_k}^{\mathcal{T}} \tag{8.36}$$

with $x_k^0(2) = R_{x_k} + jI_{x_k}$ and $y_k^0(2) = R_{y_k} + jI_{y_k}$. The problem of *minimal sensitivity of a single closed-loop eigenvalue* can be formulated as:

$$\min_{P>0} \Psi_k \Longrightarrow P_k^{opt} \tag{8.37}$$

We mention that any solution to (8.34), if one exists, belongs to the solution set of (8.37) and the converse does *not* hold and that the minimal sensitivity problem may have solutions though the lower bound can not be achieved.

It can be shown (see Theorem 4.1 in [2]) that the problem of the same form as (8.37) with Ψ_k given by (8.35) has a unique solution if Q_k and W_k are nonsingular. Here, as seen from (8.36), W_k and Q_k have at most a rank of two and hence the corresponding (8.37), generally speaking, has many solutions. Due to the limited space, we will not discuss the issue of analytical solutions to (8.37).

8.4.3 Computing Optimal Controller Structures

In this subsection, we discuss how to solve the optimal controller structure problem defined in (8.29).

A. Achieving all lower bounds simultaneously

The following theorem presents the solutions to the optimal controller structure problem defined by (8.29).

Theorem 8.3. *Suppose that there exist controller realisations such that* all *pole-sensitivity lower bounds are achieved simultaneously. The corresponding similarity transformations are given by and only by:*

$$T_{simul} = (M_{x_2} D_{\alpha\beta} M_{x_2}^{\mathcal{H}})^{1/2} V \tag{8.38}$$

where V is an arbitrary orthogonal matrix, and:

$$\mu_2(T_{simul}) = \max_{T:detT \neq 0} \mu_2(T) = \min_k \left\{ \sqrt{\frac{m_k^2}{N_c \Psi_k^{min}}} \right\} \tag{8.39}$$

The corresponding structures, denoted by X_{simul}, can be computed easily with T_{simul} and the initial structure X_0.

Proof. First, from Theorem 8.2 we know that Ψ_k achieves its minimum if and only if $y_k(2) = c_k(\beta_k/\alpha_k)x_k(2)$. Let T be a similarity transformation that yields a controller realisation in which the closed-loop system achieves all the lower bounds *simultaneously.* It follows from $x_k(2) = T^{-1}x_k^0(2)$, $y_k(2) = T^{\mathcal{T}} y_k^0(2)$ and (8.30) that: $TT^{\mathcal{T}} y_k^0(2) = c_k \frac{\beta_k}{\alpha_k} x_k^0(2)$. Combining these constraints for all k, one has $TT^{\mathcal{T}} M_{y_2} = M_{x_2} D_{\alpha\beta}$, where $D_{\alpha\beta} \triangleq \text{diag}(c_1 \frac{\beta_1}{\alpha_1}, \ldots, c_m \frac{\beta_m}{\alpha_m})$, $M_{x_2} \triangleq \begin{pmatrix} x_1^0(2) \cdots x_m^0(2) \end{pmatrix}$ and $M_{y_2} \triangleq \begin{pmatrix} y_1^0(2) \cdots y_m^0(2) \end{pmatrix}$. Noting that $M_{y_2} M_{x_2}^{\mathcal{H}}$ is the identity matrix, one has:

$$TT^{\mathcal{T}} = M_{x_2} D_{\alpha\beta} M_{x_2}^{\mathcal{H}} \tag{8.40}$$

Note that the left of the above equation should be (real) positive-definite and that $\bar{A}$ is real. It follows from Lemma 6.3 in [2] that $c_k = 1, \forall k$ and $M_{x_2} D_{\alpha\beta} M_{x_2}^{\mathcal{H}}$ is a (real) positive-definite matrix, the latter leads to (8.38). [5] The proof is completed, noting the fact that all the realisations obtained with (8.38) yield one and the same partial pole-sensitivity measure $\Psi_k = \Psi_k^{min}, \forall k$ via (8.25). □

Remark 8.4. It should be pointed out that though T_{simul} by (8.38) always exist, they do not yield the optimal controller structures if all the lower bounds cannot be achieved *simultaneously.* It is interesting to note that the structures X_{simul} do yield a very good performance.

B. An iterative algorithm for computing optimal controller structures

The general optimal structure problem can be approached as below.

Define $F_k(P) \triangleq \frac{1}{\eta_k^2} = \frac{N_c}{m_k^2}\Psi_k$, where Ψ_k is given by (8.35). It is easy to see that the optimal realisation problem (8.29) is equivalent to:

$$\min_{P>0} \max_k F_k(P) \tag{8.41}$$

[5] Clearly, c_k does not have to be real if $x_k^0(2) = y_k^0(2) = 0$.

This problem can be attacked with any standard minimisation algorithm starting with an initial realisation, say the one given by (8.38). For example, the command `minmax.m` in MATLAB® [21] can be used to solve (8.41). It has been noted that the problem is solved more accurately and efficiently if the analytical partial derivatives of the objective functions $\{F_k(P)\}$ can be provided. According to Lemma 8.1:

$$\frac{\partial F_k(P)}{\partial P} = \frac{N_c}{m_k^2}[-\mathrm{tr}(W_k P)P^{-1}Q_k P^{-1} - \beta_k^2 P^{-1}Q_k P^{-1} + \mathrm{tr}(Q_k P^{-1})W_k + \alpha_k^2 W_k], \ \forall k \tag{8.42}$$

where $\{\alpha_k, \beta_k\}$ and $\{Q_k, W_k\}$ are given by (8.26) and (8.36), respectively.

For a given $P(i)$, let $F_{k_i}(P(i)) = \max_k F_k(P(i))$. (8.41) can be attacked with the following simple gradient based algorithm:

$$P(i+1) = P(i) - \omega \left.\frac{\partial F_{k_i}(P)}{\partial P}\right|_{P=P(i)} \tag{8.43}$$

where ω is a small positive number, called step size.

Denote P_{opt} as a solution to (8.41). The corresponding optimal similarity transformation matrices can be characterised as:

$$T_{opt} = P_{opt}^{1/2} V \tag{8.44}$$

where V is an arbitrary orthogonal matrix and $M^{1/2}$ denotes the square root matrix of (semi-) positive-definite matrix $M \geq 0$, that is $M = M^{1/2}M^{1/2}$ with $M^{1/2}$ symmetric. The optimal structure X_{opt}^f can then be obtained with (8.3) where $T = T_{opt}$.

It should be pointed out that (8.41) has solutions even if *all* the lower bounds cannot be achieved *simultaneously*.

8.5 Sparse Structures

It has been observed that the optimal realisations that maximise the stability robustness related measure yield a much better performance than the canonical forms. They, however, are generally fully parameterised. This implies that they increase the complexity of implementation and slow down the processing. The main objective in this section is to develop a remarkable algorithm that can be used for searching sparse realisations that have a very nice stability behaviour. This algorithm was initially proposed by Chan [22] and later modified by Amit and Shaked [3], where the algorithm was conceived for minimum roundoff noise gain of a digital filter.

Let X_0 be a controller structure, say the one given by (8.38) or (8.44), which yield very good stability behaviour in terms of maximising μ_2 defined in (8.17). Noting that X_0 is usually fully parameterised, we wish to find a realisation, denoted as X_{spa}, such that it is very sparse and yields almost the same stability behaviour as the one by X_0, that is $\mu_2(X_{spa}) \approx \mu_2(X_0)$.

Denote $M_\eta \triangleq \mu_2^{-2}$. One has $M_\eta = \max_k \left\{\frac{N}{m_k^2}\Psi_k\right\} = \frac{N}{m_{k_0}^2}\Psi_{k_0}$, where Ψ_k is given by (8.18) with $m_k = 1 - |\lambda_k^0|$ and k_0 a certain index.

It follows from (8.25) and (8.28) that:

$$\begin{aligned}\Psi_k = &\operatorname{tr}[T^{-1}x_k^0(2)x^{0\mathcal{H}}(2)T^{-\mathcal{T}}]\operatorname{tr}[T^{\mathcal{T}}y_k^0(2)y^{0\mathcal{H}}(2)T] \\ &+\alpha_k^2\operatorname{tr}[T^{\mathcal{T}}y_k^0(2)y^{0\mathcal{H}}(2)T] + \beta_k^2\operatorname{tr}[T^{-1}x_k^0(2)x^{0\mathcal{H}}(2)T^{-\mathcal{T}}]\end{aligned} \tag{8.45}$$

Clearly, M_η is a differentiable function with respect to similarity transformations T. To indicate this fact, we denote $M_\eta = M_\eta(T)$.

Let $\{T_j\}$ be a series of similarity transformations. Denote $X_{spa}(j)$ as the structure transformed from $X_{spa}(j-1)$ by T_j with $X_{spa}(0) = X_0$. The basic idea behind this algorithm is to choose T_j such that:

- the corresponding M_η is close to the value $\mu_2^{-2}(X_0)$,
- the trivial elements such as 0 and ± 1 in $X_{spa}(j-1)$ remain unchanged in $X_{spa}(j)$, and
- one non-trivial element in $X_{spa}(j-1)$ is transformed into a trivial value in $X_{spa}(j)$.

Now, let T be a function of a fictitious variable t. The first constraint can be achieved by setting the derivative of $M_\eta(T(t))$ to zero. In the sequel, $\dot{M}$ denotes the derivative of a matrix M with respect to t. It can be shown with Lemma 8.1 that:

$$\dot{M}_\eta(T(t)) = \operatorname{tr}[\Omega^{\mathcal{T}}(t)W(t) + W^{\mathcal{H}}(t)\Omega(t)] \tag{8.46}$$

where:

$$\dot{T}(t) = T(t)\Omega(t) \tag{8.47}$$

for some $\Omega(t) \in \mathbb{R}^{n_c \times n_c}$, and:

$$\begin{aligned}W(t) \triangleq \frac{N}{m_{k_0}^2}\Bigg[&\sum_{i,j=1} \delta(A_{ij})\left\{\operatorname{tr}(E_jH_x)H_yE_i - \operatorname{tr}(E_iH_y)E_jH_x\right\} \\ &+\sum_{i=1}^{n_c}\delta(B_i)H_3(j)E_i - \sum_{i}^{n_c}\delta(C_i)E_jH_4(i)\Bigg]\end{aligned} \tag{8.48}$$

with $E_i \triangleq e_ie_i^{\mathcal{H}}, \forall i$ and:

$$\begin{aligned}&H_x \triangleq x_{k_0}(2)x_{k_0}^{\mathcal{H}}(2), && H_y \triangleq y_{k_0}(2)y_{k_0}^{\mathcal{H}}(2) \\ &H_3(j) \triangleq \alpha_{k_0}^2 y_{k_0}(2)E_jy_{k_0}^{\mathcal{H}}(2), && H_4(i) \triangleq \beta_{k_0}^2 x_{k_0}(2)E_ix_{k_0}^{\mathcal{H}}(2)\end{aligned} \tag{8.49}$$

Clearly, $\dot{M}_\eta$ is linear in all elements of $\Omega(t)$.

The second constraint can be achieved by setting the derivative of the corresponding parameters to zero. It can be shown that the corresponding derivatives can be found from:

$$\dot{A} = A\Omega(t) - \Omega(t)A, \quad \dot{B} = -\Omega(t)B, \quad \dot{C} = C\Omega(t) \tag{8.50}$$

We mention that (8.48) and (8.50) are calculated for the realisation (A, B, C) transformed from the initial realisation with the similarity transformation $T(t)$.

We note from (8.50) that for each parameter, its derivative is linear in all elements of $\Omega(t)$. By setting the derivatives of those elements that are already trivial in the previous controller structure and $\dot{M}_\eta$ to zero, one obtains a set of linear equations in Ω:

$$L(t)\text{vec}\Omega(t) = \mathbf{0}$$

where $L(t) \in \mathbb{R}^{m\times n}$, $m < n$ is a known matrix and vec$(\cdot)$ denotes the column stacking operator. By SVD, $L(t) = U(t)\Sigma V^T(t)$ and hence $L(t)V(t) = U(t)\Sigma$. Noting $m < n$, one has $L(t)v_k(t) = 0$, $\forall k = m+1, m+2, \cdots, n$, where $V(t) = \begin{pmatrix} v_1(t) & \cdots & v_k(t) & \cdots & v_n(t) \end{pmatrix}$. Clearly, one can form vec$\Omega(t)$ with:

$$\text{vec}\Omega(t) = \sum_{k=m+1}^{n} \xi_k v_k(t) \tag{8.51}$$

with the coefficients $\{\xi_k\}$ determined in such a way that the selected non-trivial element can converge to its desired value $(0, \pm 1)$ as fast as possible. See the third constraint above. For detailed discussions, we refer to [2].

With this $\Omega(t)$, $T(j)$ can be obtained by solving the equation $\dot{T}(t) = T(t)\Omega(t)$ numerically:

$$T_{j+1}(t) = T_j(t) + \dot{T}_j(t)\Delta_j(t) \tag{8.52}$$

where $\dot{T}_j(t) = T_j(t)\Omega(t)$ with $\Delta_j(t)$ a small step size. To obtain the final T_j, the above integration may be iterated several times. Repeating this procedure one can obtain a sparse structure. Usually, a pre-specified convergence tolerance γ is given (say, $\gamma = 10^{-5}$ for example). When the difference between a parameter and its nearest target (that is 0 or ± 1) is absolutely smaller than this pre-specified tolerance, this parameter is considered as a trivial element.

8.6 A Design Example

We now present a design example to illustrate the design procedures. This example was used by Chen and Francis [23] for studying the input-output stability of sampled-data systems, where the plant is an experimental flexible beam at the University of Toronto with transfer function:

$$P(s) = \frac{1.6188s^2 - 0.1575s - 43.9425}{s^4 + 0.1736s^3 + 27.9001s^2 + 0.0186s}$$

the anti-aliasing filter is given by:

$$F(s) = \frac{1}{s+1}$$

and the stabilising controller $C_s(s)$, designed using an $\mathcal{H}_\infty$ optimisation method, is given by the following transfer function:

$$\frac{0.046s^6 + 1.5862s^5 + 3.09s^4 + 44.3s^3 + 42.7785s^2 + 0.02867s + 1.58 \times 10^{-4}}{s^6 + 3.766s^5 + 34.9509s^4 + 106.2s^3 + 179.2s^2 + 166.43s + 0.0033}$$

It was shown in [23] that the corresponding sampled-data system is stable if the sampling frequency $f_s > 1$.

With $f_s = 5$, we obtained the following transfer functions:

$$P_d(z) = \frac{0.0018z^4 + 0.0003z^3 - 0.0163z^2 + 0.0011z + 0.0016}{z^5 - 3.7860z^4 + 6.3299z^3 - 6.0926z^2 + 3.3395z - 0.7908}$$

and $C_d(z)$ is given by:

$$0.0460 + \frac{0.2119z^5 - 0.8081z^4 + 1.3570z^3 - 1.3108z^2 + 0.7220z - 0.1720}{z^6 - 4.2600z^5 + 8.2537z^4 - 9.4410z^3 + 6.6649z^2 - 2.6885z + 0.4709}$$

for the discrete time system $HP(s)F(s)S$ and the digital controller, respectively.

One can easily get the controllable realisation R_p^c for $P_d(z)$, and R_c^c for $C_d(z)$ (the corresponding system matrix is denoted by X_c). We point out that the coefficients of $P_d(z)$ and $C_d(z)$ are presented in *FORMAT SHORT* (MATLAB®) and hence only the first four significant digits in fractional part of each parameter are displayed. In the sequel, the *FORMAT SHORT* display is assumed.

Taking X_c as the initial structure, one computed a transformation T_{simul} with (8.38), where V was taken as the identity matrix. The corresponding structure, denoted as X_{simul}, is fully parameterised.

With X_{simul} as the initial structure, we have run the algorithm described in Section 8.5. The obtained sparse structure, denoted by X_{simul}^{spa}, is:

$$A_{simul}^{spa} = \begin{pmatrix} 1 & -0.8566 & 1 & 0 & 0 & 0 \\ 0.3498 & 1 & -1 & 0 & 0 & 0 \\ 0 & 0.7238 & 0 & 0 & 0.0923 & 0.1074 \\ 0 & 0 & 1 & 0.4618 & -0.1727 & 0 \\ -0.2252 & 0 & 0 & 0.4725 & 1 & 0 \\ 0 & -0.1277 & 0 & 0 & 0.1112 & 0.7981 \end{pmatrix}$$

$$B_{simul}^{spa} = \begin{pmatrix} 0 & -0.0305 & 0 & -0.2185 & 0.2119 & 0 \end{pmatrix}^T$$

$$C_{simul}^{spa} = \begin{pmatrix} -0.5405 & 0 & 0 & 0 & 1 & 0 \end{pmatrix}$$

with a tolerance of $\gamma = 10^{-5}$.

Taking X_{simul} as the initial structure, we solve the optimal stability realisation problem (8.41) using the simple gradient algorithm of type (8.43) and obtain a P_{opt}. A fully parameterised optimal controller realisation/structure, denoted as X^f_{opt}, is obtained with the optimal similarity transformation T_{opt} given by (8.44) with V assigned to the unit matrix.

Starting with X^f_{opt}, we run the algorithm discussed in the previous section. The obtained structure, denoted as X^{spa}_{opt}, is given by:

$$A^{spa}_{opt} = \begin{pmatrix} 0.5684 & -1 & 0.1763 & 0 & 0 & 0.4897 \\ 0.1236 & 1 & 0 & 0 & 0.8136 & -0.0265 \\ 0 & 0 & 0.6915 & 0 & 0 & 0.0002 \\ 0 & 0.0348 & 1 & 1 & -0.9571 & 0 \\ 0 & 0 & 0.0039 & 1 & 0 & 0 \\ 0 & 0 & 0 & -1 & 1 & 1 \end{pmatrix}$$

$$B^{spa}_{opt} = \begin{pmatrix} 0 & -0.2093 & 0.0250 & 0 & 0 & 0 \end{pmatrix}^T$$
$$C^{spa}_{opt} = \begin{pmatrix} -0.4633 & -1 & 0.1016 & -1 & 1 & 0.4937 \end{pmatrix}$$

with a tolerance $\gamma = 10^{-5}$.

By computer simulation, one can find the B_c^{min} as well as μ_2 for X_c, X_{simul}, X^{spa}_{simul}, X^f_{opt} and X^{spa}_{opt}. The results are presented in the following table, where N is the number of non-trivial parameters in the corresponding structure. [6] One can see that the minimum word-length for X^f_{opt}, X_{simul}, X^{spa}_{opt} and X^{spa}_{simul} is much less than that for X_c. It is interesting to note that X_{simul} indeed yields a very good stability performance. Starting with this structure, the fully parameterised X^f_{opt} obtained with the proposed gradient algorithm is much better. In fact, simulations show that X^f_{opt} and X^{spa}_{opt} can still ensure stability of the sampled-data control system when implemented with $B_c = 13$ bits though the pole deviations get large. Compared with X^f_{opt} (X_{simul}), X^{spa}_{opt} (X^{spa}_{simul}) obtained using the proposed algorithm for sparseness not only keep almost the same stability behaviour, but also possesses just 17 non-trivial parameters. This makes it more suitable for implementation.

Table 8.1. Comparison of five structures

	X_c	X_{simul}	X^{spa}_{simul}	X^f_{opt}	X^{spa}_{opt}
μ_2	2.41×10^{-10}	2.58×10^{-6}	2.51×10^{-6}	6.73×10^{-6}	1.62×10^{-5}
B_c^{min}	28	16	16	14	14
N	13	49	17	49	17

[6] In computing μ_2, the contribution of pole-sensitivities with respect to the trivial parameters such as $0, \pm 1$ subject to a tolerance 10^{-5} is not taken into account.

References

1. R.A. Roberts and C.T. Mullis, *Digital Signal Processing.* Addison Wesley, 1987.
2. M. Gevers and G. Li, *Parametrizations in Control, Estimation and Filtering Problems: Accuracy Aspects.* Springer Verlag, London, Communication and Control Engineering Series, 1993.
3. G. Ami and U. Shaked, 'Small roundoff realization of fixed-point digital filters and controllers,' *IEEE Trans. Acoustics, Speech, and Signal Processing,* Vol.ASSP-36, No.6, pp.880–891, June 1988.
4. G. Li and M. Gevers, 'Optimal finite precision implementation of a state-estimate feedback controller,' *IEEE Trans. Circuits and Systems,* Vol.CAS-38, No.12, pp.1487–1499, Dec. 1990.
5. P. Moroney, A.S. Willsky and P.K. Houpt, 'The digital implementation of control compensators: the coefficient wordlength issue,' *IEEE Trans. Automatic Control,* Vol.AC-25, No.4, pp.621–630, August 1980.
6. D. Williamson and K. Kadiman, 'Optimal finite wordlength linear quadratic regulation,' *IEEE Trans. Automatic Control,* Vol.34, No.12, pp.1218–1228, 1989.
7. K. Liu, R. Skelton and K. Grigoriadis, 'Optimal controllers for finite wordlength implementation,' *IEEE Trans. Automatic Control,* Vol.37, pp.1294–1304, 1992.
8. A.G. Madievski, B.D.O. Anderson and M. Gevers, 'Optimum realizations of sampled-data controllers for FWL sensitivity minimization,' *Automatica,* Vol.31, No.3, pp.367–379, 1995.
9. R.H. Middleton and G.C. Goodwin, *Digital Control and Estimation: A Unified Approach,* Englewood Cliffs, NJ: Prentice-Hall, 1990.
10. T. Chen and B. A. Francis, 'Sampled data optimal design and robust stabilization,' *J. Dynamic Systems, Measurement and Control,* Dec. 1992.
11. T. Chen and B. A. Francis, 'H_2-Optimal Sampled-Data Control' , *IEEE Trans. on Automatic Control,* Vol. 36, No.4. pp. 387-397, April 1991.
12. L. Qiu, B. Bernhardsson, A. Rantzer, E.J. Davison, P.M. Young and J.C. Doyle, 'On the real structured stability radius', *in Proc. of 12th World IFAC Congress,* Vol. 8, pp. 71-78, July 1993, Sydney, Australia.
13. I.J. Fialho and T.T. Georgiou, 'On stability and performance of sampled-data systems subject to wordlength constraint,' *IEEE Trans. Automatic Control,* Vol.39, pp.2476–2481, Dec. 1994.
14. G. Li, 'On the structure of digital controllers with finite word length consideration,' *IEEE Trans. Automatic Control,* Vol.43, No.5, pp.689–693, May 1998.
15. S. Chen, J. Wu, R.H. Istepanian and J. Chu, 'Optimizing stability bounds of finite-precision PID controller structures,' *IEEE Trans. Automatic Control,* Vol.44, No.11, pp.2149–2153, 1999.
16. J. Wu, S. Chen, G. Li and J. Chu, 'Optimal finite-precision state-estimate feedback controller realizations of discrete-time systems,' *IEEE Trans. Automatic Control,* Vol.45, No.8, pp.1550-1554, Aug. 2000.
17. G.W. Stewart, *Introduction to Matrix Computations,* Academic Press, 1973.
18. G.W. Stewart and Ji-guan Sun, *Matrix Perturbation Theory,* Academic Press, 1990.

19. J.H. Wilkinson, *The Algebraic Eigenvalue Problem*, Oxford, U.K.: Clarendon, 1965.
20. J. Demmel, 'Nearest defective matrices and the geometry of ill-conditioning,' pp. 35-56 in *Reliable Numerical Computation*, edited by M.G. Cox and S. Hammarling, Clarendon Press, Oxford, 1990.
21. A. Grace, *Optimization TOOLBOX: For Use with MATLAB*, The MathWorks, Inc., 1992.
22. D. S. K. Chan, 'Constrained minimization of roundoff noise in fixed-point digital filters,' *Proc. 1979 IEEE Int. Conf. on Acoust., Speech, and Signal Processing*, pp. 335-339, April 1979.
23. T. Chen and B. A. Francis, 'Input-output stability of sampled-data systems', *IEEE Trans. on Automatic Control*, Vol. 36, No.1. pp. 50-58, January 1991.

CHAPTER 9

AN EVOLUTIONARY ALGORITHM APPROACH TO THE DESIGN OF FINITE WORD-LENGTH CONTROLLER STRUCTURES

James F. Whidborne and Robert S.H. Istepanian

Abstract. The design of the implementation of a digital controller is a multi-objective problem, in that the performance degradation needs to be minimised with a realisation that has minimal cost and complexity in terms of word-length and number of computations. An approach, which utilises evolutionary computation for solving this multi-objective problem for second-order digital controllers, is presented. Use is made of a linear system equivalence completion problem to reduce the size of the search space. The effectiveness of the approach is shown with an example.

9.1 Introduction

There are, in general, a number of different considerations in choosing the structure and the implementation architecture of a digital control scheme. The chosen word-length is an important consideration, since this will impact upon the accuracy of the implementation and the subsequent closed-loop performance and stability margins. Another important consideration is the computational complexity, in particular the number of multiplication operations, as this affects the overall computation time. Now, it is well known that the sensitivity of the system to implementation inaccuracies resulting from controller parameter rounding is dependent upon the controller structure. Hence, the problem is one of optimising the controller structure to obtain the most accurate implementation whilst simultaneously minimising the number of multiplication operations and the word-length.

These objectives are in fact conflicting. Clearly, a very accurate implementation can be obtained if a very long word-length is used. Removing the restriction on the number of multiplication operations can also be used to improve the accuracy of the implementation. This is less obvious, and is shown here by example. Assume that we wish to implement the operation $y = 3.25 \times x$, but are restricted by the choice of word-length to multiplication by factors of $1/2$. Thus, multiplication by 3.25 can be obtained using two multiplication operations by $y = 5/2 \times 3/2 \times x$. Clearly, arbitrary accuracy for the controller parameters can be obtained by repeated multiplication, but at the expense of computation time (and increased rounding noise in the variable). The optimisation problem is thus a multi-objective problem, in that there are several conflicting objectives that need to be simultaneously minimised.

Not only is the problem multi-objective, but in addition, the objective functions are of a mixed discrete/continuous nature, and the variables are in a discretised space. Evolutionary algorithms (EA) [1,2], and in particular multi-objective genetic algorithms (MOGA) [3], provide a set of robust and powerful optimisation search procedures that are able to deal with problems of this type [4–6]. Thus, this chapter describes an evolutionary algorithm approach to this problem.

Basically, the approach is to find a finite word-length (FWL) controller that is near to the originally designed controller such that the robust stability and closed-loop performance degradation, the word-length and the computational complexity (in terms of the number of multiplication operations) are simultaneously minimised. The approach is based on the generation of near-equivalent FWL controller representations by means of the solution to a linear system equivalence completion problem followed by a rounding operation. A MOGA is then used to find a set of Pareto-optimal near-equivalent FWL controllers. This allows the designer to trade-off FWL implementation stability and performance against both the word-length and the computational complexity by enabling the designer to simply choose the most appropriate design from the set. The essence of the approach was originally proposed in [7] and [8].

In the next section, the problem of multi-objective optimisation is introduced. In 9.3, evolutionary algorithms and the multi-objective genetic algorithm are described. In Section 9.4, a system equivalence completion problem is solved, and in Section 9.5, the solution is used in combination with a MOGA to provide a methodology for obtaining accurate, low word-length, low complexity digital controller implementations. In Section 9.6, the developed approach is applied to the implementation of a PID controller designed for the IFAC93 benchmark problem [9] by [10]. Some concluding remarks are made in the final section.

9.2 Multi-objective Optimisation

A multi-objective optimisation problem is one in which there are several objective functions, usually conflicting, which need to be simultaneously achieved. Thus, if there are p design objective functions $\{\phi_i(x) : i = 1 \dots p\}$, the problem is stated as:

$$\min_{x \in \mathcal{X}} \{\phi_i(x), \text{ for } i = 1 \dots p\} \tag{9.1}$$

where $\mathcal{X}$ denotes the set of possible values of x. In most cases, the objective functions are in conflict, so the reduction of one objective function leads to the increase in another. Subsequently, the result of the multi-objective optimisation is known as a Pareto-optimal solution [11]. A Pareto-optimal solution has the property that it is not possible to reduce any of the objective functions without increasing at least one of the other objective functions.

A point $x^* \in \mathcal{X}$ is defined as being Pareto-optimal if and only if there exists no other point $x \in \mathcal{X}$ such that $\phi_i(x) \leq \phi_i(x^*)$ $\forall$ $i = 1, \ldots, p$ and $\phi_j(x) < \phi_j(x^*)$ for at least one j.

The Pareto-optimal set is illustrated in Figure 9.1 for the case where there are two objective functions, *i.e.*, $p = 2$. A point lying in the interior of the attainable set is sub optimal, since both ϕ_1 and ϕ_2 can be reduced simultaneously. A point lying on the boundary of the set, *i.e.*, in the Pareto-optimal set, requires ϕ_1 to be increased if ϕ_2 is to be decreased and vice versa. A solution to a multi-objective optimisation problem must lie on this boundary.

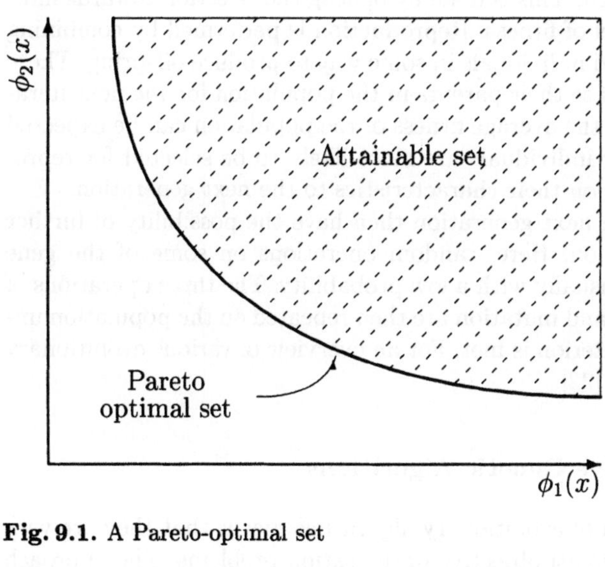

Fig. 9.1. A Pareto-optimal set

The general multi-objective optimisation problem has been studied in great detail, particularly within the operational research community (e.g., [21]), but also within the control community (e.g., [12–19], for a review, see [20]). In the next section, an evolutionary computation approach to multi-objective optimisation is described.

9.3 Evolutionary Algorithms and the Multi-objective Genetic Algorithm

9.3.1 Evolutionary Algorithms

Evolutionary algorithms are based upon the evolution of species according to the Darwinian principle of "survival of the fittest". As such, they can be used as a general optimisation method in a variety of settings. In order to mimic the evolutionary process in nature, three main operators perform the fundamental processes of *selection*, *reproduction* and *mutation*. These operators act upon a set (or population) of potential solutions (or individuals), and cause the population to evolve towards the desired solution.

The individuals are encoded as strings (the genotype), frequently binary strings, although other schemes are possible. A random initial population is created. After the independent variables (the phenotype) are decoded from the genotype, the objective function for each individual is then evaluated. The objective function determines the strength of the individual's ability to survive in the environment (fitness), and is hence used as the basis for selection for reproduction. This is done by biasing the selection towards individuals with a high level of fitness. Reproduction is performed by combining the genotypes of selected individuals in some way to produce offspring. These offspring may then replace their parents in the population for the next iteration (generation). Thus the average fitness of the population can be expected to increase as the fitter individuals are more likely to be selected for reproduction and hence pass on their characteristics to the next generation.

The offspring in the next generation then have the possibility of further perturbation by mutation. Here, random operations on some of the gene strings are performed, usually with a low probability. The three operations of selection, reproduction and mutation are then repeated on the population until some termination criterion is met. For an overview of various evolutionary algorithm schemes, see [22].

9.3.2 Multi-objective Genetic Algorithms

The natural parallelism of evolutionary algorithms means that they are very well-suited for solving multi-objective optimisation problems. The approach used in this chapter is the multi-objective genetic algorithm (MOGA) [23–25], which is an extension of an idea by Goldberg [1]. Traditionally, for multi-objective optimisation, a single Pareto-optimal solution is obtained, its location on the Pareto-optimal hyper-surface being determined by the weightings or other formulation parameters. The idea behind the MOGA is, instead of finding a single optimal solution, to develop a set or *population* of Pareto-optimal or near Pareto-optimal solutions. This then allows the designer to observe the nature of the Pareto-optimal solution surface, and hence make decisions about the appropriateness of the various solutions in the solution set.

The set of Pareto-optimal or near Pareto-optimal solutions is obtained by finding a set of solutions that are *non-dominated.*

Definition 9.1. An individual j with a set of objective functions $\phi^j = \{\phi_1^j, \ldots, \phi_p^j\}$ is said to be non-dominated if for a population of N individuals, there are no other individuals $k = 1, \ldots, N, k \neq j$ such that $\phi_i^k \leq \phi_i^j$ for all $i = 1, \ldots, p$ and $\phi_i^k < \phi_i^j$ for at least one i.

This concept is illustrated in Figure 9.2, which shows a set of potential solutions to the problem $\min(\phi_1, \phi_2)$. From this figure, it can be seen that solutions a, b, c and d are non-dominated, b dominates e, f is dominated by c but not by b and g is dominated by a, b, c, d, e and f.

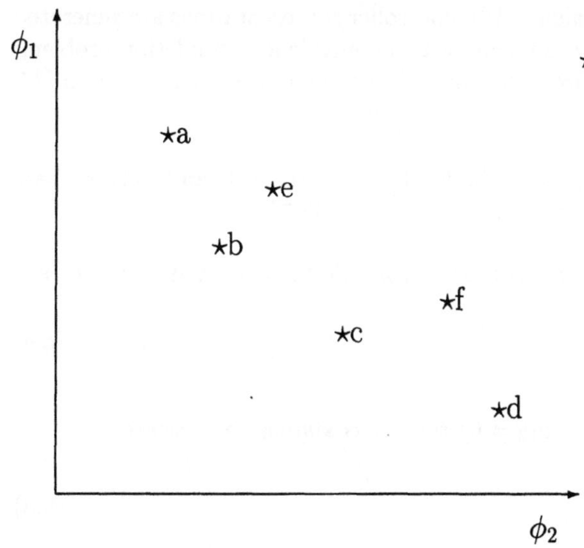

Fig. 9.2. Potential solutions to the problem $\min(\phi_1, \phi_2)$

With the MOGA, non-dominated individuals are given the greatest fitness, and individuals that are dominated by many other individuals are given a small fitness. Using this mechanism, the population evolves towards a set of non-dominated, near Pareto-optimal individuals. Details of this mechanism are given in [25,26].

9.3.3 Fitness sharing

In addition to finding a set of near Pareto-optimal individuals, clearly, it is desirable that the sample of the whole Pareto-optimal set given by the set of non-dominated individuals is fairly uniform. A phenomenon that can occur using evolutionary algorithms is *genetic drift.* Even though all non-dominated individuals in the population are assigned the same fitness, the stochastic nature of the selection mechanism means that one individual will be selected more than another equally fit individual in any one generation. This individual will then produce more offspring closely related to the parent. The result is that the influence of this individual will accumulate over generations and produce an imbalance in the population. There are mechanisms in nature that reduce genetic drift. In evolutionary algorithms, a common mechanism to prevent genetic drifts is *fitness sharing* [25,26], which works by reducing the fitness of individuals that are genetically close to each other.

9.4 A Linear System Equivalence Completion Problem

In the method, near-equivalent FWL controller representations are generated by means of the solution to a linear system equivalence completion problem; some preliminary results are first required. The definition below is from [27, p 154].

Definition 9.2. A matrix A is similar to a matrix $\tilde{A}$ if and only if there exists a non-singular matrix T such that $\tilde{A} = TAT^{-1}$.

Lemma 9.1. *Given a matrix* $A \in \mathbb{R}^{2\times 2}$ *such that* $A \neq \alpha I \; \forall \, \alpha \in \mathbb{R}$, *where:*

$$A = \begin{bmatrix} a_{11} & a_{12} \\ a_{21} & a_{22} \end{bmatrix} \tag{9.2}$$

and given the real pair $(\tilde{a}_{11}, \tilde{a}_{12} \neq 0)$ *then* A *is similar to* $\tilde{A}$ *where:*

$$\tilde{A} = \begin{bmatrix} \tilde{a}_{11} & \tilde{a}_{12} \\ \tilde{a}_{21} & \tilde{a}_{22} \end{bmatrix} \tag{9.3}$$

and where $\tilde{a}_{22} = a_{11} + a_{22} - \tilde{a}_{11}$ *and* $\tilde{a}_{21} = (\tilde{a}_{11}\tilde{a}_{22} - \det A)/\tilde{a}_{12}$.

Proof. To prove this Lemma, we use the property that if A and $\tilde{A}$ have the same minimal polynomial and the same characteristic polynomial, and their minimal polynomial is the same as their characteristic polynomial, then A and $\tilde{A}$ are similar [28, p 150]. The characteristic equations of A and $\tilde{A}$ are $P_A(t) = \det(tI - A) = t^2 - (a_{11} + a_{22})t + (a_{11}a_{22} - a_{21}a_{12})$ and $P_{\tilde{A}}(t) = \det(tI - \tilde{A}) = t^2 - (\tilde{a}_{11} + \tilde{a}_{22})t + (\tilde{a}_{11}\tilde{a}_{22} - \tilde{a}_{21}\tilde{a}_{12})$ respectively. By equating coefficients we obtain the expressions for $\tilde{a}_{21}$ and $\tilde{a}_{22}$. It remains to be shown that the minimal polynomial of A is the same as the characteristic polynomial. For the case where the eigenvalues of A are unique, then it is clear that the the minimal polynomial is the same as the characteristic polynomial. For the case where the eigenvalues of A are repeated and A is non-derogatory, then the Jordan canonical form of A is:

$$J_A = \begin{bmatrix} \lambda & 1 \\ 0 & \lambda \end{bmatrix} \tag{9.4}$$

where λ is the repeated eigenvalue and the minimal polynomial is the same as the characteristic polynomial [29, p 6]. However, if A is diagonal with repeated eigenvalues, that is $A = \lambda I$, then $A = T^{-1}AT$ for all non-singular T. Thus the Jordan canonical form equals A and A is derogatory and so the minimal polynomial is of lower order than the characteristic polynomial. This form is explicitly excluded. □

The derogatory case excluded in the above lemma is shown to be impractical for controller implementation by the following lemma.

Lemma 9.2. *For a two state SISO LTI system $F(z)$, where:*

$$F \stackrel{s}{=} \left[\begin{array}{c|c} A & B \\ \hline C & D \end{array}\right] \tag{9.5}$$

$$\stackrel{s}{=} \left[\begin{array}{cc|c} a_{11} & a_{12} & b_1 \\ a_{21} & a_{22} & b_2 \\ \hline c_1 & c_2 & d \end{array}\right] \tag{9.6}$$

then the system is uncontrollable if $A = \alpha I \; \forall \; \alpha \in \mathbb{R}$.

Proof. If $A = \alpha I$, the controllability matrix:

$$\mathcal{C} = [B \; AB] \tag{9.7}$$

is given by:

$$\mathcal{C} = \begin{bmatrix} b_1 & \alpha b_1 \\ b_2 & \alpha b_2 \end{bmatrix} \tag{9.8}$$

which is clearly rank deficient, hence the system is uncontrollable. □

The main theorem in which a linear system equivalence completion problem is solved can now be stated.

Theorem 9.1. *Given an controllable two state SISO LTI system $F(z)$, where:*

$$F \stackrel{s}{=} \left[\begin{array}{c|c} A & B \\ \hline C & D \end{array}\right] \tag{9.9}$$

$$\stackrel{s}{=} \left[\begin{array}{cc|c} a_{11} & a_{12} & b_1 \\ a_{21} & a_{22} & b_2 \\ \hline c_1 & c_2 & d \end{array}\right] \tag{9.10}$$

then given $(\tilde{a}_{11}, \tilde{a}_{12} \neq 0, \tilde{b}_1, \tilde{b}_2)$ *there exists an equivalent state space representation* $\tilde{F}$ *where:*

$$\tilde{F} \stackrel{s}{=} \left[\begin{array}{c|c} \tilde{A} & \tilde{B} \\ \hline \tilde{C} & \tilde{D} \end{array}\right] \tag{9.11}$$

$$\stackrel{s}{=} \left[\begin{array}{cc|c} \tilde{a}_{11} & \tilde{a}_{12} & \tilde{b}_1 \\ \tilde{a}_{21} & \tilde{a}_{22} & \tilde{b}_2 \\ \hline \tilde{c}_1 & \tilde{c}_2 & \tilde{d} \end{array}\right] \tag{9.12}$$

such that $\tilde{F}(z) = F(z)$ *where* $F(z) = C\left[zI - A\right]^{-1} B + D$ *and:*

$$\tilde{F}(z) = CT^{-1}\left[zI - TAT^{-1}\right]^{-1} TB + D \tag{9.13}$$

where:

$$T = \begin{bmatrix} t_{11} & t_{12} \\ t_{21} & t_{22} \end{bmatrix} \tag{9.14}$$

is non-singular if $(\tilde{A}, \tilde{B})$ *is controllable.*

Proof. The proof is by construction. From (9.13),

$$\tilde{A} = TAT^{-1} \tag{9.15}$$

From Definition 9.2, and from Lemma 9.2, since the system is controllable, the case $A = \alpha I \ \forall \ \alpha \in \mathbb{R}$ can be excluded. Hence from Lemma 9.1, $\tilde{a}_{22}$ is given by $\tilde{a}_{22} = a_{11} + a_{22} - \tilde{a}_{11}$ and $\tilde{a}_{21}$ is given by $\tilde{a}_{21} = (\tilde{a}_{11}\tilde{a}_{22} - \det A)/\tilde{a}_{12}$. From (9.15), $TA - \tilde{A}T = 0$, which gives [30, p 255]:

$$\left[I \otimes \tilde{A} - A^T \otimes I\right]\mathbf{t} = 0 \tag{9.16}$$

where $\mathbf{t} = [t_{11}, t_{21}, t_{12}, t_{22}]^T$ and $[I \otimes \tilde{A} - A^T \otimes I]$ is rank 2. Now, from (9.13), $TB = \tilde{B}$. Hence, given $\tilde{b}_1, \tilde{b}_2$:

$$\begin{bmatrix} b_1 & 0 & b_2 & 0 \\ 0 & b_1 & 0 & b_2 \end{bmatrix} \mathbf{t} = \begin{bmatrix} \tilde{b}_1 \\ \tilde{b}_2 \end{bmatrix} \tag{9.17}$$

An $X \in \mathbb{R}^{4\times 4}$ can be constructed from (9.16) and (9.17) where $X\mathbf{t} = z$, where X is non-singular rank 4 and where $z \in \mathbb{R}^4$ is a column vector with two elements of $\tilde{B}$ and two zero elements. Hence $\mathbf{t}$ can be calculated and T obtained. Since F is controllable, then the controllability matrix $\mathcal{C} = \begin{bmatrix} B & AB \end{bmatrix}$ is rank 2. Since the pair $(\tilde{A}, \tilde{B})$ is required to be controllable, the controllability matrix $\tilde{\mathcal{C}} = \begin{bmatrix} \tilde{B} & \tilde{A}\tilde{B} \end{bmatrix}$ is rank 2 and since $\begin{bmatrix} \tilde{B} & \tilde{A}\tilde{B} \end{bmatrix} = \begin{bmatrix} TB & TAB \end{bmatrix}$ then clearly T must be non-singular. Thus $\tilde{C}$ can be calculated. From (9.13), $\tilde{d} = d$. □

Note that by redefining the transformation matrix as T^{-1}, the problem given $\tilde{C}$ instead of $\tilde{B}$ can be solved [8]. The only restrictions on the problem are that the original system and the equivalent system are controllable, and that certain canonical realisations (*i.e.*, diagonal and lower triangular) are excluded by the constraint that a_{12} is not zero. These realisations can easily be programmed into the method.

9.5 FWL Controller Structure Design using Evolutionary Computation

The approach utilises the solution to the linear system equivalence completion problem described in the last section to generate FWL controller representations that are approximately equivalent to the originally designed controller. The MOGA is used to evolve these representations so that the implementation accuracy, in terms of some metric, and the implementation cost/complexity in terms of the word-length and the number of multiplication operations are simultaneously minimised. The approach provides a set of near Pareto-optimal solutions, the most appropriate can then be chosen by the designer.

The particular FWL representation used in this study is given next. The encoding of the solution space is described in Section 9.5.2, followed by a description of the procedure.

9.5.1 FWL representation

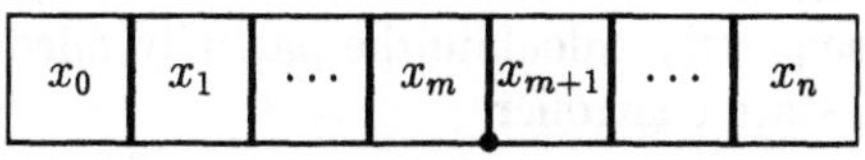

Fig. 9.3. FWL fixed-point representation

A typical two's complement FWL fixed-point representation of a number $q(x)$ is shown in Figure 9.3. The number $q(x)$ is represented by an $m+n+1$ binary string x where:

$$x = [x_0, x_1 \ldots, x_m, x_{m+1}, \ldots, x_{m+n}] \tag{9.18}$$

$x_i \in \{0,1\}$, $m \in \{0,1,2,\ldots\}$ and $n \in \{0,1,2,\ldots\}$. The value $q(x)$ is given by:

$$q(x) = -x_0 2^m + \sum_{i=1}^{m} x_i 2^{i-1} + \sum_{i=m+1}^{m+n} x_i 2^{m-i} \tag{9.19}$$

The set of possible values that can be taken by an FWL variable represented by a $m+n+1$ binary string x is defined as $\mathcal{Q}_{n,m}$ and is given by:

$$\mathcal{Q}_{n,m} := \{q : q = q(x), x_i \in \{0,1\} \; \forall \, i\} \tag{9.20}$$

9.5.2 Encoding of solution space

In this chapter, it is assumed there is a maximum possible word-length of 16 bits, however, it is a trivial matter to change this to use another maximum word-length.

In order to generate the partially filled parameterisations of K given by (9.21), the genotype of each individual consists of a 71-bit binary string $[x_1, \ldots, x_{71}], x_i \in \{0, 1\}$. The bit length of the integer part of the parameters' representation $m \in 0, \ldots, 7$ is represented by $[x_1, x_2, x_3]$, and is given by $m = \sum_{i=1}^{3} x_i 2^{i-1}$. The bit length of the fractional part of the parameters' representation $n \in 0, \ldots, 15$ is represented by $[x_4, \ldots, x_7]$, and is given by $n = \min(\sum_{i=4}^{7} x_i 2^{i-4}, 15 - m)$. The four $m + n + 1$ word-length parameters $q_j \in \mathcal{Q}_{n,m}, j = 1, \ldots, 4$, where $(q_1, q_2, q_3, q_4) = (\tilde{a}_{11}, \tilde{a}_{12}, \tilde{b}_1, \tilde{b}_2)$ respectively, are represented by $[x_{8+16(j-1)}, \ldots, x_{8+m+n+16(j-1)}]$. Thus not all the bits in x are necessarily active. The values of q_j are calculated by (9.19).

9.5.3 Solution Procedure

For each individual in the population, the following procedure is used to generate a possible solution candidate.

1. Randomly generate the binary string x, which describes $(\tilde{a}_{11}, \tilde{a}_{12}, \tilde{b}_1, \tilde{b}_2)$ as described in Section 9.5.2. Subsequently calculate the partially filled FWL parameterisations of the two state controller:

$$\tilde{K} \stackrel{s}{=} \left[\begin{array}{cc|c} q_1 & q_2 & q_3 \\ ? & ? & q_4 \\ \hline ? & ? & D_K \end{array}\right] \tag{9.21}$$

   where $q_j \in \mathcal{Q}_{n,m}, j = 1, \ldots, 4$, *i.e.*, each q_j is FWL.
2. Using Theorem 9.1, solve the linear system equivalence completion problem such that $\tilde{K}(z) = K(z)$.
3. Obtain an FWL controller that is an approximation of the designed controller realisation, $K_q \approx \tilde{K}$, by rounding the non-FWL parameters in $\tilde{K}$ so they are FWL, *i.e.*,

$$K_q \stackrel{s}{=} \left[\begin{array}{cc|c} q_1 & q_2 & q_3 \\ q_5 & q_6 & q_4 \\ \hline q_7 & q_8 & q_9 \end{array}\right] \tag{9.22}$$

   where $q_j \in \mathcal{Q}_{n,m}, j = 1, \ldots, 9$.
4. Calculate as ϕ_1 the metric on the $\mathcal{H}_\infty$-norm topology measuring the difference between the original nominal closed-loop system and the closed-loop system using the FWL controller. That is, if both the implemented controller and the closed-loop system are stable, ϕ_1 is defined as:

$$\phi_1 = ||R - R_q||_\infty \tag{9.23}$$

where:

$$R = \frac{GK}{1+GK} \tag{9.24}$$

and:

$$R_q = \frac{GK_q}{1+GK_q} \tag{9.25}$$

This is also a measure of the robust stability/performance degradation.
5. Set ϕ_2 to be the required word-length given by $m+n+1$.
6. Calculate computational complexity index, ϕ_3, as the number of multiplications operations required in evaluating u. Note that a parameter which is 2^i, where i is any integer, can be implemented using a simple shift operation. Thus a parameter which is $1, 0, -1$ or a power of 2 does not require a multiplication operation for implementation.

9.5.4 Fitness Sharing

As seen in Section 9.5.2, not all the bits within a candidate solution bit string are necessarily active. Thus, two individuals may have the same genotype, but different gene strings. Thus to implement fitness sharing, multiple copies of genotypes are simply removed from the population.

9.6 Application Example

The approach is illustrated by application to the PID controller [10] for the IFAC93 benchmark problem [9] designed for operation at "stress level 2". In this problem, the nominal plant is provided along with perturbation ranges for five of the plant parameters. The PID controller has been optimally tuned for a set of robust performance criteria. For further details, see [10].

The continuous time nominal plant, $G(s)$, is given as:

$$G(s) = \frac{25(1-0.4s)}{(s^2+3s+25)(5s+1)} \tag{9.26}$$

The plant, $G(s)$, is discretised with sampling period of $t_s = 0.05$ seconds. A PID controller is designed as [10]:

$$K(s) = 1.311 + 0.431/s + 1.048s/(s+12.92) \tag{9.27}$$

and discretised using the bilinear transform. The initial realisation K^0 is set to:

$$K^0 \stackrel{s}{=} \left[\begin{array}{cc|c} 0 & -0.51172 & 1 \\ 1 & 1.5117 & 0 \\ \hline -0.36524 & -0.17638 & 2.1139 \end{array}\right] \tag{9.28}$$

The MOGA is implemented in MATLAB® using the GA Toolbox [31]. An elitism strategy is used whereby all non-dominated individuals propagate through to the next population. Selection is performed using stochastic universal sampling with a fitness determined by the number of dominating individuals. Single point cross-over is used with a probability of 0.7. Each bit has a probability of mutation of 0.00933.

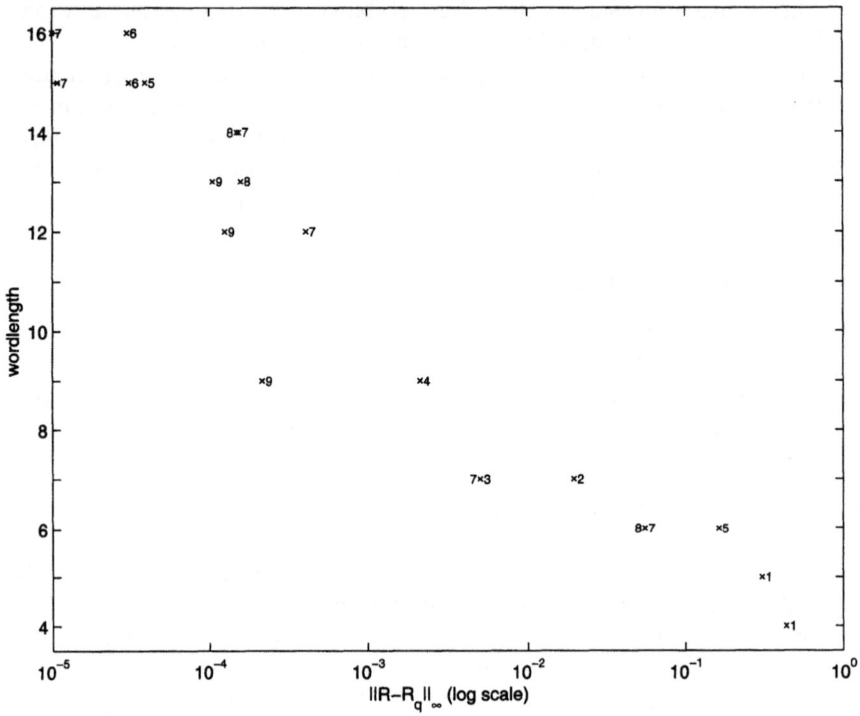

Fig. 9.4. Non-dominated solution set showing implementation accuracy, ϕ_1, against word-length, ϕ_2 — the solutions are labelled with the complexity index, ϕ_3

The MOGA is run with a population of 150. After 750 generations (which takes about four hours on a 450 MHz Pentium II), a set of non-dominated solutions is obtained. Figure 9.4 shows the implementation accuracy, ϕ_1, against the word-length, ϕ_2, for the set of non-dominated solutions. In addition, each solution is labelled with the value of the computational complexity index, ϕ_3. Note that some solutions appear almost superimposed upon each other, hence some solutions appear in the figure to have two values of ϕ_3.

It is clear from Figure 9.4 that there is a trade-off between ϕ_1 and ϕ_2. Figure 9.5 shows the design index trade-off plot of the normalised indices against index number for the non-dominated solution set. The indices have

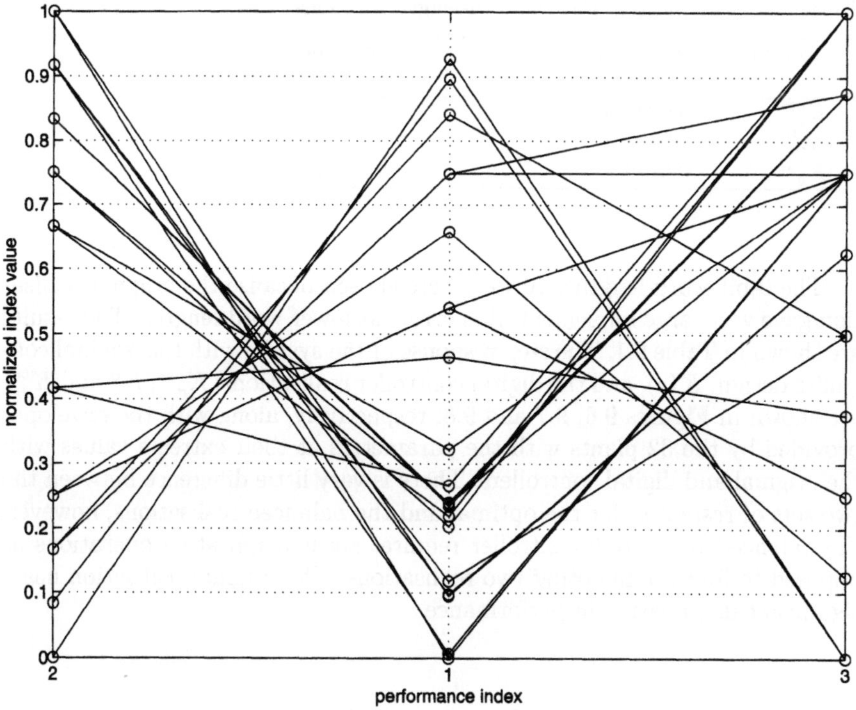

Fig. 9.5. Design index trade-offs

been re-ordered to demonstrate the trade-off between ϕ_2 and ϕ_1, and the less obvious trade-off between ϕ_1 and ϕ_3.

For the sake of comparison, the 9-bit word-length, four multiplication operations non-dominated solution realisation:

$$K_q^{opt} \stackrel{s}{=} \left[\begin{array}{cc|c} 2^{-1} & -1 & -0.796875 \\ 0 & 1 & 2^{-3} \\ \hline 0.375 & 0.578125 & 2.109375 \end{array}\right] \tag{9.29}$$

is chosen to compare with a 9-bit word-length implementation of the original canonical realisation, K^0, given by

$$K_q^{0} \stackrel{s}{=} \left[\begin{array}{cc|c} 0 & -0.515625 & 1 \\ 1 & 1.515625 & 0 \\ \hline -0.359375 & -0.171875 & 2.109375 \end{array}\right] \tag{9.30}$$

and with a 9-bit word-length implementation of a balanced realisation, [32, p 72-78]:

$$K_q^{bal} \stackrel{s}{=} \left[\begin{array}{cc|c} 1 & 0 & -0.15625 \\ 0 & 0.515625 & -0.625 \\ \hline -0.140625 & 0.609375 & 2.109375 \end{array}\right] \tag{9.31}$$

Table 9.1. Comparison of 9-bit word-length solutions

Realisation	$\|R - R_q\|$ ϕ_1	word-length ϕ_2	complexity ϕ_3
K_q^{opt}	0.002121	9	4
K_q^0	0.175467	9	5
K_q^{bal}	0.011904	9	6

The 9-bit word-length solutions were chosen because the original realisation gave very large implementation errors at lower word-lengths. The results are shown in Table 9.1. The step response of the system with the original controller design, $K(s)$, and the digital controller realisations, K_q^{opt}, K_q^0 and K_q^{bal} are shown in Figures 9.6, 9.7 and 9.8, respectively, along with the envelopes provided by the 32 plants with the parameters at their extreme values with the original and digital controllers. There is very little difference between the two sets of responses for the optimal and the balanced realisations, however, the balanced realisation controller requires six multiplication operations as opposed to four for the other two realisations. The original realisation has a significant degradation in performance.

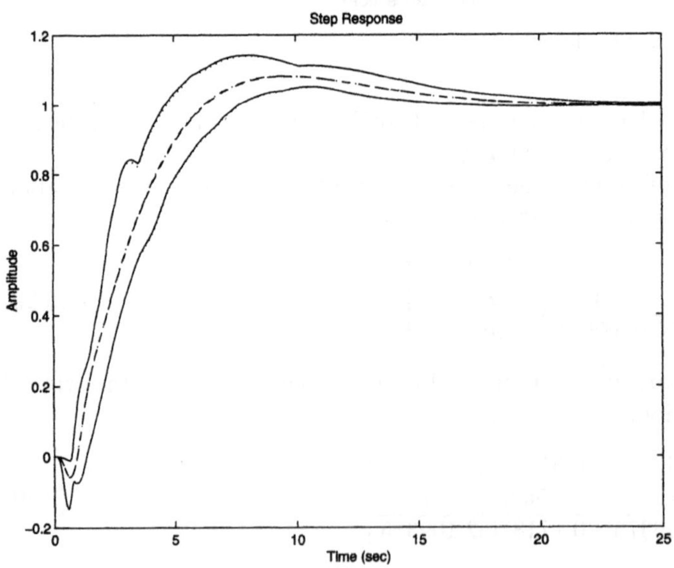

Fig. 9.6. Step response for original design, K, with nominal plant (· - · -) and envelope of extreme plants (···) and for controller, K_q^{opt}, with nominal plant (- - -) and envelope of extreme plants (—)

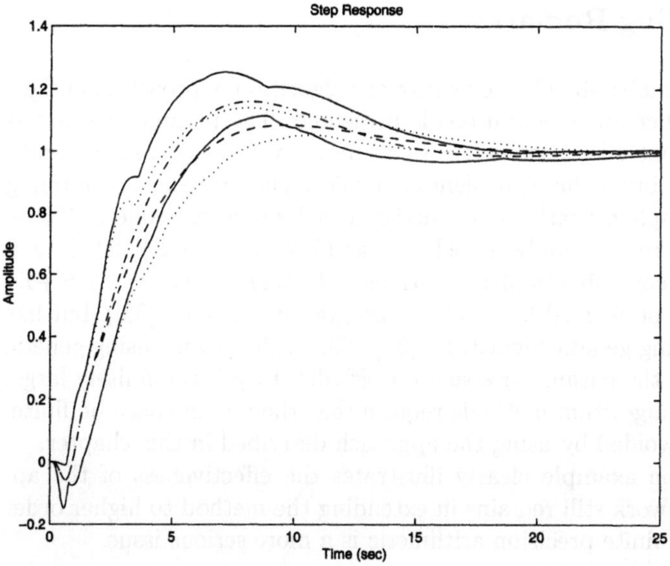

Fig. 9.7. Step response for original design, K, with nominal plant (· - - · -) and envelope of extreme plants (· · ·) and for controller, K_q^0, with nominal plant (- - -) and envelope of extreme plants (—)

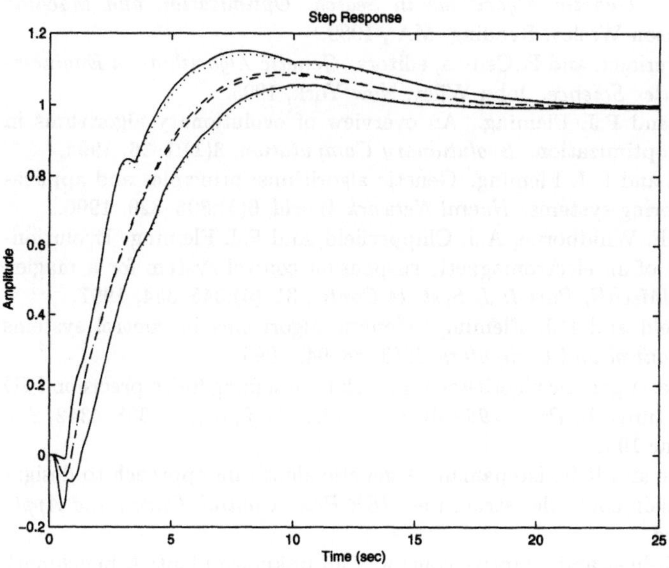

Fig. 9.8. Step response for original design, K, with nominal plant (· - - · -) and envelope of extreme plants (· · ·) and for controller, K_q^{bal}, with nominal plant (- - -) and envelope of extreme plants (—)

9.7 Concluding Remarks

In this chapter, a multi-objective evolutionary algorithm approach to designing digital controller implementations is presented. The parameter space is a subset of the controller parameters — a linear system equivalence problem is utilised to obtain the equivalent controller. The problem of obtaining good controller implementations of a designed nth-order controller, $C(zI - A)^{-1}B + D$, is more commonly solved by searching over $T \in \mathbb{R}^2, \det(T) \neq 0$ in order to obtain controller implementations, $CT^{-1}(zI - TAT^{-1})^{-1}TB + D$. This is commonly performed by nonlinear programming, (e.g., [33]), but has also been tried using genetic algorithms [34]. The problem with using genetic algorithms is that the parameter space $T \in \mathbb{R}^2, \det(T) \neq 0$ is infinitely large, and most genetic algorithm methods require that the search space be finite. This problem is avoided by using the approach described in this chapter.

The application example clearly illustrates the effectiveness of the approach. However, work still remains in extending the method to higher order controllers, where finite-precision arithmetic is a more serious issue.

References

1. D.E. Goldberg. *Genetic Algorithms in Search, Optimization and Machine Learning.* Addison-Wesley, Reading, MA., 1989.
2. G. Winter, J. Periaux, and P. Ceusta, editors. *Genetic Algorithms in Engineering and Computer Science.* John Wiley, New York, 1995.
3. C.M. Fonseca and P.J. Fleming. An overview of evolutionary algorithms in multiobjective optimization. *Evolutionary Computation*, 3(1):1–16, 1995.
4. A.M.S. Zalzala and P.J. Fleming. Genetic algorithms: principles and applications in engineering systems. *Neural Network World*, 6(5):803–820, 1996.
5. N.V. Dakev, J.F. Whidborne, A.J. Chipperfield, and P.J. Fleming. Evolutionary H_∞ design of an electromagnetic suspension control system for a maglev vehicle. *Proc. IMechE, Part I: J. Syst. & Contr.*, 311(4):345–354, 1997.
6. A.J. Chipperfield and P.J. Fleming. Genetic algorithms in control systems engineering. *Control and Computers*, 23(3):88–94, 1995.
7. J.F. Whidborne. A genetic algorithm approach to designing finite-precision PID controller structures. In *Proc. 1999 Amer. Contr. Conf.*, pages 4338–4342, San Diego, CA, June 1999.
8. J.F. Whidborne and R.H. Istepanian. A genetic algorithm approach to designing finite-precision controller structures. *IEE Proc. Control Theory and Appl.*, 2000. submitted.
9. S.F. Graebe. Robust and adaptive control of an unknown plant: A benchmark of new format. *Automatica*, 30(4):567–576, 1994.
10. J.F. Whidborne, G. Murad, D.-W. Gu, and I. Postlethwaite. Robust control of an unknown plant – the IFAC 93 benchmark. *Int. J. Control*, 61(3):589–640, 1995.
11. V. Pareto. *Manuale di Economia Politica.* Societa Editrice Libraria, Milan, Italy, 1906.

12. L.A. Zadeh. Optimality and non-scalar-valued performance criteria. *IEEE Trans. Autom. Control*, 8(1):59–60, 1963.
13. V. Zakian and U. Al-Naib. Design of dynamical and control systems by the method of inequalities. *Proceedings of the Institution of Electrical Engineers*, 120(11):1421–1427, 1973.
14. F.W. Gembicki and Y.Y. Haimes. Approach to performance and sensitivity multiobjective optimization: the goal attainment method. *IEEE Trans. Autom. Control*, 20(8):821–830, 1975.
15. J.G. Lin. Mutiple-objective problems: Pareto-optimal soultions by method of proper equality constraints. *IEEE Trans. Autom. Control*, 21(5):641–650, 1976.
16. J.G. Lin. Proper inequality constraints and maximization of index vectors. *J. Optim. Theory Applic.*, 21:505–521, 1977.
17. D.P. Giesy. Calculation of Pareto-optimal solutions to multiple-objective problems using threshold-of-acceptability constraints. *IEEE Trans. Autom. Control*, 23(6):1114–1115, 1978.
18. D. Tabak, A.A. Schy, D.P. Giesy, and K.G. Johnson. Application of multiobjective optimization in aircraft control systems design. *Automatica*, 15:595–600, 1979.
19. N. Golpalsami and C.K. Sanathanan. Satisfactory solutions approach to parameter optimization of dynamical systems with vector performance index. *J. Optim. Theory Applic.*, 47(3):301–319, 1985.
20. W.Y. Ng. *Interactive Multi-Objective Programming as a Framework for Computer-Aided Control System Design*, volume 132 of *Lect. Notes Control & Inf. Sci.* Springer-Verlag, Berlin, 1989.
21. J.L. Cohon. *Multiobjective Programming and Planning.* Academic Press, New York, 1978.
22. W.M. Spears, K.A. De Jong, T. Bäck, D.B. Fogel, and H. De Garis. An overview of evolutionary computation. In *Machine Learning: ECML-93 European Conference on Machine Learning*, volume 667 of *Lecture notes in Artificial Intelligence*, pages 442–459. Springer-Verlag, 1993.
23. C.M. Fonseca and P.J. Fleming. Genetic algorithms for multiobjective optimization: formulation, discussion and generalization. In *Genetic Algorithms: Proceeding of the Fifth International Conference*, pages 416–423, San Mateo, CA, 1993.
24. C.M. Fonseca and P.J. Fleming. Multiobjective genetic algorithms. In *IEE Colloquium on Genetic Algorithms for Control Systems Engineering*, number 1993/130, pages 6/1–6/5, London, U.K., 1993.
25. C.M. Fonseca and P.J. Fleming. Multiobjective optimization and multiple constraint handling with evolutionary algorithms – part I: a unified formulation. *IEEE Trans. Syst. Man & Cybernetics – A*, 28(1):26–37, 1995.
26. A.J. Chipperfield, J.F. Whidborne, and P.J. Fleming. Evolutionary algorithms and simulated annealing for MCDM. In T. Gal, T.J. Stewart, and T. Hanne, editors, *Multicriteria Decision Making - Advances in MCDM Models, Algorithms, Theory, and Applications*, pages 16.1–16.32. Kluwer, Boston, 1999.
27. I.N. Hernstein and D.J. Winter. *Matrix Theory and Linear Algebra.* Macmillan, New York, 1988.
28. R.A. Horn and C.R. Johnson. *Matrix Analysis.* Cambridge University Press, Cambridge, U.K., 1985.
29. S. Barnett. *Matrices in Control Theory - with applications to linear programming.* Van Nostrand Reinhold, London, U.K., 1971.

30. R.A. Horn and C.R. Johnson. *Topics in Matrix Analysis.* Cambridge University Press, Cambridge, U.K., 1991.
31. A. Chipperfield, P.J. Fleming, H. Pohlheim, and C.M. Fonseca. *Genetic Algorithm Toolbox: User's Guide.* Dept. Automatic Control and Systems Engineering, University of Sheffield, U.K., 1994.
32. K. Zhou, J.C. Doyle, and K. Glover. *Robust and Optimal Control.* Prentice Hall, Upper Saddle River, NJ., 1996.
33. J.F. Whidborne, J. Wu, and R.H. Istepanian. Finite word length stability issues in an ℓ_1 framework. *Int. J. Control*, 73(2):166–176, 2000.
34. E.G. Collins, Jr. and Y. Zhao. Design of H_2-optimal, finite word length, PID controllers. In *Proc. 2000 Amer. Contr. Conf.*, pages 902 – 903, Chicago, 2000.

CHAPTER 10
NON-FRAGILE ROBUST CONTROLLER DESIGN

Pertti M. Mäkilä and Juha Paattilammi

Abstract. This chapter studies how to obtain non-fragile controller designs via modern robust $\mathcal{H}_\infty$ control theory. It is shown that robustness conditions against both plant and controller uncertainty can be interpreted as $\mathcal{H}_\infty$-norm conditions on closed-loop transfer functions. It is emphasised that, in general, neither non-fragile nor robust designs are guaranteed by designing the $\mathcal{H}_\infty$-norm of the nominal sensitivity function and/or of the nominal complementary sensitivity function to be small. What complicates a detailed controller uncertainty analysis is the fact that the controller can be implemented in a large number of different parameterisations and in most parameterisations the controller uncertainty in the frequency domain typically depends in a complex nonlinear manner on the controller parameters. It is important to choose a parameterisation that allows realistic parametric perturbations without violating the controller frequency domain uncertainty conditions used in the design.

10.1 Introduction

Recently, a widely noticed observation [16] has been that several modern robust and optimal design methods may give designs that are sensitive, or fragile, to small or modest inaccuracies in controller implementation. These observations are elaborated upon in several papers in the 1998-1999 American Control Conferences (see e.g., [5]). It is the purpose of the present chapter to put these observations into a wider perspective. It is demonstrated here that modern robust control theory can be used to produce non-fragile designs for small and medium-sized problems.

It should be emphasised that the controller fragility issue has been earlier studied in the literature under the terminology controller sensitivity [1,12,19,25]. It should also be noted that non-fragile controller design has been studied already in the 1970s via nonlinear programming methods, see e.g., [15]. For additional references on non-fragile controller design see [5]. It is somewhat surprising that controller fragility issues have received less attention within modern robust control theory. Note that filter sensitivity and parameterisation issues have always played an important role in signal processing theory and practice (and in analogue control devices) (see e.g., [3,11]).

In classical controller design, hardly any distinction was made between plant and controller uncertainty. For single-input single-output (SISO) control loops, plant and controller uncertainty appear in a symmetrical manner in the loop transfer function PC, where P denotes the plant and C the controller. Hence one can typically include uncertainty in C as additional uncertainty in P and in this manner often obtain non-fragile controller designs, if the controller is realised sensibly. This is probably the intuitive explanation as to why it is often accepted that the most important robustness issue is related to modelling uncertainty and system perturbations [32,33]. However, also robustness to controller uncertainty has been addressed in stability analysis within modern robust control theory [10,22,23,27,29,30]. This chapter will demonstrate that much more can be said about non-fragile controller design via modern robust control theory.

An important issue in controller implementation is the choice of a proper controller parameterisation. Different realisations are, however, not considered in detail in Keel and Bhattacharyya [16,17], see also [18,20]. A well-known observation from practical control engineering is as follows: the standard expanded form transfer function realisation is often very bad and should be avoided especially for controller orders higher than about three. In [21,26], it is shown that this canonical controller form can indeed be extremely sensitive to coefficient uncertainty. Other recent work on non-fragile design and finite word-length issues is described in [8] and [31].

In the present chapter, due to its tutorial style, the emphasis is on basic concepts for understanding model/controller parameterisation sensitivity issues and for obtaining non-fragile controller designs via modern robust control theory. Hence we have left out many advanced technical concepts, such as metrics for unstable systems [2,10,23,30,32] and structured uncertainty (μ) techniques [33].

The rest of this chapter is organised as follows. Section 10.2 deals with robustness and fragility analysis via coprime factorisations. Section 10.3 explains why minimisation of the supremum norm ($\mathcal{H}_\infty$-norm) of the complementary sensitivity function, although appearing in a natural way in robustness analysis, carries with it certain dangers that can result in unrealistic designs. Section 10.4 studies a particular controller parameterisation, the factored form, that is generally considered to have desirable properties. First-order perturbation analysis indeed supports these views. In Section 10.5 another natural controller parameterisation is studied in order to get further insight about how to include controller non-fragility issues directly in modern robust controller design. Some conclusions are presented in Section 10.6.

10.2 Robustness and Fragility Analysis

There are several good references on robust $\mathcal{H}_\infty$ control from the point of view of optimal synthesis. Seven recent text books that also include many of the latest developments are [4,14,28,33,32]. A standard reference on coprime factorisations and on their use in robustness analysis is the monograph [29]. It is convenient to discuss the single-input single-output (SISO) case here. (Most of the ideas discussed here hold also in the multivariable (MIMO) case.)

10.2.1 Internal Stability

Consider the standard relationships:

$$y = P(u+v) + d \tag{10.1}$$

$$u = C(r-y) \tag{10.2}$$

where y is the output, P is the plant, u is the input, v and d are disturbances, C is the controller, and r is the reference signal. These relationships correspond to the standard one-degree of freedom feedback configuration. They can be interpreted either as transfer domain (frequency domain) relationships or as time domain relationships (through convolution operators, say) depending on, e.g., which signal spaces are considered. For our purposes it suffices to let P and C be linear, causal, time-invariant, finite-dimensional, systems. It is assumed that P and C satisfy $\overline{P(\bar{s})} = P(s)$ and $\overline{C(\bar{s})} = C(s)$, so that it is possible to represent P and C with transfer functions having real-valued coefficients only.

Then the closed-loop relationships can be readily written as:

$$y = Tr + SPv + Sd \tag{10.3}$$

$$u = CSr - Tv - CSd \tag{10.4}$$

where $S = 1/(1+PC)$ is the sensitivity function and $T = PC/(1+PC) = 1 - S$ is the complementary sensitivity function. Similarly one can derive closed-loop relationships for internal signals in the loop such as for the error $e = r - y$.

Let $1+PC$ be not identical to zero and furthermore let S, T, SP, and CS be proper transfer functions. The closed-loop system is then said to be well-posed. The standard conclusion is that four closed-loop operators (transfer functions) appear in this case in ***internal stability analysis.*** (A well-posed closed-loop system is said to be internally stable if the internal signals and the output from the system are bounded signals for any bounded external signals entering the loop, *i.e.*, for any bounded r, v and d. What is meant by a bounded signal here depends on the signal size measure used.) This means that all the four closed-loop operators (transfer functions) S, T, SP and CS should be stable for internal stability.

We shall in the sequel use the notation LTI for linear time-invariant. Let F be a stable finite-dimensional (causal) LTI system (transfer function). We shall denote $\|F\|_\infty \equiv \sup_{Re\, s \geq 0} |F(s)|$. That is, $\|F\|_\infty$ is the so-called $\mathcal{H}_\infty$-norm of F. Note that $\|F\|_\infty = \sup_\omega |F(j\omega)|$. That is, the $\mathcal{H}_\infty$-norm of F can be evaluated as the supremum of the distance from any point on the Nyquist plot of F to the origin. Note that we could similarly define the $\mathcal{H}_\infty$-norm for delay systems of the form $e^{-hs}F(s)$, and even for much more general stable infinite-dimensional systems. Similarly, although we are using continuous time systems here, an analogous treatment applies to discrete time systems.

Early work in robust $\mathcal{H}_\infty$ control concentrated on the transfer functions S and T (sensitivity and mixed sensitivity minimisation). In fact, their peak frequency response (amplitude) values $M_S = \|S\|_\infty$ and $M_T = \|T\|_\infty$ are often given rules of thumb [28], such as $M_S < 2$ and $M_T < 1.5$ (depending on the source these rules of thumb are stated with slightly different numerical values and usually for continuous time systems). Unfortunately, in general, such rules of thumb on S and T are only crude guidelines that do not guarantee satisfactory closed-loop performance and robustness properties [26].

To understand in more detail why this is so, let us consider the sensitivity of S to changes in the plant P and the controller C. To make the argument as simple as possible let P and C be both stable. Furthermore, let us assume that P' and C' are stable perturbations of P and C, respectively. Let $S' = 1/(1+P'C')$. Denote $\Delta P = P' - P$ and $\Delta C = C' - C$. Then:

$$S' = \frac{S}{1 + CS\Delta P + SP\Delta C + S\Delta P\Delta C} \tag{10.5}$$

This gives the following fact.

Theorem 10.1. *Let P and C be stable finite-dimensional (causal) LTI systems. Furthermore, let C internally stabilise P. Let P' and C' be stable finite-dimensional (causal) LTI perturbations of P and C, respectively. Let $\|P' - P\|_\infty \leq \rho_P$ and $\|C' - C\|_\infty \leq \rho_C$, where $\rho_P \geq 0$ and $\rho_C \geq 0$. Then C' internally stabilises P' if:*

$$\|CS\|_\infty \rho_P + \|SP\|_\infty \rho_C + \|S\|_\infty \rho_P \rho_C < 1 \tag{10.6}$$

Furthermore, C' internally stabilises the nominal plant P if:

$$\rho_C < \frac{1}{\|SP\|_\infty} \tag{10.7}$$

So we see that the closed-loop operators CS and SP appear naturally in robustness analysis of closed-loop stability and performance. If the design leads to large values for these operators, then even small perturbations of P and C can deteriorate the true closed-loop performance considerably (and even lead to instability). Thus, in general, it is important not only to keep $\|S\|_\infty$ and $\|T\|_\infty$ reasonably small, but also to avoid too-large values for $\|SP\|_\infty$ and $\|CS\|_\infty$. (There are also other reasons for keeping these quantities reasonably small.)

A classical popular class of stable controllers are lead-lag networks of the form:

$$C_{ll}(s) = k_C \frac{s+\alpha_1}{s+\beta_1} \cdots \frac{s+\alpha_n}{s+\beta_n} \tag{10.8}$$

where $\alpha_k > 0$, $\beta_k > 0$ for all k. Let $C'_{ll}(s)$ denote a perturbation of $C_{ll}(s)$ of the form:

$$C'_{ll}(s) = k'_C \frac{s+\alpha'_1}{s+\beta'_1} \cdots \frac{s+\alpha'_n}{s+\beta'_n} \tag{10.9}$$

where $\alpha'_k > 0$ and $\beta'_k > 0$ for all k. Clearly $C'_{ll}(s)$ is a stable perturbation of $C_{ll}(s)$ and $\|C'_{ll} - C_{ll}\|_\infty \to 0$ when $k_c \to k_c$, $\alpha'_k \to \alpha_k$, and $\beta'_k \to \beta_k$ for all k. By (10.7) the smaller we design $\|SP\|_\infty$, the less accurately we need to implement the designed controller $C_{ll}(s)$ to preserve closed-loop stability.

10.2.2 Coprime Factorisations and Robustness

We shall illustrate the use of coprime factorisations in robustness analysis (which includes controller fragility analysis). Let $P = N/M$ and $C = X/Y$ be such coprime factorisations of the plant model P and the designed stabilising controller C that the Bezout identity $NX + MY = 1$ is satisfied. $N, M \not\equiv 0$, X and $Y \not\equiv 0$ are here (causal) stable operators (transfer functions).

What if the true plant and the implemented controller are not exactly P and C but instead P' and C'? Let S' and T' denote the sensitivity function and the complementary sensitivity function associated with P' and C'. This situation has been analysed, e.g., in [23] and [29]. Denote $\Delta S = S' - S$. Then:

$$\Delta S = \frac{M\Delta Y + \Delta MY + \Delta M\Delta Y - S\Delta Z}{1+\Delta Z} \tag{10.10}$$

where:

$$\Delta Z = N\Delta X + \Delta NX + M\Delta Y + \Delta MY + \Delta N\Delta X + \Delta M\Delta Y \tag{10.11}$$

and $\Delta N = N' - N$, $\Delta M = M' - M$, $\Delta X = X' - X$, and $\Delta Y = Y' - Y$.

Denote $\Delta T = T' - T$, $\Delta(SP) = S'P' - SP$, and $\Delta(CS) = C'S' - CS$. Then:

$$\Delta T = \frac{N\Delta X + \Delta NX + \Delta N\Delta X - T\Delta Z}{1+\Delta Z} = -\Delta S \tag{10.12}$$

$$\Delta(SP) = \frac{N\Delta Y + \Delta NY + \Delta N\Delta Y - SP\Delta Z}{1+\Delta Z} \tag{10.13}$$

$$\Delta(CS) = \frac{M\Delta X + \Delta MX + \Delta M\Delta X - CS\Delta Z}{1+\Delta Z} \tag{10.14}$$

We see that it makes good sense to design the controller so that $\|X\|_\infty$ and $\|Y\|_\infty$ are small. This guarantees good robustness properties against uncertainty in the plant coprime factors. *What about uncertainty in the controller coprime factors?* Unfortunately, the level of uncertainty depends heavily on the actual parameterisation of the implemented controller. Hence, it makes sense to take the standard approach and study how the controller should be parameterised to tolerate controller uncertainty after an otherwise realistic design has been made. This separation procedure has the advantage that it can be analysed more readily.

Let X_0 and Y_0 denote any stable transfer functions such that $NX_0 + MY_0 = 1$. Then any (linear, time-invariant, causal) controller C stabilising P can be written as:

$$C = \frac{X_0 + MQ}{Y_0 - NQ} \tag{10.15}$$

for some stable transfer function Q, that is $\|Q\|_\infty < \infty$. Denote $X = X_0 + MQ$ and $Y = Y_0 - NQ$. Clearly:

$$S = M(Y_0 - NQ), \quad T = N(X_0 + MQ) \tag{10.16}$$

$$CS = M(X_0 + MQ), \quad SP = N(Y_0 - NQ) \tag{10.17}$$

Therefore:

$$\|S\|_\infty \leq \|M\|_\infty \|Y_0 - NQ\|_\infty, \; \|T\|_\infty \leq \|N\|_\infty \|X_0 + MQ\|_\infty \tag{10.18}$$

$$\|CS\|_\infty \leq \|M\|_\infty \|X_0 + MQ\|_\infty, \; \|SP\|_\infty \leq \|N\|_\infty \|Y_0 - NQ\|_\infty \tag{10.19}$$

Hence by choosing Q so that $\|X_0 + MQ\|_\infty$ and $\|Y_0 - NQ\|_\infty$ are reasonably small, then not only robustness against plant coprime factor uncertainty is optimised but also the nominal design values for $\|S\|_\infty$, $\|T\|_\infty$, $\|CS\|_\infty$ and $\|SP\|_\infty$ are kept under control.

Unfortunately, robustness optimisation against uncertainty does not need to result in satisfactory designs from a performance point of view. To illustrate this let P be a stable strictly proper plant. (Hence $P(j\infty) = 0$.) As $P \times 0 + 1 \times 1 = 1$, we can take $N = P$, $M = 1$, $X_0 = 0$, and $Y_0 = 1$. Then $S = 1 - PQ$, $T = PQ$, $CS = Q$, and $SP = P(1 - PQ)$, and $C = Q/(1 - PQ)$. Hence the quantity:

$$\kappa(N, M) = \inf_{Q \; stable} \max\{\|X_0 + MQ\|_\infty, \|Y_0 - NQ\|_\infty\} \tag{10.20}$$

is minimised by $Q = 0$ (as $\kappa(N, M) = 1$), and so we get as a robustness maximising controller the zero controller $C = 0$ (no feedback). This is quite natural if we want to maximise robustness under the a priori information that the real plant is stable! So clearly we need to include some technique to not only control the norms of the closed-loop operators S, T, CS, and SP, but in addition to shape at least the frequency response of S. That is, we need to make $|S(j\omega)|$ small for small frequencies ω. Fortunately, it is possible to utilise classical loop-shaping ideas in combination with robustness optimisation to design for both satisfactory performance and robustness [7,24,32].

The concept of normalised coprime factorisation (ncf) is useful [29]. The coprime factorisation (N, M) of $P = N/M$ is said to be an ncf of P if:

$$|N(j\omega)|^2 + |M(j\omega)|^2 = 1 \text{ for all } \omega \tag{10.21}$$

An ncf (N, M) of P is unique within multiplication by ± 1, and satisfies $\|N\|_\infty \leq 1$ and $\|M\|_\infty \leq 1$. One could also normalise coprime factorisations in other ways. Normalisation gives a useful way to introduce several metrics for measuring the distance between two linear systems [2,10,22,23,29,30,32], which can be used to introduce various robustness optimisation techniques. We shall in this tutorial use the normalisation as defined above together with the robustness quantity κ.

Clearly (10.10-10.14) give then directly the following fact.

Theorem 10.2. *Let P be a finite-dimensional (causal) LTI system. Let $P = N/M$, where (N, M) is a normalised coprime factorisation of P and $NX_0 + MY_0 = 1$. Here N, M, X_0, and Y_0 are stable finite-dimensional (causal) LTI systems. Let:*

$$C = \frac{X_0 + MQ_\kappa}{Y_0 - NQ_\kappa} \tag{10.22}$$

where Q_κ is a solution to the minimisation problem in:

$$\kappa_P = \inf_{Q \text{ stable}} \max\{\|X_0 + MQ\|_\infty, \|Y_0 - NQ\|_\infty\} \tag{10.23}$$

Denote $X = X_0 + MQ_\kappa$ and $Y = Y_0 - NQ_\kappa$. Let N', M', X', and Y' denote stable finite-dimensional perturbations of N, M, X and Y, respectively. Let S', T', $C'S'$, and $S'P'$ denote the corresponding perturbed closed-loop operators. (Here $P' = N'/M'$ and $C' = X'/Y'$.) Finally let $\Delta N = N' - N$, $\Delta M = M' - M$, $\Delta X = X' - X$, and $\Delta Y = Y' - Y$, $\Delta S = S' - S$, $\Delta T = T' - T$, $\Delta(SP) = S'P' - SP$, and $\Delta(CS) = C'S' - CS$. Let:

$$\max\{\|\Delta N\|_\infty, \|\Delta M\|_\infty\} \leq \lambda_P \tag{10.24}$$

$$\max\{\|\Delta X\|_\infty, \|\Delta Y\|_\infty\} \leq \lambda_C \tag{10.25}$$

where $\lambda_P \geq 0$ and $\lambda_C \geq 0$ are given numbers satisfying:

$$\lambda_P \kappa_P + \lambda_C + \lambda_P \lambda_C < \frac{1}{2} \tag{10.26}$$

Then C' stabilises P' internally and:

$$\|\Delta G\|_\infty \leq \frac{\lambda_C + \lambda_P \lambda_C + \kappa_P[\lambda_P + 2(\lambda_P \kappa_P + \lambda_C + \lambda_P \lambda_C)]}{1 - 2(\lambda_P \kappa_P + \lambda_C + \lambda_P \lambda_C)} \tag{10.27}$$

where $\Delta G = \Delta S, \Delta T, \Delta(CS)$, or $\Delta(SP)$.

Therefore we come to the conclusion that by choosing the controller that maximises robustness against plant coprime factor uncertainty, it gives certain robustness against both plant and controller uncertainty.

It remains to illustrate in an elementary manner that one can indeed even design for good disturbance rejection by combining loop-shaping ideas with robustness optimisation as in the method of McFarlane-Glover [24,33]. Thus let W denote a pre-compensator for shaping the frequency response of WP to a desired form. Often W can be chosen to be a proportional (P) or a proportional-integral (PI) pre-compensator.

Let (N_W, M_W) denote an ncf of WP. Thus as $|N_W(j\omega)|^2 + |M_W(j\omega)|^2 = 1$, it follows that:

$$|M_W(j\omega)|^2 = \frac{1}{1 + |W(j\omega)P(j\omega)|^2} \tag{10.28}$$

Hence we can shape $|M_W(j\omega)|$ to be small for small frequencies by choosing $|W(j\omega)|$ large enough for small frequencies if $|P(j\omega)|$ is bounded away from zero for small frequencies. Let us assume for convenience that the plant P has this property. Let us now design a controller C_W for the product WP using robustness optimisation. Thus let X_W and Y_W denote any stable transfer functions such that $N_W X_W + M_W Y_W = 1$. Then C_W is given by:

$$C_W = \frac{X_W + M_W Q_W}{Y_W - N_W Q_W} \tag{10.29}$$

where Q_W is a solution to the minimisation problem in:

$$\kappa_W = \inf_{Q \text{ stable}} \max\{\|X_W + M_W Q\|_\infty, \|Y_W - N_W Q\|_\infty\} \tag{10.30}$$

The sensitivity function $S = 1/(1 + WPC_W)$ can be expressed as $S = M_W(Y_W - N_W Q_W)$, so that:

$$|S(j\omega)| \leq \kappa_W |M_W(j\omega)| \tag{10.31}$$

Hence W should be chosen so that $\kappa_W |M_W(j\omega)|$ is $<< 1$ for small frequencies (for good disturbance rejection properties.) A good choice of W then results in a total controller $C = WC_W$ that gives a nice trade-off between performance and robustness properties.

Let us illustrate this with a simple analytical example.

Example 10.1. Consider the nominal plant:

$$P(s) = \frac{1}{s+1} \tag{10.32}$$

Let $W(s) = k$, where k is a constant. Then (N_W, M_W) is an ncf of kP, where:

$$N_W(s) = \frac{k}{s + \sqrt{1+k^2}} \tag{10.33}$$

$$M_W(s) = \frac{s+1}{s + \sqrt{1+k^2}} \tag{10.34}$$

We can take $X_W = 0$ and $Y_W = (s + \sqrt{1+k^2})/(s+1)$. Note that $\kappa_W \geq 1$ as $\|Y_W - N_W Q\| \geq |Y_W(j\infty)| = 1$ for any stable Q. Take $Q = a \times M_W^{-1}$, where a is a real number. Then $X_W + M_W Q = a$ and:

$$Y_W - N_W Q = \frac{s + \sqrt{1+k^2} - ak}{s+1} \tag{10.35}$$

Hence $Q_W = Q$ as above solves the robustness optimisation problem for κ_W if a and k satisfy:

$$-1 \leq a \leq 1 \tag{10.36}$$

$$-1 \leq \sqrt{1+k^2} - ak \leq 1 \tag{10.37}$$

Clearly with the above Q:

$$S(s) = \frac{s + \sqrt{1+k^2} - ak}{s + \sqrt{1+k^2}} \tag{10.38}$$

Thus taking e.g., $k = 5$ and $a = 1$ gives the controller:

$$C_W(s) = \frac{s+1}{s + \sqrt{26} - 5} \tag{10.39}$$

This results in $S(0) \approx 0.02$ and $|S(j)| \approx 0.2$, so that indeed the total controller $C(s) = 5C_W(s)$ attenuates well low frequency output disturbances for the nominal plant $P(s)$. The transfer function CS is now given by:

$$C(s)S(s) = 5\frac{s+1}{s + \sqrt{26}} \tag{10.40}$$

so that $\|CS\|_\infty = 5$. Note that if we were to design a proportional controller $C(s) = k_C$ for $P(s)$ so that $S(0) \approx 0.02$, this would require $k_C \approx 50$ and thus result in $\|CS\|_\infty \approx 50$. Such a P-controller would, however, have very small robustness margins against gain and phase uncertainties in the nominal model.

The reason why we have chosen the particular robustness optimisation problem as discussed above, is that it is very easy to illustrate the design procedure with minimum amount of difficult mathematical concepts. In fact, we see that only rather elementary technical tools are required to explain many of the fine points of modern robust control design. The McFarlane-Glover loop-shaping and robustness optimisation procedure differs in technical details from the above procedure and can be implemented with efficient numerical techniques also for multivariable systems. The McFarlane-Glover technique seems to be presently the best candidate for a realistic, easy-to-use, robust control design method [9,24,33].

We should remark that robustness optimisation as described above is closely related to designing the controller so that an appropriate norm, say the $\mathcal{H}_\infty$-norm or the $\mathcal{L}_1/\ell_1$-norm, of the transfer matrix:

$$H(P,C) = \begin{pmatrix} T & SP \\ CS & S \end{pmatrix} \tag{10.41}$$

is minimised over all internally stabilising controllers. (Actually these design procedures are essentially equivalent [32,33].) More generally, one can minimise a frequency-weighted version of $H(P,C)$.

To understand different robustness and performance trade-offs more fully, let us next study robustness against a class of multiplicative uncertainties.

10.3 Another View on Robustness and Fragility

Consider the uncertain plant $P'(s) = k_P P(s)$, where $P(s) = N(s)/M(s)$ is the nominal plant, $k_P = 1 + \Delta_P$ and Δ_P denotes a multiplicative (relative) uncertainty. That is, Δ_P is a stable transfer function satisfying $\|\Delta_P\|_\infty \leq \rho_P < 1$. (We could include a frequency dependent weighting in Δ_P, but the above suffices for our purposes of illustration here).

Similarly let $C' = k_C(X_0 + MQ)/(Y_0 - NQ)$ denote the uncertain controller, where $C = (X_0 + MQ)/(Y_0 - NQ)$ is the nominal controller and $k_C = 1 + \Delta_C$, where Δ_C is a stable transfer function satisfying $\|\Delta_C\|_\infty \leq \rho_C < 1$. Let C stabilise internally the nominal plant P. Note that the notation is the same as in the previous sections.

The perturbed sensitivity function $S' = 1/(1 + P'C')$ can then be written as:

$$S' = \frac{M(Y_0 - NQ)}{1 + (\Delta_P + \Delta_C + \Delta_P\Delta_C)N(X_0 + MQ)} \tag{10.42}$$

Similarly the perturbed complementary sensitivity function T', $C'S'$ and $S'P'$ become

$$T' = \frac{(1 + \Delta_P + \Delta_C + \Delta_P\Delta_C)N(X_0 + MQ)}{1 + (\Delta_P + \Delta_C + \Delta_P\Delta_C)N(X_0 + MQ)} \tag{10.43}$$

$$C'S' = \frac{(1 + \Delta_C)M(X_0 + MQ)}{1 + (\Delta_P + \Delta_C + \Delta_P\Delta_C)N(X_0 + MQ)} \tag{10.44}$$

$$S'P' = \frac{(1 + \Delta_P)N(Y_0 - NQ)}{1 + (\Delta_P + \Delta_C + \Delta_P\Delta_C)N(X_0 + MQ)} \tag{10.45}$$

Recall that $T = N(X_0 + MQ)$. It is seen that the closed-loop system is stable if:

$$\rho_P + \rho_C + \rho_P\rho_C < 1/\|T\|_\infty \tag{10.46}$$

If $\rho_P < 1/||T||_\infty$ then closed-loop stability is preserved for a relative controller uncertainty satisfying:

$$||\Delta_C||_\infty < \frac{||T||_\infty^{-1} - \rho_P}{1 + \rho_P} \tag{10.47}$$

Hence, it makes sense, for the multiplicative uncertainties considered here, to choose the controller C so that the peak value of the nominal complementary sensitivity function T, that is $||T||_\infty$, is not too far from its smallest possible value. In fact, choosing the internally stabilising controller that minimises $||T||_\infty$, maximises the stability margin of the closed loop against multiplicative plant and controller uncertainties.

But it is typically not a good idea to try to minimise $||T||_\infty$ fully as this can result in a controller with an associated closed-loop system that has very small robustness margins against general coprime factor uncertainty in the plant and the controller. Furthermore, there are certain other problems with this idea. Note that general coprime factor uncertainty is a more general type of uncertainty than multiplicative uncertainty.

10.4 Factored Controller Form

We still need to study the controller parameterisation issue to give the finishing touch to a non-fragile design procedure via modern robust control theory. This will be done in this section by studying one particular controller parameterisation, the factored form, that allows a fairly complete controller uncertainty analysis. Another controller parameterisation will be studied in Section 10.5.

We start by recalling the robustness analysis in Section 10.2. By that analysis, internal stability of the closed-loop stability is preserved if the quantity ΔZ defined in (10.11) satisfies:

$$||\Delta Z||_\infty < 1 \tag{10.48}$$

and for convenience we restate the definition of ΔZ below:

$$\Delta Z = N\Delta X + \Delta N X + M\Delta Y + \Delta M Y + \Delta N\Delta X + \Delta M\Delta Y \tag{10.49}$$

Hence we see that stability is preserved if:

$$\begin{aligned} &||N\Delta X||_\infty + ||M\Delta Y||_\infty + ||\Delta N||_\infty ||X||_\infty + ||\Delta M||_\infty ||Y||_\infty \\ &+ ||\Delta N||_\infty ||\Delta X||_\infty + ||\Delta M||_\infty ||\Delta Y||_\infty < 1 \end{aligned} \tag{10.50}$$

This condition means that it makes sense to choose the controller parameterisation so that:

$$J(\Delta X, \Delta Y) \equiv ||N\Delta X||_\infty + ||M\Delta Y||_\infty \tag{10.51}$$

is kept small. To show how this can be utilised let us analyse the factored controller parameterisation:

$$C(s) = k_C \frac{\prod_{k=1}^{m}(s - z_k)}{\prod_{\ell=1}^{n}(s - p_\ell)} \tag{10.52}$$

where n denotes the controller degree (order) and $n \geq m$. It is assumed that there are no common factors in the nominator and denominator of $C(s)$. Here $\{z_k\}$ and $\{p_\ell\}$ denote the zeros and the poles of $C(s)$, respectively. This is a classical controller parameterisation, reminiscent of a lead-lag network, that is generally considered to be a well-behaved parameterisation (see e.g., [1,11]).

Remark 10.1. It should be noted that the factored form of the controller would be realised in practice a bit differently. A transfer function with real-valued coefficients would be factored so that complex conjugate zero, or pole, pairs would be written as second-order factors. In addition, an unstable factor in the denominator would not be implemented in practice via an unstable filter, but instead such a factor would be multiplied in $C(s)$. To illustrate what this means let us consider the controller $C_0(s)$ given earlier in Section 10.4 as:

$$C_0(s) = 9\frac{s+1}{s-5} \tag{10.53}$$

As $u(s) = C_0(s)(r(s) - y(s))$ we can write:

$$\frac{s-5}{s+1}u(s) = 9(r(s) - y(s)) \tag{10.54}$$

Hence:

$$u(s) = 9(r(s) - y(s)) + \frac{6}{s+1}u(s) \tag{10.55}$$

That is, we could realise the control signal as a proportional feedback of $r(s) - y(s)$ plus a stably filtered feedback of the history of the input signal.

We prefer to use in the sequel the standard form (10.52) as this allows a systematic analysis that suffices for our purposes.

Hence we need to analyse the terms $||N\Delta X||_\infty$ and $||M\Delta Y||_\infty$. The first problem is that the controller (10.52) is not given in the coprime factor form that we need for our analysis. We know that $C(s) = X(s)/Y(s)$ and that $C(s)$ satisfies (10.52). Note that for any stable and stably invertable transfer function $V(s)$, (XV, YV) provides also a coprime factorisation of $C(s)$. Similarly, (NV^{-1}, MV^{-1}) provides a coprime factorisation of the (nominal) plant P. As $NX + MY = 1$ it follows that also $(NV^{-1})(XV) + (MV^{-1})(YV) = 1$. Consider V as a fixed, exactly known, auxiliary quantity.

Then clearly $(NV^{-1})\Delta(XV) = N\Delta X$ and $(MV^{-1})\Delta(YV) = M\Delta Y$. (Also $\Delta(NV^{-1})\Delta(XV) = \Delta N\Delta X$ and $\Delta(MV^{-1})\Delta(YV) = \Delta M\Delta Y$.) Hence, whatever coprime factorisation we choose for the controller $C(s)$, it does not change any of the terms we need to estimate.

Hence we choose:

$$\tilde{X}(s) = k_C \frac{\prod_{k=1}^{m}(s - z_k)}{(s+1)^n} \tag{10.56}$$

$$\tilde{Y}(s) = \frac{\prod_{\ell=1}^{n}(s - p_\ell)}{(s+1)^n} \tag{10.57}$$

Clearly $\tilde{X}/\tilde{Y}$ provides a coprime factorisation of C. Let $\tilde{N}/\tilde{M}$ be the corresponding coprime factorisation of the nominal plant P, so that $\tilde{N}\tilde{X}+\tilde{M}\tilde{Y} = 1$ and $P = \tilde{N}/\tilde{M}$. Then, as noted above, $N\Delta X = \tilde{N}\Delta\tilde{X}$. Hence we can estimate:

$$\begin{aligned} N(s)\Delta X(s) &= \tilde{N}(s)\left[(k_C + \Delta k_C)\frac{\prod_{k=1}^{m}(s-z_k-\Delta z_k)}{(s+1)^n} - k_C\frac{\prod_{k=1}^{m}(s-z_k)}{(s+1)^n}\right] \\ &= \frac{\Delta k_C}{k_C}N(s)X(s) - \tilde{N}(s)k_C\frac{\sum_{\ell=1}^{m}\left[\prod_{k\neq\ell}(s-z_k)\right]\Delta z_\ell}{(s+1)^n} + R(\Delta k_C, \Delta z) \end{aligned} \tag{10.58}$$

where R is the residual term collecting terms that depend on higher order terms in the perturbations of the controller gain k_C and the controller zeros $\{z_k\}$. As in first-order perturbation analysis of planetary orbital stability *etc.*, we shall concentrate here on the first-order perturbation terms. (The residual R becomes vanishingly small compared to the first-order terms when $|\Delta k_C| \to 0$ and $|\Delta z_k| \to 0$ for all k.) Let ΔX_1 and $\Delta\tilde{X}_1$ denote the first-order perturbation terms in ΔX and $\Delta\tilde{X}$, respectively.

Recall that $P(s)$ and $C(s)$ are defined over the real field. That is, all transfer functions are taken to have real-valued coefficients.

Assumption A. Let $k_C \neq 0$, $z_k \neq 0$ for $k = 1, \ldots, m$, and $p_\ell \neq 0$ for $\ell = 1, \ldots, n$. Let all zeros, $\{z_k\}$, and poles, $\{p_\ell\}$, of the controller $C(s)$ be real. Furthermore, let $z_k \neq p_\ell$ for all k, ℓ.

Remark 10.2. The assumption about real-valued controller zeros and poles is an important case that includes classical lead-lag networks and many other controller structures.

We shall assume now that Assumption A holds. Clearly:

$$\|N\Delta X_1\|_\infty \leq \frac{|\Delta k_C|}{|k_C|}\|NX\|_\infty + \sum_{\ell=1}^{m}\left\|\tilde{N}(s)k_C\frac{\prod_{k\neq\ell}(s-z_k)}{(s+1)^n}\right\|_\infty |\Delta z_\ell| \tag{10.59}$$

Now:

$$\left\| \tilde{N}(s) k_C \frac{\prod_{k \neq \ell}(s - z_k)}{(s+1)^n} \right\|_\infty = \left\| \frac{\tilde{N}(s)\tilde{X}(s)}{s - z_\ell} \right\|_\infty \leq \frac{\|NX\|_\infty}{|z_\ell|} \tag{10.60}$$

if z_ℓ is a stable real zero (that is if $z_\ell < 0$). Similarly:

$$\begin{aligned} \left\| \tilde{N}(s) k_C \frac{\prod_{k \neq \ell}(s - z_k)}{(s+1)^n} \right\|_\infty &= \left\| \tilde{N}(s) \frac{s - z_\ell}{s + z_\ell} k_C \frac{\prod_{k \neq \ell}(s - z_k)}{(s+1)^n} \right\|_\infty \\ &\leq \frac{\|NX\|_\infty}{|z_\ell|} \end{aligned} \tag{10.61}$$

if z_ℓ is an unstable real zero (that is, if $z_\ell > 0$.) We can derive in a completely similar manner an upper bound for $\|M\Delta Y_1\|_\infty$, where ΔY_1 denotes the first-order perturbation terms in ΔY. Therefore under Assumption A:

$$\frac{\|N\Delta X_1\|_\infty}{\|NX\|_\infty} \leq \frac{|\Delta k_C|}{|k_C|} + \sum_{k=1}^{m} \frac{|\Delta z_k|}{|z_k|} \tag{10.62}$$

$$\frac{\|M\Delta Y_1\|_\infty}{\|MY\|_\infty} \leq \sum_{\ell=1}^{n} \frac{|\Delta p_\ell|}{|p_\ell|} \tag{10.63}$$

These bounds give a strong indication that the factored form controller parameterisation has good robustness properties against small relative coefficient perturbations/errors. Let:

$$\frac{|\Delta k_C|}{|k_C|} \leq \gamma_{k_C} \tag{10.64}$$

$$\frac{|\Delta z_k|}{|z_k|} \leq \gamma_z,\ k = 1, \ldots, m \tag{10.65}$$

$$\frac{|\Delta p_\ell|}{|p_\ell|} \leq \gamma_p,\ \ell = 1, \ldots, n \tag{10.66}$$

where $\gamma_{k_C} \geq 0$, $\gamma_z \geq 0$, and $\gamma_p \geq 0$ are given numbers. Recall that $T = NX$ and $S = MY$. Hence under Assumption A:

$$J(\Delta X_1, \Delta Y_1) \leq (\gamma_{k_C} + m\gamma_z)\|T\|_\infty + n\gamma_p\|S\|_\infty \tag{10.67}$$

Hence if we design $\|T\|_\infty$ and $\|S\|_\infty$ to be both small then we get non-fragility of the controller for free in this parameterisation! But this is precisely what can be done with modern robust control systems design. Hence non-fragility does not require a significant change of robust design philosophy in accordance with [20]. Note that this conclusion depends on using a good controller parameterisation. By the above bounds we see that, loosely speaking, controller fragility, or sensitivity, only increases linearly in the number of uncertain controller coefficients. The expanded transfer function parameterisation has often much worse sensitivity properties [21,26].

Indeed we can now understand that non-fragility of a controller for a given nominal plant does not guarantee good overall robustness properties of the controller. This is due to the fact that non-fragility does not guarantee anything with regard to $||CS||_\infty$, and so a non-fragile controller can have arbitrarily poor robustness margins against plant coprime factor uncertainty.

The question still remains as to the fragility analysis for the factored form controller when the nominal (designed) controller has complex valued zeros and/or poles. We shall next reflect on this briefly.

Assumption B. Let $k_C \neq 0$, $z_k \neq 0$ for $k = 1, \ldots, m$, and $p_\ell \neq 0$ for $\ell = 1, \ldots, n$. Let $z_k \neq p_\ell$ for all k, ℓ. Let all unstable zeros and all unstable poles of the controller $C(s)$ be real. (That is, if $\operatorname{Re} z_k \geq 0$ then $\operatorname{Im} z_k = 0$. Similarly, if $\operatorname{Re} p_\ell \geq 0$ then $\operatorname{Im} p_\ell = 0$.) Furthermore, for stable complex zeros and poles let $|\operatorname{Re} z_k| \geq (1 - \epsilon_k)|z_k|$ and $|\operatorname{Re} p_\ell| \geq (1 - \alpha_\ell)|p_\ell|$, respectively. Here $0 < \epsilon_k < 1$ and $0 < \alpha_\ell < 1$.

We get readily by repeating the earlier analysis but now under Assumption B that:

$$\frac{||N\Delta X_1||_\infty}{||NX||_\infty} \leq \frac{|\Delta k_C|}{|k_C|} + \sum_{k=1}^{m} \beta_k \frac{|\Delta z_k|}{|z_k|} \tag{10.68}$$

$$\frac{||M\Delta Y_1||_\infty}{||MY||_\infty} \leq \sum_{\ell=1}^{n} \theta_\ell \frac{|\Delta p_\ell|}{|p_\ell|} \tag{10.69}$$

where $\beta_k = 1$ if z_k is real and $\beta_k = 1/(1 - \epsilon_k)$ if z_k is stable and non-real, and $\theta_\ell = 1$ if p_ℓ is real and $\theta_\ell = 1/(1 - \alpha_\ell)$ if p_ℓ is stable and non-real.

Therefore under Assumption B, using the notation in (10.64)-(10.66):

$$J(\Delta X_1, \Delta Y_1) \leq \left(\gamma_{k_C} + \gamma_z \sum_{k=1}^{m} \beta_k \right) ||T||_\infty + \gamma_p \left(\sum_{\ell=1}^{n} \theta_\ell \right) ||S||_\infty \tag{10.70}$$

Hence if Assumption B holds and all non-real stable zeros and poles are dominantly real, that is, say, all $\beta_k \leq 2$ and $\theta_\ell \leq 2$, then designing $||T||_\infty$ and $||S||_\infty$ both to be small remains a reasonable strategy for obtaining both robust and non-fragile designs. When there are stable zeros or poles that have a dominating imaginary part, then there may be problems if such zeros or poles are not realised accurately. Thus it is probably safer to not use such zeros or poles as online tuning parameters. In contrast, the controller gain parameter k_C is a good tuning parameter.

We have arrived at the conclusion that non-real zeros and poles may cause increased fragility, or sensitivity, even in the factored form parameterisation of the controller. So designing for small $||T||_\infty$ and $||S||_\infty$ may not always suffice to guarantee non-fragile designs, especially if one insists on considerable online tuning of any controller parameter. In particular we cannot conclude that the factored form controller would be somehow uniformly the best controller parameterisation. The issue of optimal controller parameterisations is in fact a deep topic [12], and should attract more attention within modern robust control theory.

10.5 Partial Fraction Controller Form

In many applications, the plant and the perturbed plant are known to be stable. In this case it is often desirable to use a stable controller for feedback purposes. For this setup it is possible to make a simplified analysis that gives further insight into how to perform robust design so that also controller non-fragility conditions are addressed in the design itself. We shall here assume that the controller is implemented in partial fraction form (*i.e.*, as a parallel connection of first-order systems).

Thus let us consider the standard one degree of freedom control system, where there is uncertainty both in the controller and the plant.

The controller uncertainty is naturally modelled as structured uncertainty (due to implementation inaccuracies in the controller parameters). However, it is quite difficult to use this structured information in the design. It is therefore practical to look for alternative uncertainty descriptions. Here we consider a case where both the model of the plant and the controller are given as some nominal model with additive uncertainty. That is, we have:

$$C' = C + \Delta_C W_C$$

and

$$P' = P + \Delta_P W_P$$

where P and C are the stable nominal plant and the stable controller, respectively, W_C and W_P are stable weighting functions, and Δ_C and Δ_P are arbitrary (linear, time-invariant, causal, stable) perturbations (uncertainties) satisfying:

$$\|\Delta_C\|_\infty \leq 1 \text{ and } \|\Delta_P\|_\infty \leq 1 \tag{10.71}$$

Let C stabilise internally the nominal plant P.

Recall the defining expressions for the sensitivity function $S = 1/(1+PC)$ and for the complementary sensitivity function $T = PC/(1+PC)$.

The perturbed sensitivity functions S' and T' are then given as:

$$S' = \frac{S}{1 + CS\Delta_P W_P + \Delta_C W_C SP + \Delta_C \Delta_P W_C W_P S} \tag{10.72}$$

$$T' = \frac{T + CS\Delta_P W_P + \Delta_C W_C SP + \Delta_C \Delta_P W_C W_P S}{1 + CS\Delta_P W_P + \Delta_C W_C SP + \Delta_C \Delta_P W_C W_P S} \tag{10.73}$$

From these we get directly that:

$$S'P' = \frac{S(P + \Delta_P W_P)}{1 + CS\Delta_P W_P + \Delta_C W_C SP + \Delta_C \Delta_P W_C W_P S}$$

$$C'S' = \frac{(C + \Delta_C W_C)S}{1 + CS\Delta_P W_P + \Delta_C W_C SP + \Delta_C \Delta_P W_C W_P S}$$

For the closed loop to be stable under these uncertainties we must have that:

$$\|CS\Delta_P W_P + \Delta_C W_C SP + \Delta_C \Delta_P W_C W_P S\|_\infty < 1 \tag{10.74}$$

Condition (10.74) is guaranteed to hold if:

$$\sup_\omega \left[|(W_C SP)(j\omega)| + |(W_P W_C S)(j\omega)| + |(W_P CS)(j\omega)|\right] < 1 \tag{10.75}$$

Instead of condition (10.75) it is easier to deal with a more standard $\mathcal{H}_\infty$-norm condition which implies (10.75), by considering the following H_∞ controller design problem where the objective is to find a controller C such that:

$$\left\| \begin{pmatrix} W_C SP \\ W_P W_C S \\ W_P CS \end{pmatrix} \right\|_\infty < \frac{1}{\sqrt{3}} \tag{10.76}$$

One can also append to the criterion (10.76) a performance weighting on, say, the sensitivity function S resulting in the more general problem:

$$\left\| \begin{pmatrix} W_C SP \\ W_P W_C S \\ W_P CS \\ W_e S \end{pmatrix} \right\|_\infty < \frac{1}{2} \tag{10.77}$$

where $W_e(s)$ is a suitably chosen stable weighting function.

Therefore it appears that standard robust control system design methods can be used to synthesise non-fragile controllers.

Let us study a specific parameterisation of the controller, namely the partial fraction form.

Thus we assume that the controller is given in the partial fraction form:

$$C(s) = k + \sum_{i=1}^{n} \frac{a_i}{s + b_i} \tag{10.78}$$

where $a_i \neq 0$, $b_i > 0$, $i = 1, \ldots, n$. Here we have assumed (for the sake of clarity) that $C(s)$ has simple real (stable) poles only. The perturbed controller is:

$$C'(s) = k' + \sum_{i=1}^{n} \frac{a'_i}{s + b'_i} \tag{10.79}$$

Denote $k' = k + \Delta k$, $a'_i = a_i + \Delta a_i$, $b'_i = b_i + \Delta b_i$, $i = 1, \ldots, n$. Now:

$$\Delta C(s) = C'(s) - C(s)$$

$$= k' - k + \sum_{i=1}^{n} (\frac{a_i'}{s+b_i'} - \frac{a_i}{s+b_i})$$

$$= \Delta k + \sum_{i=1}^{n} \frac{a_i}{s+b_i} \left[(\frac{1+\frac{\Delta a_i}{a_i}}{1+\frac{\Delta b_i}{s+b_i}}) - 1 \right]$$

$$= \Delta k + \sum_{i=1}^{n} \frac{a_i}{s+b_i} (-\frac{\Delta b_i}{s+b_i} + \ldots) + \sum_{i=1}^{n} \frac{\Delta a_i}{s+b_i} (1 - \frac{\Delta b_i}{s+b_i} + \ldots)$$

Let us assume that:

$$\left| \frac{\Delta b_i}{b_i} \right| < 1, \quad i = 1, \ldots, n \tag{10.80}$$

We then get that for $\operatorname{Re} s \geq 0$:

$$|C'(s) - C(s)| \leq |\Delta k| + \sum_{i=1}^{n} \frac{|a_i|}{|s+b_i|} \frac{\left|\frac{\Delta b_i}{b_i}\right|}{\left|1 - \left|\frac{\Delta b_i}{b_i}\right|\right|} + \sum_{i=1}^{n} \frac{|a_i|}{|s+b_i|} \frac{\left|\frac{\Delta a_i}{a_i}\right|}{\left|1 - \left|\frac{\Delta b_i}{b_i}\right|\right|}$$
$$\leq |W_c(s)| \tag{10.81}$$

Suppose that:

$$\left| \frac{\Delta k}{k} \right| \leq \sigma, \left| \frac{\Delta b_i}{b_i} \right| \leq \sigma, \left| \frac{\Delta a_i}{a_i} \right| \leq \sigma, \text{ where } 0 \leq \sigma < 1 \tag{10.82}$$

We also assume that $|a_i| \leq A$ and that $|b_i| \geq B$ for $i = 1, \ldots, n$, where $A > 0$ and $B > 0$. Then we get from (10.81) and (10.82) that:

$$|C'(s) - C(s)| \leq \sigma |k| + \sum_{i=1}^{n} \frac{|a_i|}{|s+b_i|} \frac{2\sigma}{1-\sigma}$$
$$\leq \sigma |k| + \frac{2\sigma n}{1-\sigma} \left| \frac{A}{s+B} \right|$$
$$\leq |W_C(s)| \tag{10.83}$$

for some suitably chosen stable transfer function $W_C(s)$. For example, when $|\Delta k| = 0$, we see that (10.83) is satisfied if we choose simply:

$$W_C(s) = \frac{2\sigma n}{1-\sigma} \left(\frac{A}{s+B} \right) \tag{10.84}$$

More generally, the above analysis suggests that a first-order weighting function of the form:

$$W_C(s) = \alpha \frac{s+\beta_1}{s+\beta_2} \tag{10.85}$$

where $\beta_1 > 0$ and $\beta_2 > 0$, may often suffice for non-fragile robust controller design. This is important as the higher the order of the weighting function is, the higher will be the order of the analytically designed controller, and by the above analysis a higher controller order increases the risk of controller fragility.

It is possible to consider $W_C(s)$ as a design function to be chosen so that the designed controller can be implemented in a non-fragile manner (not necessarily in partial fraction form). The above analysis also suggests that the partial fraction form may often be a better controller realisation than the companion (or extended) form realisation [26].

10.6 Conclusions

We have re-examined the observations in [16] concerning fragility of controllers designed with modern robust control theory. It has been demonstrated here that modern robust control theory can often be used to obtain non-fragile designs that have at the same time good robustness and performance properties. Instrumental in our discussion has been the use of coprime factorisations and the factored form controller parameterisation. However, the factored form controller parameterisation that we have used here may not be actually the optimal parameterisation to use [12]. In fact, a simplified analysis indicates that e.g., the partial fraction form may often share the good non-fragility properties of the factored form.

Acknowledgements

Financial support to the authors from the Academy of Finland under grant number 40536 is gratefully acknowledged.

References

1. K.J. Åström and B. Wittenmark. *Computer-Controlled Systems*, 2nd ed., Englewood Cliffs, N.J.: Prentice Hall, 1990.
2. C. Bonnet and J.R. Partington. Robust stabilization in the BIBO gap topology. *Int. J. Robust and Nonlinear Control*, vol. 7, pp. 429-447, 1997.
3. D.B. Chester. Digital IF filter technology for 3G systems: An introduction. *IEEE Communications Magaz.*, vol. 37, pp. 102-107, 1999.
4. M.A. Dahleh and I.J. Diaz-Bobillo. *Control of Uncertain Systems: A Linear Programming Approach*, Englewood Cliffs, NJ: Prentice-Hall, 1995.
5. P. Dorato. Non-fragile controller design: An overview. *Proc. 1998 American Control Conf.*, Philadelphia, pp. 2829-2831, 1998.
6. J.C. Doyle, B.A. Francis and A.R. Tannenbaum. *Feedback Control Theory*, New York: MacMillan, 1992.

7. M.J. Englehart and M.C. Smith. A four-block problem for H_∞ design: properties and applications. *Automatica*, vol. 27, pp. 811-818, 1991.
8. D. Famularo, P. Dorato, C.T. Abdallah, W.M. Haddad and A. Jadbabaie. Robust non-fragile LQ controllers: the static state feedback case. *Int. J. Control*, vol. 73, pp. 159-165, 2000.
9. J. Feng and M.C. Smith. When is a controller optimal in the sense of H_∞ loop-shaping? *IEEE Trans. Automat. Control*, vol. 40, pp. 2026-2039, 1995.
10. T.T. Georgiou and M.C. Smith. Optimal robustness in the gap metric. *IEEE Trans. Automat. Control*, vol. 35, pp. 673-686, 1990.
11. M.S. Ghausi. Analog active filters. *IEEE Trans. Circuits Syst.*, vol. CAS-31, pp. 13-31, 1984.
12. M. Gevers and G. Li. *Parametrizations in Control, Estimation and Filtering Problems*, Springer-Verlag, 1993.
13. K. Glover and D. McFarlane. Robust stabilization of normalized coprime factor plant descriptions with H_∞ -bounded uncertainty. *IEEE Trans. Automat. Control*, vol. 34, pp. 821-830, 1989.
14. M. Green and D.J.N. Limebeer. *Linear Robust Control*, Englewood Cliffs, NJ: Prentice-Hall, 1995.
15. J.S. Karmarkar and D.D. Šiljak. A computer-aided regulator design. *Proc. 9th Annual Allerton Conf. on Circuit and System Theory*, pp. 585-594, 1971.
16. L.H. Keel and S.P. Bhattacharyya. Robust, Fragile, or Optimal? *IEEE Trans. Automat. Control*, vol. 42, pp. 1098-1105, 1997.
17. L.H. Keel and S.P. Bhattacharyya. A linear programming approach to controller design. *Proc. 36th IEEE Conf. Decision and Control*, San Diego, pp. 2139-2148, 1997.
18. L.H. Keel and S.P. Bhattacharyya. Author's reply. *IEEE Trans. Automat. Control*, vol. 43, pp. 1268, 1998.
19. A.G. Madievski, B.D.O. Anderson and M. Gevers. Optimum realizations of sampled-data controllers for FWL sensitivity minimization. *Automatica*, vol. 31, pp. 367-379, 1995.
20. P.M. Mäkilä. Comments on "Robust, Fragile, or Optimal ?". *IEEE Trans. Automat. Control*, vol. 43, pp. 1265-1267, 1998.
21. P.M. Mäkilä. Puzzles in systems and control. In *Lecture Notes in Control and Information Sciences*, vol. 245, pp. 242-257. Springer, 1999. Eds A. Garulli, A. Tesi and A. Vicino.
22. P.M. Mäkilä and J.R. Partington. Robust identification of strongly stable systems. *IEEE Trans. Automat. Control*, vol. 37, 1709-1716, 1992.
23. P.M. Mäkilä and J.R. Partington. Robust stabilization – BIBO stability, distance notions and robustness optimization. *Automatica*, vol. 23, pp. 681-693, 1993.
24. D. McFarlane and K. Glover. *Robust Controller Design using Normalized Coprime Factor Plant Descriptions*, Lecture Notes in Control and Information Sciences, vol. 138. NY: Springer, 1990.
25. P. Moroney. *Issues in the Implementation of Digital Compensators*, Cambridge, U.S.A: The MIT Press, 1983.
26. J. Paattilammi and P.M. Mäkilä. Fragility and robustness: A case study on paper machine headbox control. *IEEE Control Syst. Magaz.*, vol. 20, pp. 13-22, 2000.
27. L. Qiu and E.J. Davison. Feedback stability under simultaneous gap metric uncertainties in plant and controller. *Syst. Control Lett.*, vol. 18, pp. 9-22, 1992.

28. S. Skogestad and I. Postlethwaite. *Multivariable Feedback Control*, Chichester: Wiley, 1996.
29. M. Vidyasagar. *Control System Synthesis*, MA: MIT Press, 1985.
30. G. Vinnicombe. Frequency domain uncertainty and the graph topology. *IEEE Trans. Automat. Control*, vol. 38, pp. 1371-1383, 1993.
31. J.F. Whidborne, J. Wu and R.S.H. Istepanian. Finite word length stability issues in an ℓ_1 framework. *Int. J. Control*, vol. 73, pp. 166-176, 2000.
32. K. Zhou with J.C. Doyle. *Essentials of Robust Control*, Upper Saddle River, NJ: Prentice Hall, 1998.
33. K. Zhou, J.C. Doyle and K. Glover. *Robust and Optimal Control*, Englewood Cliffs, NJ: Prentice-Hall, 1996.

CHAPTER 11
ROBUST RESILIENT CONTROLLER DESIGN

Wassim M. Haddad and Joseph R. Corrado

Abstract. One of the fundamental problems in feedback control design is the ability of the control system to maintain stability and performance in the face of system uncertainties. To this end, elegant multivariable robust control design frameworks such as $\mathcal{H}_\infty$ control, $\mathcal{L}_1$ control, and μ synthesis have been developed to address the robust stability and performance control problem. An implicit assumption inherent in these design frameworks is that the controller will be implemented exactly. In a recent paper by Keel and Bhattacharyya, it was shown that even though such frameworks are robust with respect to system uncertainty, they are extremely fragile with respect to errors in the controller coefficients. In this chapter, we extend the robust fixed-structure controller synthesis approach to develop controllers which are robust to system uncertainties and non-fragile or *resilient* to controller gain variations.

11.1 Introduction

It is well known that unavoidable discrepancies between mathematical models and real-world systems can result in the degradation of control system performance, including instability. Thus, it is not surprising that a considerable amount of research over the past two decades has concentrated on analysis and synthesis of feedback controllers that guarantee robustness with respect to system uncertainties in the design model (see [1] and the numerous references therein). These robust controller synthesis frameworks include the Youla parameterisation of all stabilising controllers [2], $\mathcal{H}_2$ and $\mathcal{H}_\infty$ (including desired weighting functions for loop shaping) synthesis [3–5], $\mathcal{L}_1$ control design [6], μ synthesis for structured real and complex uncertainty [5], and robust fixed-structure controller synthesis [7]. Almost all of these techniques yield very high-order controllers in relation to the original system order. A notable exception is the fixed-structure controller design methodology [7], which directly accounts for controller complexity constraints, including controller order, within the control system design process. Furthermore, an implicit assumption inherent in *all* of the above mentioned design frameworks is that the resulting robust controller will be implemented *exactly*. However, in most applications and in particular, aerospace applications, reduction in size and cost of digital control hardware results in limitations in available computer memory and word-length capabilities of the digital processor and the

A/D and D/A converters. This further results in roundoff errors in numerical computations leading to controller implementation imprecision. Hence, any controller that is part of a feedback system must be insensitive to *some* amount of error with respect to its gains.

Within the context of robust controller synthesis, the above issues were first pointed out in the enlightening and very interesting paper entitled, "Robust, Fragile, or Optimal?" [8]. Specifically, [8] very elegantly point out that the powerful (weighted) $\mathcal{H}_2$, (weighted) $\mathcal{H}_\infty$, $\mathcal{L}_1$, and μ controller design approaches, even though quite robust with respect to system uncertainty, are surprisingly very sensitive with respect to errors in the controller coefficients resulting in vanishingly small stability margins. Of course, since the control system is part of the overall closed-loop system, [8] show through a series of examples that most of the elegant multivariable robust control frameworks discussed above destabilise the closed-loop system for extremely small perturbations in the controller coefficients. Hence, even though these controllers are robust (with respect to plant uncertainty) and in some cases optimal, they are extremely fragile! This further implies that the resulting controllers preclude the control system designer from tuning the controller gains around a designed nominal controller, which has been the creed of practising control engineers to capture performance requirements that are not directly addressed within the original design problem. Finally, it is interesting to note that numerical experiments seem to indicate that the fragility or brittleness of the controller is exacerbated with increasing controller order.

In this chapter, the robust fixed-structure guaranteed cost controller synthesis framework of Bernstein and Haddad [9,10], and Haddad and Bernstein [11] for systems with structured parametric uncertainty is extended to address the design of non-fragile or robust *resilient* fixed-order (*i.e.*, full- and reduced-order) dynamic compensation. For flexibility in controller synthesis, we adopt the approach of fixed-structure controller design, which allows consideration of arbitrary controller structure, including order, internal structure, and decentralisation. Specifically, using quadratic Lyapunov bounds, a rigorous development of sufficient conditions for robust stability and worst-case $\mathcal{H}_2$ performance via fixed-order dynamic compensation is presented for uncertain feedback systems wherein the controller can tolerate multiplicative and additive gain variations with respect to its nominal coefficients. These sufficient conditions are in the form of a coupled system of algebraic Riccati equations that characterise robust resilient reduced-order controllers. Hence, the proposed robust resilient controllers guarantee robust stability and robust performance in the face of both system uncertainty *and* controller errors. The proposed approach is applied on several numerical examples that clearly demonstrate the need for robust resilient control.

Notation

$\mathbb{R}$, $\mathbb{R}^n$, $\mathbb{R}^{n\times m}$	real numbers, $n \times 1$ real column vectors, $n \times m$ real matrices
$\mathbb{S}^n$	$n \times n$ symmetric matrices
$\mathbb{N}^n$	$n \times n$ nonnegative definite matrices
$\mathbb{E}$, I_n	expectation, $n \times n$ identity

11.2 Robust Stability and Performance

In this section we state the robust stability and performance problem. This problem involves a set $\mathcal{U} \subset \mathbb{R}^{n\times n}$ of uncertain plant perturbations ΔA of the nominal dynamics A and a set $\mathcal{U}_\mathrm{c} \subset \mathbb{R}^{n_\mathrm{c}\times n_\mathrm{c}} \times \mathbb{R}^{n_\mathrm{c}\times l} \times \mathbb{R}^{m\times n_\mathrm{c}}$ of uncertain controller perturbations $(\Delta A_\mathrm{c}, \Delta B_\mathrm{c}, \Delta C_\mathrm{c})$ of the nominal controller gain matrices $(A_\mathrm{c}, B_\mathrm{c}, C_\mathrm{c})$. The objective of this problem is to determine a fixed-order, strictly proper dynamic compensator $(A_\mathrm{c}, B_\mathrm{c}, C_\mathrm{c})$ that stabilises the plant for all variations in $\mathcal{U} \times \mathcal{U}_\mathrm{c}$ and minimises the worst-case $\mathcal{H}_2$-norm of the closed-loop system. In this section and the following section no explicit structure is assumed for the elements of $\mathcal{U} \times \mathcal{U}_\mathrm{c}$. In Section 11.4 and Section 11.7, two specific structures of the variations in $\mathcal{U} \times \mathcal{U}_\mathrm{c}$ will be introduced.

Robust Stability and Performance Problem. Given the nth-order stabilisable and detectable uncertain plant:

$$\dot{x}(t) = (A + \Delta A)x(t) + Bu(t) + D_1 w(t), \quad t \in [0, \infty) \tag{11.1}$$

$$y(t) = Cx(t) + D_2 w(t) \tag{11.2}$$

determine an $n_\mathrm{c}^{\mathrm{th}}$-order robust resilient dynamic compensator $(A_\mathrm{c}, B_\mathrm{c}, C_\mathrm{c})$ such that the closed-loop system consisting of (11.1), (11.2), and controller dynamics:

$$\dot{x}_\mathrm{c}(t) = (A_\mathrm{c} + \Delta A_\mathrm{c})x_\mathrm{c}(t) + (B_\mathrm{c} + \Delta B_\mathrm{c})y(t) \tag{11.3}$$

$$u(t) = (C_\mathrm{c} + \Delta C_\mathrm{c})x_\mathrm{c}(t) \tag{11.4}$$

is asymptotically stable for all plant and controller gain variations $(\Delta A, \Delta A_\mathrm{c}, \Delta B_\mathrm{c}, \Delta C_\mathrm{c}) \in \mathcal{U} \times \mathcal{U}_\mathrm{c}$ and the performance criterion:

$$J(A_\mathrm{c}, B_\mathrm{c}, C_\mathrm{c}) \triangleq \sup_{(\Delta A, \Delta A_\mathrm{c}, \Delta B_\mathrm{c}, \Delta C_\mathrm{c}) \in \mathcal{U}\times\mathcal{U}_\mathrm{c}} \limsup_{t\to\infty} \frac{1}{t}\mathbb{E}\int_0^t \left[x^\mathrm{T}(s)R_1 x(s) + u^\mathrm{T}(s)R_2 u(s)\right] \mathrm{d}s \tag{11.5}$$

is minimised.

For each uncertain plant and controller variation $(\Delta A, \Delta A_\mathrm{c}, \Delta B_\mathrm{c}, \Delta C_\mathrm{c}) \in \mathcal{U} \times \mathcal{U}_\mathrm{c}$, the closed-loop system (11.1)–(11.4) can be written as:

$$\dot{\tilde{x}}(t) = (\tilde{A} + \Delta\tilde{A})\tilde{x}(t) + (\tilde{D} + \Delta\tilde{D})w(t), \quad t \in [0, \infty) \tag{11.6}$$

where:

$$\tilde{x}(t) \triangleq \begin{bmatrix} x(t) \\ x_c(t) \end{bmatrix}, \quad \tilde{A} \triangleq \begin{bmatrix} A & BC_c \\ B_cC & A_c \end{bmatrix}, \quad \Delta\tilde{A} \triangleq \begin{bmatrix} \Delta A & B\Delta C_c \\ \Delta B_cC & \Delta A_c \end{bmatrix}$$

$$\tilde{D} \triangleq \begin{bmatrix} D_1 \\ B_cD_2 \end{bmatrix}, \quad \Delta\tilde{D} \triangleq \begin{bmatrix} 0 \\ \Delta B_cD_2 \end{bmatrix}$$

and where the closed-loop disturbance $(\tilde{D} + \Delta\tilde{D})w(t)$ has non-negative-definite intensity:

$$\tilde{V}_\Delta \triangleq (\tilde{D} + \Delta\tilde{D})(\tilde{D} + \Delta\tilde{D})^{\mathrm{T}} \tag{11.7}$$

$$= \begin{bmatrix} V_1 & 0 \\ 0 & B_cV_2B_c^{\mathrm{T}} + \Delta B_cV_2B_c^{\mathrm{T}} + B_cV_2\Delta B_c^{\mathrm{T}} + \Delta B_cV_2\Delta B_c^{\mathrm{T}} \end{bmatrix} \tag{11.8}$$

where $V_1 \triangleq D_1D_1^{\mathrm{T}}$, $V_2 \triangleq D_2D_2^{\mathrm{T}} > 0$, and $V_{12} \triangleq D_1D_2^{\mathrm{T}} = 0$.

11.3 Sufficient Conditions for Robust Stability and Performance

In practice, steady-state performance is only of interest when the undisturbed closed-loop system is robustly stable over $\mathcal{U} \times \mathcal{U}_c$. The following result is immediate. For convenience, define:

$$\tilde{R}_\Delta \triangleq (\tilde{E} + \Delta\tilde{E})^{\mathrm{T}}(\tilde{E} + \Delta\tilde{E})$$

$$= \begin{bmatrix} R_1 & 0 \\ 0 & C_c^{\mathrm{T}}R_2C_c + \Delta C_c^{\mathrm{T}}R_2C_c + C_c^{\mathrm{T}}R_2\Delta C_c + \Delta C_c^{\mathrm{T}}R_2\Delta C_c \end{bmatrix}$$

$$\tilde{E} \triangleq [E_1 \quad E_2C_c], \quad \Delta\tilde{E} \triangleq [0 \quad E_2\Delta C_c]$$

$$R_1 \triangleq E_1^{\mathrm{T}}E_1, \; R_2 \triangleq E_2^{\mathrm{T}}E_2 > 0, \; R_{12} \triangleq E_1^{\mathrm{T}}E_2 = 0$$

Lemma 11.1. *Let (A_c, B_c, C_c) be given and assume that $\tilde{A} + \Delta\tilde{A}$ is asymptotically stable for all plant and controller gain variations $(\Delta A, \Delta A_c, \Delta B_c, \Delta C_c) \in \mathcal{U} \times \mathcal{U}_c$. Then:*

$$J(A_c, B_c, C_c) = \sup_{(\Delta A, \Delta A_c, \Delta B_c, \Delta C_c) \in \mathcal{U} \times \mathcal{U}_c} \operatorname{tr} \tilde{Q}_\Delta \tilde{R}_\Delta \tag{11.9}$$

where $\tilde{Q}_\Delta \in \mathbb{R}^{\tilde{n} \times \tilde{n}}$, $\tilde{n} \triangleq n + n_c$, is the unique, non-negative-definite solution to:

$$0 = (\tilde{A} + \Delta\tilde{A})\tilde{Q}_\Delta + \tilde{Q}_\Delta(\tilde{A} + \Delta\tilde{A})^{\mathrm{T}} + \tilde{V}_\Delta \tag{11.10}$$

The key step in guaranteeing robust stability and performance is to bound the uncertain terms $\tilde{R}_\Delta$ in the cost function (11.9) and $\Delta\tilde{A}\tilde{Q}_\Delta + \tilde{Q}_\Delta\Delta\tilde{A}^{\mathrm{T}}$ and $\tilde{V}_\Delta$ in the Lyapunov equation (11.10) by smooth bounding functions $\Omega_i(\cdot)$, $i = 1, 2, 3$. For the statement of the next result, define the notation:

$$\tilde{R} \triangleq \tilde{E}^{\mathrm{T}}\tilde{E} = \begin{bmatrix} R_1 & 0 \\ 0 & C_{\mathrm{c}}^{\mathrm{T}} R_2 C_{\mathrm{c}} \end{bmatrix}, \quad \tilde{V} \triangleq \tilde{D}\tilde{D}^{\mathrm{T}} = \begin{bmatrix} V_1 & 0 \\ 0 & B_{\mathrm{c}} V_2 B_{\mathrm{c}}^{\mathrm{T}} \end{bmatrix}$$

$$\Delta\tilde{R} \triangleq \tilde{R}_\Delta - \tilde{R} = \begin{bmatrix} 0 & 0 \\ 0 & \Delta C_{\mathrm{c}}^{\mathrm{T}} R_2 C_{\mathrm{c}} + C_{\mathrm{c}}^{\mathrm{T}} R_2 \Delta C_{\mathrm{c}} + \Delta C_{\mathrm{c}}^{\mathrm{T}} R_2 \Delta C_{\mathrm{c}} \end{bmatrix}$$

$$\Delta\tilde{V} \triangleq \tilde{V}_\Delta - \tilde{V} = \begin{bmatrix} 0 & 0 \\ 0 & \Delta B_{\mathrm{c}} V_2 B_{\mathrm{c}}^{\mathrm{T}} + B_{\mathrm{c}} V_2 \Delta B_{\mathrm{c}}^{\mathrm{T}} + \Delta B_{\mathrm{c}} V_2 \Delta B_{\mathrm{c}}^{\mathrm{T}} \end{bmatrix}$$

Theorem 11.1. *Let* $(A_{\mathrm{c}}, B_{\mathrm{c}}, C_{\mathrm{c}})$ *be given and let* $\Omega_1 : \mathbb{N}^{\tilde{n}} \times \mathbb{R}^{n_{\mathrm{c}} \times n_{\mathrm{c}}} \times \mathbb{R}^{n_{\mathrm{c}} \times l} \times \mathbb{R}^{m \times n_{\mathrm{c}}} \to \mathbb{S}^{\tilde{n}}$, $\Omega_2 : \mathbb{R}^{n_{\mathrm{c}} \times l} \to \mathbb{S}^{\tilde{n}}$, *and* $\Omega_3 : \mathbb{R}^{m \times n_{\mathrm{c}}} \to \mathbb{S}^{\tilde{n}}$, *be such that:*

$$\Delta\tilde{A}\tilde{Q} + \tilde{Q}\Delta\tilde{A}^{\mathrm{T}} \leq \Omega_1(\tilde{Q}, A_{\mathrm{c}}, B_{\mathrm{c}}, C_{\mathrm{c}}) \tag{11.11}$$

$$\Delta\tilde{V} \leq \Omega_2(B_{\mathrm{c}}) \tag{11.12}$$

$$\Delta\tilde{R} \leq \Omega_3(C_{\mathrm{c}}) \tag{11.13}$$

$$(\Delta A, \Delta A_{\mathrm{c}}, \Delta B_{\mathrm{c}}, \Delta C_{\mathrm{c}}) \in \mathcal{U} \times \mathcal{U}_{\mathrm{c}}$$

$$\left(\tilde{Q}, A_{\mathrm{c}}, B_{\mathrm{c}}, C_{\mathrm{c}}\right) \in \mathbb{N}^{\tilde{n}} \times \mathbb{R}^{n_{\mathrm{c}} \times n_{\mathrm{c}}} \times \mathbb{R}^{n_{\mathrm{c}} \times l} \times \mathbb{R}^{m \times n_{\mathrm{c}}}$$

and suppose there exists $\tilde{Q} \in \mathbb{N}^{\tilde{n}}$ *satisfying:*

$$0 = \tilde{A}\tilde{Q} + \tilde{Q}\tilde{A}^{\mathrm{T}} + \Omega_1(\tilde{Q}, A_{\mathrm{c}}, B_{\mathrm{c}}, C_{\mathrm{c}}) + \tilde{V} + \Omega_2(B_{\mathrm{c}}) \tag{11.14}$$

Then:

$$(\tilde{A} + \Delta\tilde{A}, \tilde{D} + \Delta\tilde{D}) \text{ is stabilisable,} \tag{11.15}$$

for all $(\Delta A, \Delta A_{\mathrm{c}}, \Delta B_{\mathrm{c}}, \Delta C_{\mathrm{c}}) \in \mathcal{U} \times \mathcal{U}_{\mathrm{c}}$, *if and only if* $\tilde{A} + \Delta\tilde{A}$ *is asymptotically stable for all* $(\Delta A, \Delta A_{\mathrm{c}}, \Delta B_{\mathrm{c}}, \Delta C_{\mathrm{c}}) \in \mathcal{U} \times \mathcal{U}_{\mathrm{c}}$. *In this case:*

$$\tilde{Q}_\Delta \leq \tilde{Q}, \qquad (\Delta A, \Delta A_{\mathrm{c}}, \Delta B_{\mathrm{c}}, \Delta C_{\mathrm{c}}) \in \mathcal{U} \times \mathcal{U}_{\mathrm{c}} \tag{11.16}$$

where $\tilde{Q}_\Delta$ *is given by (11.10), and:*

$$J(A_{\mathrm{c}}, B_{\mathrm{c}}, C_{\mathrm{c}}) \leq \mathcal{J}(A_{\mathrm{c}}, B_{\mathrm{c}}, C_{\mathrm{c}}) \triangleq \operatorname{tr} \tilde{Q}\left[\tilde{R} + \Omega_3(C_{\mathrm{c}})\right] \tag{11.17}$$

Proof. We stress that in (11.11), $\tilde{Q}$ denotes an arbitrary element of $\mathbb{N}^{\tilde{n}}$, whereas in (11.14), $\tilde{Q}$ denotes a specific solution of the modified Lyapunov equation (11.14). This minor abuse of notation considerably simplifies the presentation. Now, for $(\Delta A, \Delta A_{\mathrm{c}}, \Delta B_{\mathrm{c}}, \Delta C_{\mathrm{c}}) \in \mathcal{U} \times \mathcal{U}_{\mathrm{c}}$, (11.14) is equivalent to:

$$\begin{aligned} 0 = {} & (\tilde{A} + \Delta\tilde{A})\tilde{Q} + \tilde{Q}(\tilde{A} + \Delta\tilde{A})^{\mathrm{T}} + \Omega_1(\tilde{Q}, A_{\mathrm{c}}, B_{\mathrm{c}}, C_{\mathrm{c}}) \\ & -(\Delta\tilde{A}\tilde{Q} + \tilde{Q}\Delta\tilde{A}^{\mathrm{T}}) + \tilde{V}_\Delta + \Omega_2(B_{\mathrm{c}}) - \Delta\tilde{V} \end{aligned} \tag{11.18}$$

Hence, by assumption, (11.18) has a solution $\tilde{Q} \in \mathbb{N}^{\tilde{n}}$ for all $(\Delta A, \Delta A_c, \Delta B_c, \Delta C_c) \in \mathcal{U} \times \mathcal{U}_c$ and, by (11.11) and (11.12), $\Omega_1(\tilde{Q}, A_c, B_c, C_c) - (\Delta\tilde{A}\tilde{Q} + \tilde{Q}\Delta\tilde{A}^{\mathrm{T}})$ and $\Omega_2(B_c) - \Delta\tilde{V}$ are non-negative-definite. Now, if the stabilisability condition (11.15) holds for all $(\Delta A, \Delta A_c, \Delta B_c, \Delta C_c) \in \mathcal{U} \times \mathcal{U}_c$, it follows from Theorem 3.6 of [12] that:

$$\left(\tilde{A}+\Delta\tilde{A}, \left[\tilde{V}_\Delta+\Omega_1(\tilde{Q}, A_c, B_c, C_c)-(\Delta\tilde{A}\tilde{Q}+\tilde{Q}\Delta\tilde{A}^{\mathrm{T}})+\Omega_2(B_c)-\Delta\tilde{V}\right]^{1/2}\right)$$

is stabilisable for all $(\Delta A, \Delta A_c, \Delta B_c, \Delta C_c) \in \mathcal{U} \times \mathcal{U}_c$. It now follows from (11.18) and Lemma 12.2 of [12] that $\tilde{A} + \Delta\tilde{A}$ is asymptotically stable for all $(\Delta A, \Delta A_c, \Delta B_c, \Delta C_c) \in \mathcal{U} \times \mathcal{U}_c$. Conversely, if $\tilde{A} + \Delta\tilde{A}$ is asymptotically stable for all $(\Delta A, \Delta A_c, \Delta B_c, \Delta C_c) \in \mathcal{U} \times \mathcal{U}_c$, then (11.15) holds. Next, subtracting (11.10) from (11.18) yields:

$$\begin{aligned} 0 &= (\tilde{A} + \Delta\tilde{A})(\tilde{Q} - \tilde{Q}_\Delta) + (\tilde{Q} - \tilde{Q}_\Delta)(\tilde{A} + \Delta\tilde{A})^{\mathrm{T}} + \Omega_1(\tilde{Q}, A_c, B_c, C_c) \\ &\quad -(\Delta\tilde{A}\tilde{Q} + \tilde{Q}\Delta\tilde{A}^{\mathrm{T}}) + \Omega_2(B_c) - \Delta\tilde{V} \end{aligned}$$

or, equivalently, since $\tilde{A} + \Delta\tilde{A}$ is asymptotically stable for all $(\Delta A, \Delta A_c, \Delta B_c, \Delta C_c) \in \mathcal{U} \times \mathcal{U}_c$:

$$\begin{aligned} \tilde{Q} - \tilde{Q}_\Delta &= \int_0^\infty e^{(\tilde{A}+\Delta\tilde{A})t}\Big[\Omega_1(\tilde{Q}, A_c, B_c, C_c) - (\Delta\tilde{A}\tilde{Q} + \tilde{Q}\Delta\tilde{A}^{\mathrm{T}}) \\ &\qquad + \Omega_2(B_c) - \Delta\tilde{V}\Big] e^{(\tilde{A}+\Delta\tilde{A})^{\mathrm{T}}t}\,\mathrm{d}t \\ &\geq 0 \end{aligned}$$

which implies (11.16). The performance bound (11.17) is now an immediate consequence of (11.16) and (11.13) by noting that:

$$\begin{aligned} J(A_c, B_c, C_c) &= \sup_{\Delta\tilde{A} \in \mathcal{U}\times\mathcal{U}_c} \operatorname{tr} \tilde{Q}_\Delta \tilde{R}_\Delta \leq \sup_{\Delta\tilde{A} \in \mathcal{U}\times\mathcal{U}_c} \operatorname{tr} \tilde{Q}\tilde{R}_\Delta \\ &= \sup_{\Delta\tilde{A} \in \mathcal{U}\times\mathcal{U}_c} \operatorname{tr} \tilde{Q}(\tilde{R} + \Delta\tilde{R}) \leq \operatorname{tr} \tilde{Q}(\tilde{R} + \Omega_3(C_c)) \end{aligned}$$

□

Remark 11.1. In applying Theorem 11.1 it may be convenient to replace Condition (11.15) with the stronger condition:

$$\left(\tilde{A} + \Delta\tilde{A}, \left[\tilde{V}_\Delta + \Omega_1(\tilde{Q}, A_c, B_c, C_c) - (\Delta\tilde{A}\tilde{Q} + \tilde{Q}\Delta\tilde{A}^{\mathrm{T}}) + \Omega_2(B_c) - \Delta\tilde{V}\right]^{1/2}\right) \text{ is stabilisable,} \tag{11.19}$$

for all $(\Delta A, \Delta A_c, \Delta B_c, \Delta C_c) \in \mathcal{U} \times \mathcal{U}_c$, which is easier to verify in practice. Clearly, (11.19) is satisfied if $\tilde{V}_\Delta + \Omega_1(\tilde{Q}, A_c, B_c, C_c) - (\Delta\tilde{A}\tilde{Q} + \tilde{Q}\Delta\tilde{A}^{\mathrm{T}}) + \Omega_2(B_c) - \Delta\tilde{V}$ is positive-definite for all $(\Delta A, \Delta A_c, \Delta B_c, \Delta C_c) \in \mathcal{U} \times \mathcal{U}_c$. This will be the case, e.g., if either $\tilde{V}$ is positive-definite or strict inequalities hold in (11.11) or (11.12).

11.4 Multiplicative Controller Uncertainty Structure and Guaranteed Cost Bound

Having established the theoretical basis for our approach, we now assign an explicit structure to the sets $\mathcal{U}$ and $\mathcal{U}_c$ and the bounding functions $\Omega_1(\tilde{Q}, A_c, B_c, C_c)$, $\Omega_2(B_c)$, and $\Omega_3(C_c)$. Specifically, the uncertainty set $\mathcal{U}$ capturing parametric plant uncertainty and the uncertainty set $\mathcal{U}_c$ capturing multiplicative (relative) controller gain variations are defined by:

$$\mathcal{U} \triangleq \{\Delta A : \Delta A = B_0 F C_0,\ F^{\mathrm{T}} F \leq \gamma^{-2} I_r\} \tag{11.20}$$

$$\begin{aligned}\mathcal{U}_c \triangleq \{(\Delta A_c, \Delta B_c, \Delta C_c) : \Delta A_c = \delta A_c,\ \Delta B_c = \delta B_c,\\ \Delta C_c = \delta C_c,\ |\, \delta \,| \leq \gamma_c^{-1}\}\end{aligned} \tag{11.21}$$

where $B_0 \in \mathbb{R}^{n \times s}$, $C_0 \in \mathbb{R}^{r \times n}$ are fixed matrices denoting the structure of the plant uncertainty, $F \in \mathbb{R}^{s \times r}$ is an uncertain matrix, δ is an uncertain real parameter, and γ, γ_c are given positive numbers. With this uncertainty characterisation, the closed-loop system (11.6) has structured uncertainty of the form:

$$\Delta\tilde{A} = \tilde{B}_1 F \tilde{C}_1 + \delta \tilde{B}_2 \tilde{C}_2 \tag{11.22}$$

where:

$$\tilde{B}_1 \triangleq \begin{bmatrix} B_0 \\ 0 \end{bmatrix}, \quad \tilde{C}_1 \triangleq [C_0\ 0] \tag{11.23}$$

$$\tilde{B}_2 \triangleq \begin{bmatrix} 0 & 0 & B \\ I_{n_c} & I_{n_c} & 0 \end{bmatrix}, \quad \tilde{C}_2 \triangleq \begin{bmatrix} 0 & A_c \\ B_c C & 0 \\ 0 & C_c \end{bmatrix} \tag{11.24}$$

For the structure of $\mathcal{U}$ and $\mathcal{U}_c$ as specified by (11.20) and (11.21), the bounding functions $\Omega_1(\tilde{Q}, A_c, B_c, C_c)$, $\Omega_2(B_c)$, and $\Omega_3(C_c)$ can now be given a concrete form. For the statement of the next result, define $\tilde{R}_{c_1} \triangleq \tilde{E}_{c_1}^{\mathrm{T}} \tilde{E}_{c_1}$ and $\tilde{V}_{c_1} \triangleq \tilde{D}_{c_1} \tilde{D}_{c_1}^{\mathrm{T}}$, where:

$$\tilde{E}_{c_1} \triangleq [0\ E_2 C_c], \quad \tilde{D}_{c_1} \triangleq \begin{bmatrix} 0 \\ B_c D_2 \end{bmatrix}$$

Proposition 11.1. *Let $\alpha > 0, \alpha_c > 0$. Furthermore, let $\mathcal{U}$ and $\mathcal{U}_c$ be defined by (11.20) and (11.21), respectively, and define $\Omega_1(\cdot)$, $\Omega_2(\cdot)$, and $\Omega_3(\cdot)$ by:*

$$\begin{aligned}\Omega_1(\tilde{Q}, A_c, B_c, C_c) = \alpha \tilde{B}_1 \tilde{B}_1^{\mathrm{T}} + \alpha_c \tilde{B}_2 \tilde{B}_2^{\mathrm{T}} + \tilde{Q}(\alpha^{-1}\gamma^{-2}\tilde{C}_1^{\mathrm{T}}\tilde{C}_1\\ +\alpha_c^{-1}\gamma_c^{-2}\tilde{C}_2^{\mathrm{T}}\tilde{C}_2)\tilde{Q}\end{aligned} \tag{11.25}$$

$$\Omega_2(B_c) = (\gamma_c^{-2} + 2\gamma_c^{-1})\tilde{V}_{c_1} \tag{11.26}$$

$$\Omega_3(C_c) = (\gamma_c^{-2} + 2\gamma_c^{-1})\tilde{R}_{c_1} \tag{11.27}$$

Then (11.11)–(11.13) are satisfied.

Proof. Note that:

$$
\begin{aligned}
0 \leq & \left[\alpha^{\frac{1}{2}} \tilde{B}_1 - \alpha^{-\frac{1}{2}} \tilde{Q}\tilde{C}_1^{\mathrm{T}} F^{\mathrm{T}}\right]\left[\alpha^{\frac{1}{2}} \tilde{B}_1 - \alpha^{-\frac{1}{2}} \tilde{Q}\tilde{C}_1^{\mathrm{T}} F^{\mathrm{T}}\right]^{\mathrm{T}} \\
& + \left[\alpha_{\mathrm{c}}^{\frac{1}{2}} \tilde{B}_2 - \alpha_{\mathrm{c}}^{-\frac{1}{2}} \delta \tilde{Q}\tilde{C}_2^{\mathrm{T}}\right]\left[\alpha_{\mathrm{c}}^{\frac{1}{2}} \tilde{B}_2 - \alpha_{\mathrm{c}}^{-\frac{1}{2}} \delta \tilde{Q}\tilde{C}_2^{\mathrm{T}}\right]^{\mathrm{T}} \\
\leq & \ \alpha \tilde{B}_1 \tilde{B}_1^{\mathrm{T}} + \alpha_{\mathrm{c}} \tilde{B}_2 \tilde{B}_2^{\mathrm{T}} + \tilde{Q}(\alpha^{-1}\gamma^{-2}\tilde{C}_1^{\mathrm{T}}\tilde{C}_1 \\
& + \alpha_{\mathrm{c}}^{-1}\gamma_{\mathrm{c}}^{-2}\tilde{C}_2^{\mathrm{T}}\tilde{C}_2)\tilde{Q} - (\Delta\tilde{A}\tilde{Q} + \tilde{Q}\Delta\tilde{A}^{\mathrm{T}})
\end{aligned}
$$

which proves (11.11) with $\mathcal{U}$ and $\mathcal{U}_{\mathrm{c}}$ given by (11.20) and (11.21), respectively. Next, note that:

$$
\Delta\tilde{V} = \begin{bmatrix} 0 & 0 \\ 0 & (\delta^2 + 2\delta)B_{\mathrm{c}}V_2B_{\mathrm{c}}^{\mathrm{T}} \end{bmatrix} \leq \begin{bmatrix} 0 & 0 \\ 0 & (\gamma_{\mathrm{c}}^{-2} + 2\gamma_{\mathrm{c}}^{-1})B_{\mathrm{c}}V_2B_{\mathrm{c}}^{\mathrm{T}} \end{bmatrix} = \Omega_2(B_{\mathrm{c}})
$$

which proves (11.12) with $\mathcal{U}_{\mathrm{c}}$ given by (11.21). Finally, a similar construction proves (11.13). □

Next, using Theorem 11.1 and Proposition 11.1 we have the following immediate result.

Theorem 11.2. *Let $\alpha > 0, \alpha_{\mathrm{c}} > 0$, and suppose there exists an $\tilde{n} \times \tilde{n}$ non-negative-definite matrix $\tilde{Q}$ satisfying:*

$$
\begin{aligned}
0 = & \ \tilde{A}\tilde{Q} + \tilde{Q}\tilde{A}^{\mathrm{T}} + \tilde{Q}(\alpha^{-1}\gamma^{-2}\tilde{C}_1^{\mathrm{T}}\tilde{C}_1 + \alpha_{\mathrm{c}}^{-1}\gamma_{\mathrm{c}}^{-2}\tilde{C}_2^{\mathrm{T}}\tilde{C}_2)\tilde{Q} + \tilde{V} \\
& + \alpha\tilde{B}_1\tilde{B}_1^{\mathrm{T}} + \alpha_{\mathrm{c}}\tilde{B}_2\tilde{B}_2^{\mathrm{T}} + (\gamma_{\mathrm{c}}^{-2} + 2\gamma_{\mathrm{c}}^{-1})\tilde{V}_{\mathrm{c}_1}
\end{aligned} \tag{11.28}
$$

Then $(\tilde{A}+\Delta\tilde{A}, \tilde{D}+\Delta\tilde{D})$ is stabilisable for all $(\Delta A, \Delta A_{\mathrm{c}}, \Delta B_{\mathrm{c}}, \Delta C_{\mathrm{c}}) \in \mathcal{U} \times \mathcal{U}_{\mathrm{c}}$ if and only if $\tilde{A}+\Delta\tilde{A}$ is asymptotically stable for all $(\Delta A, \Delta A_{\mathrm{c}}, \Delta B_{\mathrm{c}}, \Delta C_{\mathrm{c}}) \in \mathcal{U} \times \mathcal{U}_{\mathrm{c}}$. In this case:

$$
\tilde{Q}_{\Delta} \leq \tilde{Q}, \qquad (\Delta A, \Delta A_{\mathrm{c}}, \Delta B_{\mathrm{c}}, \Delta C_{\mathrm{c}}) \in \mathcal{U} \times \mathcal{U}_{\mathrm{c}} \tag{11.29}
$$

where $\tilde{Q}_{\Delta}$ satisfies (11.10), and:

$$
J(A_{\mathrm{c}}, B_{\mathrm{c}}, C_{\mathrm{c}}) \leq \operatorname{tr} \tilde{Q}\left[\tilde{R} + (\gamma_{\mathrm{c}}^{-2} + 2\gamma_{\mathrm{c}}^{-1})\tilde{R}_{\mathrm{c}_1}\right] \tag{11.30}
$$

11.5 Decentralised Static Output Feedback Formulation

In this section we use the fixed-structure control framework of [13] and [14] to transform the Robust Stability and Performance Problem to a decentralised static output feedback setting. Specifically, note that for every dynamic controller (11.3), (11.4) with gain variations $(\Delta A_c, \Delta B_c, \Delta C_c) \in \mathcal{U}_c$ given by (11.21), the uncertain closed-loop system (11.1)-(11.4) can be written as:

$$\begin{bmatrix} \dot{x}(t) \\ \dot{x}_c(t) \end{bmatrix} = \begin{bmatrix} A + \Delta A & (1+\delta)BC_c \\ (1+\delta)B_cC & (1+\delta)A_c \end{bmatrix} \begin{bmatrix} x(t) \\ x_c(t) \end{bmatrix} + \begin{bmatrix} D_1 \\ (1+\delta)B_cD_2 \end{bmatrix} w(t) \tag{11.31}$$

Furthermore, by treating A_c, B_c, and C_c as decentralised static output feedback gains with multiplicative (relative) uncertainty as shown in Figure 11.1, we can pull the uncertainty into the plant model obtaining:

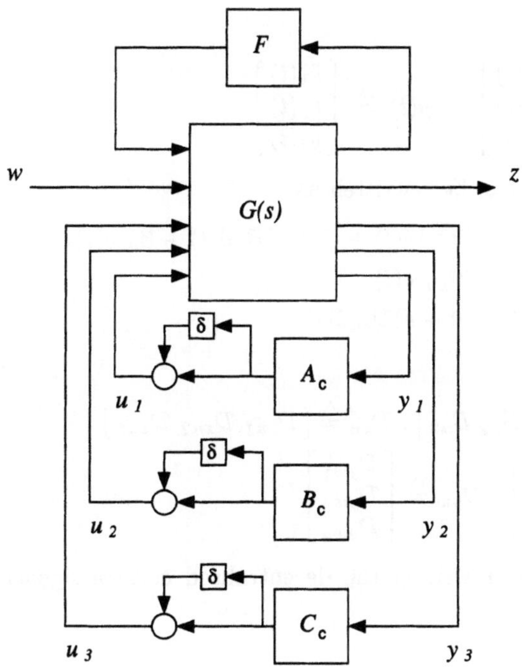

Fig. 11.1. Decentralised static output feedback: multiplicative controller uncertainty

$$\dot{\tilde{x}}(t) = (\mathcal{A} + \tilde{B}_1 F \tilde{C}_1)\tilde{x}(t) + (1+\delta)\sum_{i=1}^{3} \mathcal{B}_{ui}\hat{u}_i(t) + \mathcal{B}_w w(t) \tag{11.32}$$

$$\hat{y}_i(t) = \mathcal{C}_{y_i}\tilde{x}(t) + \mathcal{D}_{yw_i} w(t), \quad i = 1,2,3 \tag{11.33}$$

$$z(t) = \mathcal{C}_z \tilde{x}(t) + (1+\delta)\sum_{i=1}^{3} \mathcal{D}_{zui}\hat{u}_i(t) \tag{11.34}$$

$$\hat{u}_1(t) = A_c\hat{y}_1(t), \quad \hat{u}_2(t) = B_c\hat{y}_2(t), \quad \hat{u}_3(t) = C_c\hat{y}_3(t) \tag{11.35}$$

where:

$$\tilde{x}(t) \triangleq \begin{bmatrix} x(t) \\ x_c(t) \end{bmatrix}, \quad \mathcal{A} \triangleq \begin{bmatrix} A & 0 \\ 0 & 0 \end{bmatrix}$$

$$\mathcal{B}_{u1} \triangleq \begin{bmatrix} 0 \\ I_{n_c} \end{bmatrix}, \quad \mathcal{B}_{u2} \triangleq \begin{bmatrix} 0 \\ I_{n_c} \end{bmatrix}, \quad \mathcal{B}_{u3} \triangleq \begin{bmatrix} B \\ 0 \end{bmatrix}$$

$$\mathcal{C}_{y_1} \triangleq \begin{bmatrix} 0 & I_{n_c} \end{bmatrix}, \mathcal{C}_{y_2} \triangleq \begin{bmatrix} C & 0 \end{bmatrix}, \mathcal{C}_{y_3} \triangleq \begin{bmatrix} 0 & I_{n_c} \end{bmatrix}$$

$$\mathcal{B}_w \triangleq \begin{bmatrix} D_1 \\ 0 \end{bmatrix}, \mathcal{D}_{yw_1} \triangleq 0, \mathcal{D}_{yw_2} \triangleq D_2, \mathcal{D}_{yw_3} \triangleq 0$$

$$\mathcal{C}_z \triangleq \begin{bmatrix} E_1 & 0 \end{bmatrix}, \mathcal{D}_{zu1} \triangleq 0, \mathcal{D}_{zu2} \triangleq 0, \mathcal{D}_{zu3} \triangleq E_2$$

Next, defining:

$$\hat{u}(t) \triangleq \begin{bmatrix} \hat{u}_1(t) \\ \hat{u}_2(t) \\ \hat{u}_3(t) \end{bmatrix}, \quad \hat{y}(t) \triangleq \begin{bmatrix} \hat{y}_1(t) \\ \hat{y}_2(t) \\ \hat{y}_3(t) \end{bmatrix}$$

(11.32)–(11.34) can be rewritten as

$$\dot{\tilde{x}}(t) = (\mathcal{A} + \tilde{B}_1 F \tilde{C}_1)\tilde{x}(t) + (1+\delta)\mathcal{B}_u\hat{u}(t) + \mathcal{B}_w w(t) \tag{11.36}$$

$$\hat{y}(t) = \mathcal{C}_y\tilde{x}(t) + \mathcal{D}_{yw}w(t) \tag{11.37}$$

$$z(t) = \mathcal{C}_z\tilde{x}(t) + (1+\delta)\mathcal{D}_{zu}\hat{u}(t) \tag{11.38}$$

where:

$$\mathcal{B}_u \triangleq \begin{bmatrix} \mathcal{B}_{u1} & \mathcal{B}_{u2} & \mathcal{B}_{u3} \end{bmatrix}, \mathcal{D}_{zu} \triangleq \begin{bmatrix} \mathcal{D}_{zu1} & \mathcal{D}_{zu2} & \mathcal{D}_{zu3} \end{bmatrix}$$

$$\mathcal{C}_y \triangleq \begin{bmatrix} \mathcal{C}_{y_1} \\ \mathcal{C}_{y_2} \\ \mathcal{C}_{y_3} \end{bmatrix}, \mathcal{D}_{yw} \triangleq \begin{bmatrix} \mathcal{D}_{yw_1} \\ \mathcal{D}_{yw_2} \\ \mathcal{D}_{yw_3} \end{bmatrix}$$

Furthermore, by rewriting the decentralised control signals (11.35) in the compact form:

$$\hat{u}(t) = \mathcal{K}\hat{y}(t) \tag{11.39}$$

where:

$$\mathcal{K} \triangleq \begin{bmatrix} A_c & 0 & 0 \\ 0 & B_c & 0 \\ 0 & 0 & C_c \end{bmatrix}$$

the uncertain closed-loop system is given by:

$$\dot{\tilde{x}}(t) = \left(\tilde{A} + \Delta\tilde{A}\right)\tilde{x}(t) + \left(\tilde{D} + \Delta\tilde{D}\right)w(t) \tag{11.40}$$

$$z(t) = \left(\tilde{E} + \Delta\tilde{E}\right)\tilde{x}(t) \tag{11.41}$$

where:

$$\tilde{A} \triangleq \mathcal{A} + \mathcal{B}_u\mathcal{K}\mathcal{C}_y, \quad \Delta\tilde{A} \triangleq \tilde{B}_1F\tilde{C}_1 + \delta\mathcal{B}_u\mathcal{K}\mathcal{C}_y, \quad \tilde{D} \triangleq \mathcal{B}_w + \mathcal{B}_u\mathcal{K}\mathcal{D}_{yw}$$
$$\Delta\tilde{D} \triangleq \delta\mathcal{B}_u\mathcal{K}\mathcal{D}_{yw}, \quad \tilde{E} \triangleq \mathcal{C}_z + \mathcal{D}_{zu}\mathcal{K}\mathcal{C}_y, \quad \Delta\tilde{E} \triangleq \delta\mathcal{D}_{zu}\mathcal{K}\mathcal{C}_y$$

Note that $\tilde{B}_2 = \mathcal{B}_u$ and $\tilde{C}_2 = \mathcal{K}\mathcal{C}_y$.

We can now recast the Robust Stability and Performance Problem as the following Auxiliary Optimisation Problem.

Auxiliary Optimisation Problem. For given α, $\alpha_c \in \mathbb{R}$, $\alpha > 0$, and $\alpha_c > 0$, determine $\mathcal{K} \in \mathbb{R}^{(2n_c+m)\times(2n_c+l)}$ that minimises:

$$\mathcal{J}(\mathcal{K}) = \operatorname{tr} \tilde{Q}\left[\tilde{R} + (\gamma_c^{-2} + 2\gamma_c^{-1})\tilde{R}_{c_1}\right] \tag{11.42}$$

where $\tilde{Q} \geq 0$ satisfies (11.28).

It follows from Theorem 11.2 that the satisfaction of (11.28) for $\tilde{Q} \in \mathbb{N}^{\tilde{n}}$ along with the generic stabilisability condition $(\tilde{A} + \Delta\tilde{A}, \tilde{D} + \Delta\tilde{D})$ leads to closed-loop robust stability along with robust $\mathcal{H}_2$ performance.

11.6 Sufficient Conditions for Fixed-order Resilient Compensation with Multiplicative Uncertainty

In this section we state sufficient conditions for characterising dynamic output feedback controllers guaranteeing robust stability with respect to system plant uncertainty and multiplicative controller gain variations and robust $\mathcal{H}_2$ performance.

Theorem 11.3. *Let $\alpha > 0$ and $\alpha_c > 0$. Suppose there exists $\tilde{n} \times \tilde{n}$ non-negative-definite matrices $\tilde{Q}$ and $\tilde{P}$ satisfying:*

$$\begin{aligned} 0 = \tilde{A}\tilde{Q} + \tilde{Q}\tilde{A}^{\mathrm{T}} &+ \tilde{Q}\left(\alpha^{-1}\gamma^{-2}\tilde{C}_1^{\mathrm{T}}\tilde{C}_1 + \alpha_c^{-1}\gamma_c^{-2}\tilde{C}_2^{\mathrm{T}}\tilde{C}_2\right)\tilde{Q} + \tilde{V} \\ &+ \alpha\tilde{B}_1\tilde{B}_1^{\mathrm{T}} + \alpha_c\tilde{B}_2\tilde{B}_2^{\mathrm{T}} + \left(\gamma_c^{-2} + 2\gamma_c^{-1}\right)\tilde{V}_{c_1} \end{aligned} \tag{11.43}$$

$$\begin{aligned} 0 = &\left[\tilde{A} + \tilde{Q}\left(\alpha^{-2}\gamma^{-2}\tilde{C}_1^{\mathrm{T}}\tilde{C}_1 + \alpha_c^{-1}\gamma_c^{-2}\tilde{C}_2^{\mathrm{T}}\tilde{C}_2\right)\right]^{\mathrm{T}}\tilde{P} \\ &+ \tilde{P}\left[\tilde{A} + \tilde{Q}\left(\alpha^{-1}\gamma^{-2}\tilde{C}_1^{\mathrm{T}}\tilde{C}_1 + \alpha_c^{-1}\gamma_c^{-2}\tilde{C}_2^{\mathrm{T}}\tilde{C}_2\right)\right] \\ &+ \tilde{R} + \left(\gamma_c^{-2} + 2\gamma_c^{-1}\right)\tilde{R}_{c_1} \end{aligned} \tag{11.44}$$

and let (A_c, B_c, C_c) satisfy:

$$0 = \mathcal{B}_{u_1}^{\mathrm{T}}\tilde{P}\tilde{Q}\mathcal{C}_{y_1}^{\mathrm{T}} + \alpha_c^{-1}\gamma_c^{-2}A_c\mathcal{C}_y\tilde{Q}\tilde{P}\tilde{Q}\mathcal{C}_{y_1}^{\mathrm{T}} \tag{11.45}$$

$$\begin{aligned} 0 = &\mathcal{B}_{u2}^{\mathrm{T}}\tilde{P}\tilde{D}\mathcal{D}_{yw2}^{\mathrm{T}} + \mathcal{B}_{u2}^{\mathrm{T}}\tilde{P}\tilde{Q}\mathcal{C}_{y2}^{\mathrm{T}} + \alpha_c^{-1}\gamma_c^{-2}B_c\mathcal{C}_y\tilde{Q}\tilde{P}\tilde{Q}\mathcal{C}_{y2}^{\mathrm{T}} \\ &+ \left(\gamma_c^{-2} + 2\gamma_c^{-1}\right)\mathcal{B}_{u2}^{\mathrm{T}}\tilde{P}\mathcal{B}_u\mathcal{K}\mathcal{D}_{yw}\mathcal{D}_{yw2}^{\mathrm{T}} \end{aligned} \tag{11.46}$$

$$\begin{aligned} 0 = &\mathcal{B}_{u3}^{\mathrm{T}}\tilde{P}\tilde{Q}\mathcal{C}_{y3}^{\mathrm{T}} + \mathcal{D}_{zu3}^{\mathrm{T}}\tilde{E}\tilde{Q}\mathcal{C}_{y3}^{\mathrm{T}} + \alpha_c^{-1}\gamma_c^{-2}C_c\mathcal{C}_y\tilde{Q}\tilde{P}\tilde{Q}\mathcal{C}_{y3}^{\mathrm{T}} \\ &+ \left(\gamma_c^{-2} + 2\gamma_c^{-1}\right)\mathcal{D}_{zu3}^{\mathrm{T}}\mathcal{D}_{zu}\mathcal{K}\mathcal{C}_y\tilde{Q}\mathcal{C}_y^{\mathrm{T}} \end{aligned} \tag{11.47}$$

Then $(\tilde{A}+\Delta\tilde{A}, \tilde{D}+\Delta\tilde{D})$ is stabilisable for all $(\Delta A, \Delta A_c, \Delta B_c, \Delta C_c) \in \mathcal{U} \times \mathcal{U}_c$ if and only if $\tilde{A}+\Delta\tilde{A}$ is asymptotically stable for all $(\Delta A, \Delta A_c, \Delta B_c, \Delta C_c) \in \mathcal{U} \times \mathcal{U}_c$. In this case, the worst-case $\mathcal{H}_2$ performance of the closed-loop system (11.9) satisfies the bound:

$$J(A_c, B_c, C_c) \leq \operatorname{tr} \tilde{Q}\left[\tilde{R} + \left(\gamma_c^{-2} + 2\gamma_c^{-1}\right) \tilde{R}_{c_1}\right] \tag{11.48}$$

Proof. First we obtain necessary conditions for the Auxiliary Optimisation Problem and then show, by construction, that these conditions serve as sufficient conditions for closed-loop stability and robust $\mathcal{H}_2$ performance. Thus, to optimise (11.42) subject to (11.28), form the Lagrangian:

$$\begin{aligned}\mathcal{L}(\mathcal{K}, \tilde{P}, \lambda) \triangleq \operatorname{tr} \Big\{ & \lambda \left[\tilde{Q}\tilde{R} + \left(\gamma_c^{-2} + 2\gamma_c^{-1}\right) \tilde{Q}\tilde{R}_{c_1}\right] \\ & + \tilde{P}[\tilde{A}\tilde{Q} + \tilde{Q}\tilde{A}^{\mathrm{T}} + \tilde{V} + \alpha \tilde{B}_1 \tilde{B}_1^{\mathrm{T}} + \alpha_c \tilde{B}_2 \tilde{B}_2^{\mathrm{T}} \\ & + \tilde{Q}\left(\alpha^{-1}\gamma^{-2}\tilde{C}_1^{\mathrm{T}}\tilde{C}_1 + \alpha_c^{-1}\gamma_c^{-2}\tilde{C}_2^{\mathrm{T}}\tilde{C}_2\right)\tilde{Q} \\ & + \left(\gamma_c^{-2} + 2\gamma_c^{-1}\right) \tilde{V}_{c_1}] \Big\}\end{aligned} \tag{11.49}$$

where the Lagrange multipliers $\lambda \geq 0$ and $\tilde{P} \in \mathbb{R}^{\tilde{n}\times\tilde{n}}$ are not both zero. By viewing $\mathcal{K}$ and $\tilde{Q}$ as independent variables, we obtain:

$$\begin{aligned}\frac{\partial \mathcal{L}}{\partial \tilde{Q}} = & \left[\tilde{A} + \tilde{Q}\left(\alpha^{-1}\gamma^{-2}\tilde{C}_1^{\mathrm{T}}\tilde{C}_1 + \alpha_c^{-1}\gamma_c^{-2}\tilde{C}_2^{\mathrm{T}}\tilde{C}_2\right)\right]^{\mathrm{T}} \tilde{P} \\ & + \tilde{P}\left[\tilde{A} + \tilde{Q}\left(\alpha^{-1}\gamma^{-2}\tilde{C}_1^{\mathrm{T}}\tilde{C}_1 + \alpha_c^{-1}\gamma_c^{-2}\tilde{C}_2^{\mathrm{T}}\tilde{C}_2\right)\right] \\ & + \lambda\left[\tilde{R} + \left(\gamma_c^{-2} + 2\gamma_c^{-1}\right)\tilde{R}_{c_1}\right]\end{aligned} \tag{11.50}$$

If $\tilde{A} + \tilde{Q}\left(\alpha^{-1}\gamma^{-2}\tilde{C}_1^{\mathrm{T}}\tilde{C}_1 + \alpha_c^{-1}\gamma_c^{-2}\tilde{C}_2^{\mathrm{T}}\tilde{C}_2\right)$ is Hurwitz, then $\lambda = 0$ implies $\tilde{P} = 0$. Hence, it can be assumed without loss of generality that $\lambda = 1$. Furthermore, note that $\tilde{P}$ is non-negative-definite. Thus the stationary conditions with $\lambda = 1$ are given by:

$$\begin{aligned}\frac{\partial \mathcal{L}}{\partial \tilde{Q}} = & \left[\tilde{A} + \tilde{Q}\left(\alpha^{-1}\gamma^{-2}\tilde{C}_1^{\mathrm{T}}\tilde{C}_1 + \alpha_c^{-1}\gamma_c^{-2}\tilde{C}_2^{\mathrm{T}}\tilde{C}_2\right)\right]^{\mathrm{T}} \tilde{P} \\ & + \tilde{P}\left[\tilde{A} + \tilde{Q}\left(\alpha^{-1}\gamma^{-2}\tilde{C}_1^{\mathrm{T}}\tilde{C}_1 + \alpha_c^{-1}\gamma_c^{-2}\tilde{C}_2^{\mathrm{T}}\tilde{C}_2\right)\right] \\ & + \tilde{R} + \left(\gamma_c^{-2} + 2\gamma_c^{-1}\right)\tilde{R}_{c_1} = 0\end{aligned}$$

$$\frac{\partial \mathcal{L}}{\partial A_c} = \mathcal{B}_{u_1}^{\mathrm{T}} \tilde{P}\tilde{Q}\mathcal{C}_{y_1}^{\mathrm{T}} + \alpha_c^{-1}\gamma_c^{-2} A_c \mathcal{C}_y \tilde{Q}\tilde{P}\tilde{Q}\mathcal{C}_{y_1}^{\mathrm{T}} = 0$$

$$\begin{aligned}\frac{\partial \mathcal{L}}{\partial B_c} = & \mathcal{B}_{u_2}^{\mathrm{T}} \tilde{P}\tilde{D}\mathcal{D}_{yw2}^{\mathrm{T}} + \mathcal{B}_{u2}^{\mathrm{T}}\tilde{P}\tilde{Q}\mathcal{C}_{y2}^{\mathrm{T}} + \alpha_c^{-1}\gamma_c^{-2} B_c \mathcal{C}_y \tilde{Q}\tilde{P}\tilde{Q}\mathcal{C}_{y2}^{\mathrm{T}} \\ & + \left(\gamma_c^{-2} + 2\gamma_c^{-1}\right) \mathcal{B}_{u2}^{\mathrm{T}}\tilde{P}\mathcal{B}_u \mathcal{K}\mathcal{D}_{yw}\mathcal{D}_{yw2}^{\mathrm{T}} = 0\end{aligned}$$

$$\frac{\partial \mathcal{L}}{\partial C_c} = \mathcal{B}_{u3}^{\mathrm{T}} \tilde{P} \tilde{Q} \mathcal{C}_{y3}^{\mathrm{T}} + \mathcal{D}_{zu3}^{\mathrm{T}} \tilde{E} \tilde{Q} \mathcal{C}_{y3}^{\mathrm{T}} + \alpha_c^{-1} \gamma_c^{-2} C_c \mathcal{C}_y \tilde{Q} \tilde{P} \tilde{Q} \mathcal{C}_{y3}^{\mathrm{T}}$$
$$+ \left(\gamma_c^{-2} + 2\gamma_c^{-1}\right) \mathcal{D}_{zu3}^{\mathrm{T}} \mathcal{D}_{zu} \mathcal{K} \mathcal{C}_y \tilde{Q} \mathcal{C}_y^{\mathrm{T}} = 0$$

which are equivalent to (11.44) and (11.45)–(11.46), respectively. Equation (11.43) is a restatement of (11.28). It now follows from Theorem 11.2 that the stabilisability condition $(\tilde{A} + \Delta\tilde{A}, \tilde{D} + \Delta\tilde{D})$ for all $(\Delta A, \Delta A_c, \Delta B_c, \Delta C_c) \in \mathcal{U} \times \mathcal{U}_c$ is equivalent to the stability of $\tilde{A} + \Delta\tilde{A}$ for all $(\Delta A, \Delta A_c, \Delta B_c, \Delta C_c) \in \mathcal{U} \times \mathcal{U}_c$. Finally, the $\mathcal{H}_2$ performance bound (11.48) is a restatement of (11.42). □

Equations (11.44)–(11.46) provide constructive sufficient conditions that yield dynamic controllers for robust resilient fixed-order (*i.e.*, full- and reduced-order) output feedback compensation. In the design equations (11.44)–(11.46), one can view α and α_c as free parameters and optimise the performance criterion (11.42) with respect to α and α_c. In particular, setting $\frac{\partial \mathcal{J}}{\partial \alpha} = 0$ and $\frac{\partial \mathcal{J}}{\partial \alpha_c} = 0$ yields:

$$\alpha = \frac{1}{\gamma} \left[\frac{\operatorname{tr} \tilde{P} \tilde{Q} \tilde{C}_1^{\mathrm{T}} \tilde{C}_1 \tilde{Q}}{\operatorname{tr} \tilde{P} \tilde{B}_1 \tilde{B}_1^{\mathrm{T}}} \right]^{\frac{1}{2}}, \quad \alpha_c = \frac{1}{\gamma_c} \left[\frac{\operatorname{tr} \tilde{P} \tilde{Q} \tilde{C}_2^{\mathrm{T}} \tilde{C}_2 \tilde{Q}}{\operatorname{tr} \tilde{P} \tilde{B}_2 \tilde{B}_2^{\mathrm{T}}} \right]^{\frac{1}{2}} \tag{11.51}$$

It is important to note that α and α_c given by (11.51) are implicit since $\tilde{Q}$ and $\tilde{P}$ are functions of α and α_c. However, the optimal robust reduced-order controller gains and the scaling parameters α and α_c can be determined simultaneously within a numerical optimisation algorithm using $\frac{\partial \mathcal{J}}{\partial \alpha} = 0$ and $\frac{\partial \mathcal{J}}{\partial \alpha_c} = 0$. For details of this fact, see Section 11.10.

11.7 Additive Controller Uncertainty Structure and Guaranteed Cost Bound

In this section we assign a different structure to the uncertainty set $\mathcal{U}_c$ and consequently the bounding functions $\Omega_i(\cdot), i = 1, 2, 3$. Specifically, the uncertainty set $\mathcal{U}_c$ is assumed to be of the form:

$$\mathcal{U}_c \triangleq \{(\Delta A_c, \Delta B_c, \Delta C_c) : \Delta A_c = \delta \mathcal{I}_{A_c},\ \Delta B_c = \delta \mathcal{I}_{B_c},$$
$$\Delta C_c = \delta \mathcal{I}_{C_c},\ |\, \delta \,| \leq \gamma_c^{-1}\} \tag{11.52}$$

where $\mathcal{I}_{A_c}$, $\mathcal{I}_{B_c}$, and $\mathcal{I}_{C_c}$ are ones matrices of dimension $\mathbb{R}^{n_c \times n_c}$, $\mathbb{R}^{n_c \times l}$, and $\mathbb{R}^{m \times n_c}$ respectively, δ is an uncertain real parameter, and γ_c is a given positive number. Note that, unlike the multiplicative uncertainty characterisation addressed in Section 11.4, the uncertainty characterisation given by (11.52) can capture controller gain variations with zero entries in the nominal gain matrices (A_c, B_c, C_c). With this additive (absolute) uncertainty characterisation, the closed-loop system (11.6) has structured uncertainty of the form:

$$\Delta\tilde{A} = \tilde{B}_0 \tilde{F} \tilde{C}_0 \tag{11.53}$$

where:

$$\tilde{B}_0 \triangleq \begin{bmatrix} B_0 & 0 & 0 & B \\ 0 & I_{n_c} & \mathcal{I}_{B_c} & 0 \end{bmatrix}, \quad \tilde{F} \triangleq \begin{bmatrix} F & 0 & 0 & 0 \\ 0 & \delta I_{n_c} & 0 & 0 \\ 0 & 0 & \delta I_l & 0 \\ 0 & 0 & 0 & \delta I_m \end{bmatrix} \tag{11.54}$$

$$\tilde{C}_0 \triangleq \begin{bmatrix} C_0 & 0 \\ 0 & \mathcal{I}_{A_c} \\ C & 0 \\ 0 & \mathcal{I}_{C_c} \end{bmatrix} \tag{11.55}$$

For the structure of $\mathcal{U} \times \mathcal{U}_c$ as specified by (11.20) and (11.52), the bounding functions $\Omega_1(\tilde{Q})$, $\Omega_2(B_c)$, and $\Omega_3(C_c)$ can now be given a concrete form. For the statement of the next result, define $\tilde{R}_{c_2} \triangleq \tilde{E}_{c_2}^{\mathrm{T}} \tilde{E}_{c_2}$ and $\tilde{V}_{c_2} \triangleq \tilde{D}_{c_2}^{\mathrm{T}} \tilde{D}_{c_2}$, where:

$$\tilde{E}_{c_2} \triangleq \begin{bmatrix} 0 & E_2 \mathcal{I}_{C_c} \end{bmatrix}, \quad \tilde{D}_{c_2} \triangleq \begin{bmatrix} 0 \\ \mathcal{I}_{B_c} D_2 \end{bmatrix}$$

Furthermore, to enforce the block-structure of the uncertainty matrix $\tilde{F}$, define the set of compatible scaling matrices $\mathcal{D}$ by:

$$\mathcal{D} \triangleq \{\tilde{D}_0 > 0 : \tilde{F}\tilde{D}_0 = \tilde{D}_0\tilde{F},\ \tilde{F}^{\mathrm{T}}\tilde{F} \leq \tilde{N}\}$$

where:

$$\tilde{N} \triangleq \begin{bmatrix} \gamma^2 I_r & 0 \\ 0 & \gamma_c^2 I_{(n_c+l+m)} \end{bmatrix}$$

The condition $\tilde{F}\tilde{D}_0 = \tilde{D}_0\tilde{F}$ in $\mathcal{D}$ is analogous to the commuting assumption between the D-scales and Δ blocks in μ-analysis and synthesis, which accounts for the structure in the uncertainty $\tilde{F}$. It is easy to see that there *always exists* such a matrix $\tilde{D}_0$ even if $\tilde{F}$ is *neither diagonal nor symmetric.* For example, if $F = fI_r$, where f is a scalar uncertainty, then $\tilde{D}_0$ can be an arbitrary positive definite matrix. Alternatively, if $F \in \mathbb{R}^{r\times r}$ is nondiagonal, then one can always choose $\tilde{D}_0 =$ block-diag$[dI_r, D_0]$, where d is a scalar and $D_0 \in \mathbb{R}^{(n_c+l+m)\times(n_c+l+m)}$ is an arbitrary invertible matrix. Of course, $\tilde{F}$ and $\tilde{D}_0$ may have more intricate structure, e.g., they may be block-diagonal with commuting blocks situated on the diagonal.

Proposition 11.2. *Let $\alpha_1 > 0, \alpha_2 > 0$, and let $\tilde{D}_0 \in \mathcal{D}$. Furthermore, let $\mathcal{U}$ and $\mathcal{U}_c$ be defined by (11.20) and (11.52), respectively, and define $\Omega_1(\cdot)$, $\Omega_2(\cdot)$, and $\Omega_3(\cdot)$ by:*

$$\Omega_1(\tilde{Q}) = \tilde{Q}\tilde{C}_0^{\mathrm{T}}\tilde{D}_0\tilde{N}\tilde{D}_0\tilde{C}_0\tilde{Q} + \tilde{B}_0\tilde{D}_0^{-2}\tilde{B}_0^{\mathrm{T}} \tag{11.56}$$

$$\Omega_2(B_c) = \alpha_1^{-1}\gamma_c^{-2}\tilde{V}_{c_1} + (\alpha_1 + \gamma_c^{-2})\tilde{V}_{c_2} \tag{11.57}$$

$$\Omega_3(C_c) = \alpha_2^{-1}\gamma_c^{-2}\tilde{R}_{c_1} + (\alpha_2 + \gamma_c^{-2})\tilde{R}_{c_2} \tag{11.58}$$

Then (11.11)–(11.13) are satisfied.

Proof. Note that with $\tilde{D}_0 \in \mathcal{D}$:

$$0 \leq \left[\tilde{B}_0\tilde{D}_0^{-1} - \tilde{Q}\tilde{C}_0^{\mathrm{T}}\tilde{D}_0\tilde{F}^{\mathrm{T}}\right]\left[\tilde{B}_0\tilde{D}_0^{-1} - \tilde{Q}\tilde{C}_0^{\mathrm{T}}\tilde{D}_0\tilde{F}^{\mathrm{T}}\right]^{\mathrm{T}}$$
$$\leq \tilde{B}_0\tilde{D}_0^{-2}\tilde{B}_0^{\mathrm{T}} + \tilde{Q}\tilde{C}_0^{\mathrm{T}}\tilde{D}_0\tilde{N}\tilde{D}_0\tilde{C}_0\tilde{Q} - (\Delta\tilde{A}\tilde{Q} + \tilde{Q}\Delta\tilde{A}^{\mathrm{T}})$$

which proves (11.11) with $\mathcal{U}$ and $\mathcal{U}_{\mathrm{c}}$ given by (11.20) and (11.52), respectively. Next, note that:

$$\Delta\tilde{V} = \delta\tilde{D}_{\mathrm{c}_1}\tilde{D}_{\mathrm{c}_2}^{\mathrm{T}} + \delta\tilde{D}_{\mathrm{c}_2}\tilde{D}_{\mathrm{c}_1}^{\mathrm{T}} + \delta^2\tilde{D}_{\mathrm{c}_2}\tilde{D}_{\mathrm{c}_2}^{\mathrm{T}}$$

Now, since $\delta^2\tilde{D}_{\mathrm{c}_2}\tilde{D}_{\mathrm{c}_2}^{\mathrm{T}} \leq \gamma_{\mathrm{c}}^{-2}\tilde{D}_{\mathrm{c}_2}\tilde{D}_{\mathrm{c}_2}^{\mathrm{T}}$ and:

$$0 \leq \left[\alpha_1^{\frac{1}{2}}\tilde{D}_{\mathrm{c}_2} - \alpha_1^{-\frac{1}{2}}\delta\tilde{D}_{\mathrm{c}_1}\right]\left[\alpha_1^{\frac{1}{2}}\tilde{D}_{\mathrm{c}_2} - \alpha_1^{-\frac{1}{2}}\delta\tilde{D}_{\mathrm{c}_1}\right]^{\mathrm{T}}$$
$$\leq \alpha_1\tilde{D}_{\mathrm{c}_2}\tilde{D}_{\mathrm{c}_2}^{\mathrm{T}} + \alpha_1^{-1}\gamma_{\mathrm{c}}^{-2}\tilde{D}_{\mathrm{c}_1}\tilde{D}_{\mathrm{c}_1}^{\mathrm{T}} - \delta(\tilde{D}_{\mathrm{c}_1}\tilde{D}_{\mathrm{c}_2}^{\mathrm{T}} + \tilde{D}_{\mathrm{c}_2}\tilde{D}_{\mathrm{c}_1}^{\mathrm{T}})$$

it follows that:

$$\Delta\tilde{V} \leq (\alpha_1 + \gamma_{\mathrm{c}}^{-2})\tilde{V}_{\mathrm{c}_2} + \alpha_1^{-1}\gamma_{\mathrm{c}}^{-2}\tilde{V}_{\mathrm{c}_1}$$

which proves (11.12) with $\mathcal{U}_{\mathrm{c}}$ given by (11.52). Finally, a similar construction proves (11.13). □

Next, using Theorem 11.1 and Proposition 11.2 we have the following immediate result.

Theorem 11.4. *Let $\alpha_1 > 0$, $\alpha_2 > 0$, $\tilde{D}_0 \in \mathcal{D}$, and suppose there exists an $\tilde{n} \times \tilde{n}$ non-negative-definite matrix $\tilde{Q}$ satisfying:*

$$0 = \tilde{A}\tilde{Q} + \tilde{Q}\tilde{A}^{\mathrm{T}} + \tilde{Q}\tilde{C}_0^{\mathrm{T}}\tilde{D}_0\tilde{N}\tilde{D}_0\tilde{C}_0\tilde{Q} + \tilde{B}_0\tilde{D}_0^{-2}\tilde{B}_0^{\mathrm{T}} + \tilde{V}$$
$$+\alpha_1^{-1}\gamma_{\mathrm{c}}^{-2}\tilde{V}_{\mathrm{c}_1} + (\alpha_1 + \gamma_{\mathrm{c}}^{-2})\tilde{V}_{\mathrm{c}_2} \tag{11.59}$$

Then $(\tilde{A}+\Delta\tilde{A}, \tilde{D}+\Delta\tilde{D})$ is stabilisable for all $(\Delta A, \Delta A_{\mathrm{c}}, \Delta B_{\mathrm{c}}, \Delta C_{\mathrm{c}}) \in \mathcal{U} \times \mathcal{U}_{\mathrm{c}}$ if and only if $\tilde{A}+\Delta\tilde{A}$ is asymptotically stable for all $(\Delta A, \Delta A_{\mathrm{c}}, \Delta B_{\mathrm{c}}, \Delta C_{\mathrm{c}}) \in \mathcal{U} \times \mathcal{U}_{\mathrm{c}}$. In this case:

$$\tilde{Q}_\Delta \leq \tilde{Q}, \qquad (\Delta A, \Delta A_{\mathrm{c}}, \Delta B_{\mathrm{c}}, \Delta C_{\mathrm{c}}) \in \mathcal{U} \times \mathcal{U}_{\mathrm{c}} \tag{11.60}$$

where $\tilde{Q}_\Delta$ satisfies (11.10), and:

$$J(A_{\mathrm{c}}, B_{\mathrm{c}}, C_{\mathrm{c}}) \leq \operatorname{tr}\tilde{Q}\left[\tilde{R} + \alpha_2^{-1}\gamma_{\mathrm{c}}^{-2}\tilde{R}_{\mathrm{c}_1} + (\alpha_2 + \gamma_{\mathrm{c}}^{-2})\tilde{R}_{\mathrm{c}_2}\right] \tag{11.61}$$

11.8 Decentralised Static Output Feedback Formulation

As in Section 11.5, note that for every dynamic controller (11.3), (11.4) with gain variations $(\Delta A_c, \Delta B_c, \Delta C_c) \in \mathcal{U}_c$ given by (11.52), the closed-loop system (11.1)–(11.4) can be written as:

$$\begin{bmatrix} \dot{x}(t) \\ \dot{x}_c(t) \end{bmatrix} = \begin{bmatrix} A + B_0 F C_0 & BC_c + B\delta\mathcal{I}_{C_c} \\ B_c C + \delta\mathcal{I}_{B_c} C & A_c + \delta\mathcal{I}_{A_c} \end{bmatrix} \begin{bmatrix} x(t) \\ x_c(t) \end{bmatrix} + \begin{bmatrix} D_1 \\ B_c D_2 + \delta\mathcal{I}_{B_c} D_2 \end{bmatrix} w(t) \tag{11.62}$$

Furthermore, by treating A_c, B_c, and C_c as decentralised static output feedback gains with additive uncertainty as shown in Figure 11.2, we can pull the uncertainty into the plant model obtaining:

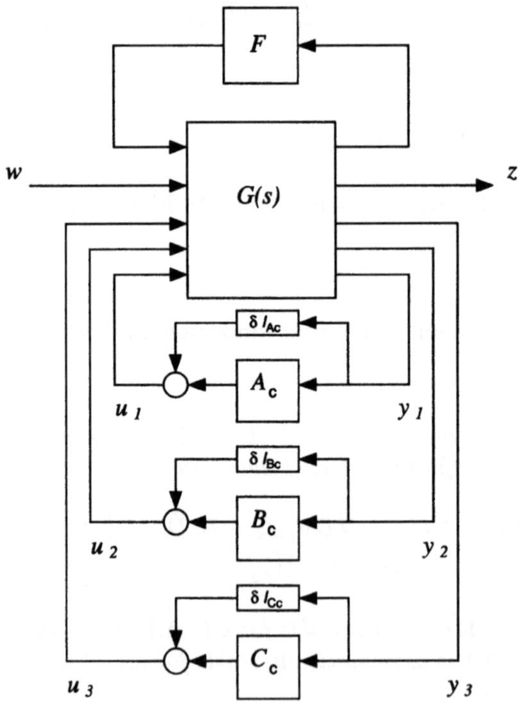

Fig. 11.2. Decentralised static output feedback: additive controller uncertainty

$$\dot{\tilde{x}}(t) = (\mathcal{A} + \Delta\mathcal{A})\tilde{x}(t) + \sum_{i=1}^{3} \mathcal{B}_{ui}\hat{u}_i(t) + (\mathcal{B}_w + \Delta\mathcal{B}_w)w(t) \tag{11.63}$$

$$\hat{y}_i(t) = \mathcal{C}_{y_i}\tilde{x}(t) + \mathcal{D}_{yw_i}w(t), \quad i = 1, 2, 3 \tag{11.64}$$

$$z(t) = (\mathcal{C}_z + \Delta\mathcal{C}_z)\tilde{x}(t) + \sum_{i=1}^{3} \mathcal{D}_{zui}\hat{u}_i(t) \tag{11.65}$$

$$\hat{u}_1(t) = A_c\hat{y}_1(t), \quad \hat{u}_2(t) = B_c\hat{y}_2(t), \quad \hat{u}_3(t) = C_c\hat{y}_3(t) \tag{11.66}$$

where:

$$\Delta\mathcal{A} \triangleq \begin{bmatrix} B_0FC_0 & \delta B\mathcal{I}_{C_c} \\ \delta\mathcal{I}_{B_c}C & \delta\mathcal{I}_{A_c} \end{bmatrix}, \quad \Delta\mathcal{B}_w \triangleq \begin{bmatrix} 0 \\ \delta\mathcal{I}_{B_c}D_2 \end{bmatrix}, \quad \Delta\mathcal{C}_z \triangleq \begin{bmatrix} 0 & \delta E_2\mathcal{I}_{C_c} \end{bmatrix}$$

The uncertain closed-loop system is now given by:

$$\dot{\tilde{x}}(t) = \left(\tilde{A} + \Delta\tilde{A}\right)\tilde{x}(t) + \left(\tilde{D} + \Delta\tilde{D}\right)w(t) \tag{11.67}$$

$$z(t) = \left(\tilde{E} + \Delta\tilde{E}\right)\tilde{x}(t) \tag{11.68}$$

where $\Delta\tilde{A} \triangleq \Delta\mathcal{A}$, $\Delta\tilde{D} \triangleq \Delta\mathcal{B}_w$, and $\Delta\tilde{E} \triangleq \Delta\mathcal{C}_z$.

Now, as in the multiplicative controller uncertainty case, we introduce an Auxiliary Optimisation Problem by considering:

$$\mathcal{J}(\mathcal{K}) = \operatorname{tr}\tilde{Q}\left[\tilde{R} + \alpha_2^{-1}\gamma_c^{-2}\tilde{R}_{c_1} + (\alpha_2 + \gamma_c^{-2})\tilde{R}_{c_2}\right] \tag{11.69}$$

with $\tilde{Q} \geq 0$ satisfying (11.59), and proceed by determining controller gains that minimise $\mathcal{J}(\mathcal{K})$.

11.9 Sufficient Conditions for Fixed-order Resilient Compensation with Additive Uncertainty

In this section we state sufficient conditions for characterising dynamic output feedback controllers guaranteeing robust stability with respect to system plant uncertainty and additive controller gain variations and robust $\mathcal{H}_2$ performance.

Theorem 11.5. *Let $\alpha_1 > 0$, $\alpha_2 > 0$, and let $\tilde{D}_0 \in \mathcal{D}$. Suppose there exists $\tilde{n} \times \tilde{n}$ non-negative-definite matrices $\tilde{Q}$ and $\tilde{P}$ satisfying:*

$$\begin{aligned} 0 &= \tilde{A}\tilde{Q} + \tilde{Q}\tilde{A}^{\mathrm{T}} + \tilde{Q}\tilde{C}_0^{\mathrm{T}}\tilde{D}_0\tilde{N}\tilde{D}_0\tilde{C}_0\tilde{Q} + \tilde{B}_0\tilde{D}_0^{-2}\tilde{B}_0^{\mathrm{T}} + \tilde{V} \\ &\quad + \alpha_1^{-1}\gamma_c^{-2}\tilde{V}_{c_1} + (\alpha_1 + \gamma_c^{-2})\tilde{V}_{c_2} \end{aligned} \tag{11.70}$$

$$\begin{aligned} 0 &= \left(\tilde{A} + \tilde{Q}\tilde{C}_0^{\mathrm{T}}\tilde{D}_0\tilde{N}\tilde{D}_0\tilde{C}_0\right)^{\mathrm{T}}\tilde{P} + \tilde{P}\left(\tilde{A} + \tilde{Q}\tilde{C}_0^{\mathrm{T}}\tilde{D}_0\tilde{N}\tilde{D}_0\tilde{C}_0\right) + \tilde{R} \\ &\quad + \alpha_2^{-1}\gamma_c^{-2}\tilde{R}_{c_1} + (\alpha_2 + \gamma_c^{-2})\tilde{R}_{c_2} \end{aligned} \tag{11.71}$$

and let (A_c, B_c, C_c) satisfy:

$$0 = \mathcal{B}_{u_1}^{\mathrm{T}}\tilde{P}\tilde{Q}\mathcal{C}_{y_1}^{\mathrm{T}} \tag{11.72}$$

$$0 = \mathcal{B}_{u2}^{\mathrm{T}}\tilde{P}\tilde{D}\mathcal{D}_{yw2}^{\mathrm{T}} + \mathcal{B}_{u2}^{\mathrm{T}}\tilde{P}\tilde{Q}\mathcal{C}_{y2}^{\mathrm{T}} + \alpha_1^{-1}\gamma_c^{-2}\mathcal{B}_{u2}^{\mathrm{T}}\tilde{P}\mathcal{B}_u\mathcal{K}\mathcal{D}_{yw}\mathcal{D}_{yw2}^{\mathrm{T}} \tag{11.73}$$

$$0 = \mathcal{B}_{u3}^{\mathrm{T}}\tilde{P}\tilde{Q}\mathcal{C}_{y3}^{\mathrm{T}} + \mathcal{D}_{zu3}^{\mathrm{T}}\tilde{E}\tilde{Q}\mathcal{C}_{y3}^{\mathrm{T}} + \alpha_2^{-1}\gamma_c^{-2}\mathcal{D}_{zu3}^{\mathrm{T}}\mathcal{D}_{zu}\mathcal{K}\mathcal{C}_y\tilde{Q}\mathcal{C}_{y3}^{\mathrm{T}} \tag{11.74}$$

Then $(\tilde{A}+\Delta\tilde{A}, \tilde{D}+\Delta\tilde{D})$ is stabilisable for all $(\Delta A, \Delta A_c, \Delta B_c, \Delta C_c) \in \mathcal{U} \times \mathcal{U}_c$ if and only if $\tilde{A}+\Delta\tilde{A}$ is asymptotically stable for all $(\Delta A, \Delta A_c, \Delta B_c, \Delta C_c) \in \mathcal{U} \times \mathcal{U}_c$. In this case, the worst-case $\mathcal{H}_2$ performance of the closed-loop system (11.9) satisfies the bound:

$$J(A_c, B_c, C_c) \le \operatorname{tr} \tilde{Q}\left[\tilde{R} + \alpha_2^{-1}\gamma_c^{-2}\tilde{R}_{c_1} + (\alpha_2 + \gamma_c^{-2})\tilde{R}_{c_2}\right] \tag{11.75}$$

Proof. The proof is similar to the proof of Theorem 11.3. □

As in Section 11.6, one can view α_i, $i = 1, 2$, and $\tilde{D}_0$ as free parameters and optimise the performance criterion $\mathcal{J}(\mathcal{K})$ given by (11.69) with respect to α_i, $i = 1, 2$, and $\tilde{D}_0$. In particular, setting $\frac{\partial \mathcal{J}}{\partial \alpha_i} = 0$, $i = 1, 2$, and $\frac{\partial \mathcal{J}}{\partial \tilde{D}_0} = 0$ yields, respectively:

$$\alpha_1 = \frac{1}{\gamma_c}\left[\frac{\operatorname{tr} \tilde{P}\tilde{V}_{c_1}}{\operatorname{tr} \tilde{P}\tilde{V}_{c_2}}\right]^{\frac{1}{2}}, \quad \alpha_2 = \frac{1}{\gamma_c}\left[\frac{\operatorname{tr} \tilde{Q}\tilde{R}_{c_1}}{\operatorname{tr} \tilde{Q}\tilde{R}_{c_2}}\right]^{\frac{1}{2}} \tag{11.76}$$

$$0 = \tilde{N}\tilde{D}_0\tilde{C}_0\tilde{Q}\tilde{P}\tilde{Q}\tilde{C}_0^{\mathrm{T}} - \tilde{D}_0^{-1}\tilde{B}_0^{\mathrm{T}}\tilde{P}\tilde{B}_0\tilde{D}_0^{-2} \tag{11.77}$$

11.10 Quasi-Newton Optimisation Algorithm

To solve the Robust Stability and Performance Problems posed in Sections 11.5 and 11.8, a general-purpose BFGS quasi-Newton algorithm [15] is used. The line-search portions of the algorithm were modified to include a constraint-checking subroutine, which decreases the length of the search direction vector until it lies entirely within the set of parameters that yield a stable closed-loop system. This modification ensures that the cost function $\mathcal{J}$ remains defined at every point in the line-search process. Numerical experience indicates that this subroutine is usually invoked only during the first few iterations of a synthesis problem.

One requirement of gradient-based optimisation algorithms is an initial stabilising design. For full-order controller design, the algorithm was initialised with an LQG controller, while for reduced-order control, the algorithm was initialised with a balanced truncated LQG controller. A large value was chosen for γ and then a feasible value of γ_c was calculated. The quasi-Newton optimisation algorithm was used to find the controller gains A_c, B_c, and C_c. After each iteration, γ_c was decreased and the current values of the controller gains (A_c, B_c, C_c) were then used as the starting point for the next iteration. When γ_c could not be decreased any further, γ was decreased and a feasible γ_c was calculated for the new value of γ and the process was repeated. For details of the algorithm, see [14].

11.11 Illustrative Numerical Examples

11.11.1 Second-order Unstable System

To demonstrate the design of robust resilient controllers, consider the second-order unstable system originally used by [16] to illustrate the lack of guaranteed gain margins for LQG controllers. Specifically, the state space system is given by:

$$\dot{x} = \begin{bmatrix} 1 & 1 \\ 0 & 1 \end{bmatrix} x + \begin{bmatrix} 0 \\ 1 \end{bmatrix} u$$
$$y = \begin{bmatrix} 1 & 0 \end{bmatrix} x$$

The matrices D_1, D_2, E_1, and E_2 are chosen to be:

$$D_1 = \begin{bmatrix} \sqrt{60} & 0 \\ \sqrt{60} & 0 \end{bmatrix}, \; D_2 = \begin{bmatrix} 0 & 1 \end{bmatrix}, \; E_1 = \begin{bmatrix} \sqrt{60} & \sqrt{60} \\ 0 & 0 \end{bmatrix}, \; E_2 = \begin{bmatrix} 0 \\ 1 \end{bmatrix}$$

Here we consider uncertainty in the $(2,1)$ component of the dynamics matrix. Using the uncertainty structure given by (11.20), the actual dynamics are given by $A + B_0 f C_0$, where $B_0 = [1 \quad 0]^{\mathrm{T}}$ and $C_0 = [0 \quad 1]$.

The quasi-Newton optimisation algorithm discussed in Section 11.10 was used to compute full-order controllers ($n_c = 2$) that minimise the cost bound $\mathcal{J}$ for several values of γ and γ_c for both the multiplicative and additive uncertainty characterisations. The actual $\mathcal{H}_2$ cost was computed for a range of values of the controller error parameter δ and the plant uncertainty f. The cost dependence for the multiplicative (relative) uncertainty characterisation (11.21) is shown in Figure 11.3. As γ_c decreases, the $\mathcal{H}_2$ cost of the nominal closed-loop system increases while the $\mathcal{H}_2$ cost of the perturbed closed-loop system remains near the nominal value for a larger range of perturbations. The LQG controller stabilises the closed-loop system for only small perturbations in the controller error parameter, while the resilient controllers stabilise the closed-loop system and provide performance close to the optimal level for much larger perturbations in the controller error parameter. Hence, robust performance over a large range of the uncertain parameter is achieved for some increase in the $\mathcal{H}_2$ cost above the optimal.

The effects of both plant uncertainty and controller uncertainty can be seen in the parameter plot shown in Figure 11.4. In this case, the stability of the closed-loop system was checked over a grid of the uncertain parameters for both the LQG and a robust resilient controller. The dashed line shows the region of asymptotic stability of the LQG controller, the solid line corresponds to the robust resilient controller, and 'x' marks the point corresponding to the nominal conditions. It is evident from the figure that the robust resilient controller renders the closed-loop system more robust to perturbations in both the plant and controller.

Next, Figure 11.5 shows the Nyquist plots of the loop gain of the system for the LQG controller and three different robust resilient controller designs.

These plots clearly demonstrate the resiliency of the non-fragile controllers over the LQG controller in terms of their increased gain and phase margins. Figures 11.6–11.8 give the same plots for the case of additive (absolute) controller uncertainty. Once again, the achieved robustness of the robust resilient controllers over the LQG controller is obvious. Figure 11.6 also shows the cost dependence of an $\mathcal{H}_\infty$ robust controller which shows that although the $\mathcal{H}_\infty$ controller is robust against plant variations, it is highly fragile with respect to controller gain variations. The Nyquist plots of the loop gain of the $\mathcal{H}_\infty$ robust controller and the robust resilient controller corresponding to identical plant uncertainty levels are shown in Figure 11.9, which clearly shows the relative stability superiority of the robust resilient controller over the robust $\mathcal{H}_\infty$ controller.

11.11.2 Two-mass Benchmark Problem

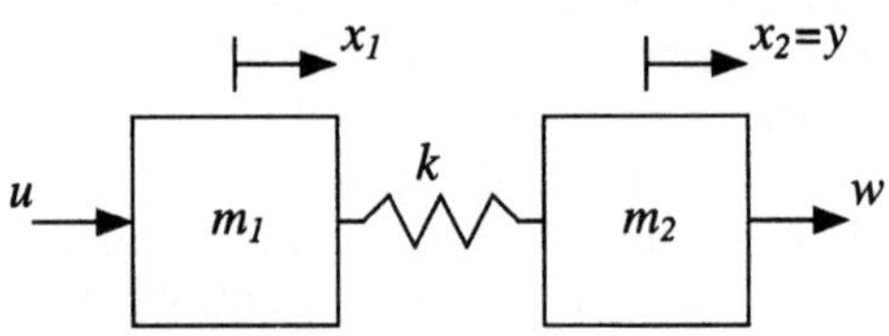

Fig. 11.10. Two-mass oscillator

Consider the two-mass system shown in Figure 11.10 with $m_1 = m_2 = 1$ and an uncertain spring stiffness k [17]. A control force acts on mass 1 and the position of mass 2 is measured, resulting in a noncolocated control problem. The nominal system dynamics with $k_{\mathrm{nom}} = 1$ and states defined in the figure is given by:

$$\dot{x} = \begin{bmatrix} 0 & 0 & 1 & 0 \\ 0 & 0 & 0 & 1 \\ -1 & 1 & 0 & 0 \\ 1 & -1 & 0 & 0 \end{bmatrix} x + \begin{bmatrix} 0 \\ 0 \\ 1 \\ 0 \end{bmatrix} u + \begin{bmatrix} 0 & 0 \\ 0 & 0 \\ 0 & 0 \\ 1 & 0 \end{bmatrix} w$$

$$y = \begin{bmatrix} 0 & 1 & 0 & 0 \end{bmatrix} x + \begin{bmatrix} 0 & 1 \end{bmatrix} w$$

Using the uncertainty structure given by (11.20), the actual dynamics are given by $A + B_0 f C_0$, where $B_0 = [0 \quad 0 \quad -1 \quad 1]^{\mathrm{T}}$ and $C_0 = [1 \quad -1 \quad 0 \quad 0]$. The matrices E_1 and E_2 are chosen to be:

$$E_1 = \begin{bmatrix} 0 & 1 & 0 & 0 \\ 0 & 0 & 0 & 0 \end{bmatrix}, \quad E_2 = \begin{bmatrix} 0 \\ \sqrt{0.001} \end{bmatrix}$$

As in Subsection 11.11.1, the quasi-Newton optimisation algorithm was used to compute full-order controllers ($n_c = 4$) that minimise the cost bound

$\mathcal{J}$. The cost dependence for the multiplicative uncertainty characterisation is shown in Figure 11.11. The Nyquist plots of the LQG controller and three robust resilient controllers are shown in Figure 11.12. For the additive uncertainty characterisation, the cost dependence is compared to the LQG controller and an $\mathcal{H}_\infty$ robust controller in Figure 11.13. The asymptotic stability regions of the LQG controller and the robust resilient controller are shown in Figure 11.14. Finally, the Nyquist plots of the LQG controller and three robust resilient controllers are shown in Figure 11.15, while the Nyquist plots of the $\mathcal{H}_\infty$ robust controller and the robust resilient controller corresponding to identical plant uncertainty levels are shown in Figure 11.16. In all cases, the robust resilient controllers are superior in their ability to tolerate plant *and* controller uncertainty as compared to the LQG and robust $\mathcal{H}_\infty$ controllers. Furthermore, the robust resilient controllers possess far superior gain and phase margins.

11.12 Conclusion

This chapter extended the robust fixed-structure guaranteed cost controller synthesis framework to synthesise robust resilient controllers for controller gain variations and system parametric uncertainty. Specifically, the guaranteed cost approach of [9] and [10] was used to develop sufficient conditions for robust stability and $\mathcal{H}_2$ performance via fixed-order dynamic compensation. A quasi-Newton optimisation algorithm was used to obtain robust controllers for two illustrative examples.

References

1. P. Dorato and R. K. Yedavalli, Eds., *Recent Advances in Robust Control.* IEEE Press, 1990.
2. M. Vidyasagar, *Control System Synthesis.* Cambridge, MA: The MIT Press, 1985.
3. B. D. O. Anderson and J. B. Moore, *Optimal Control: Linear Quadratic Methods.* Englewood Cliffs, NJ: Prentice-Hall, 1990.
4. B. A. Francis, *A Course in $\mathcal{H}_\infty$ Control Theory.* New York: Springer-Verlag, 1987.
5. K. Zhou, J. C. Doyle, and K. Glover, *Robust and Optimal Control.* Englewood Cliffs, NJ: Prentice-Hall, 1996.
6. M. A. Dahleh and I. J. Diaz-Bobillo, *Control of Uncertain Systems: A Linear Programming Approach.* Englewood Cliffs, NJ: Prentice-Hall, 1995.
7. D. S. Bernstein and D. C. Hyland, "Optimal projection approach to robust, fixed-structure control design," in *Mechanics and Control of Space Structures* (J. Junkins, ed.), pp. 237–293, AIAA, Washington, D.C., 1993.
8. L. H. Keel and S. P. Bhattacharyya, "Robust, fragile, or optimal?," *IEEE Trans. Autom. Contr.*, vol. 42, pp. 1098–1105, 1997.

9. D. S. Bernstein and W. M. Haddad, "The optimal projection equations with Petersen-Hollot bounds: Robust stability and performance via fixed-order dynamic compensation for systems with structured real-valued parameter uncertainty," *IEEE Trans. Autom. Contr.*, vol. 33, pp. 578–582, 1988.
10. D. S. Bernstein and W. M. Haddad, "Robust stability and performance via fixed-order dynamic compensation with guaranteed cost bounds," *Math. Contr. Sig. Sys.*, vol. 3, pp. 139–163, 1990.
11. W. M. Haddad and D. S. Bernstein, "Parameter-dependent Lyapunov functions and the Popov criterion in robust analysis and synthesis," *IEEE Trans. Autom. Contr.*, vol. 40, pp. 536–543, 1995.
12. W. M. Wonham, *Linear Multivariable Control: A Geometric Approach.* New York: Springer-Verlag, 2nd ed., 1979.
13. D. S. Bernstein, W. M. Haddad, and C. N. Nett, "Minimal complexity control law synthesis, part 2: Problem solution via $\mathcal{H}_2/\mathcal{H}_\infty$ optimal static output feedback," in *Proc. Amer. Contr. Conf.*, (Pittsburgh, PA), pp. 2506–2511, June 1989.
14. R. S. Erwin, A. G. Sparks, and D. S. Bernstein, "Decentralized real structured singular value synthesis," in *Proc. 13th IFAC World Congress*, vol. C: Control Design I, (San Francisco, CA), pp. 79–84, July 1996.
15. J. E. Dennis, Jr. and R. B. Schnabel, *Numerical Methods for Unconstrained Optimization and Nonlinear Equations.* Englewood Cliffs, NJ: Prentice-Hall, 1983.
16. J. C. Doyle, "Guaranteed margins for LQG regulators," *IEEE Trans. Autom. Contr.*, vol. 23, pp. 756–757, 1978.
17. B. Wei and D. S. Bernstein, "Benchmark problems for robust control design," *AIAA J. Guid. Contr. Dyn.*, vol. 15, pp. 1057–1059, 1992.

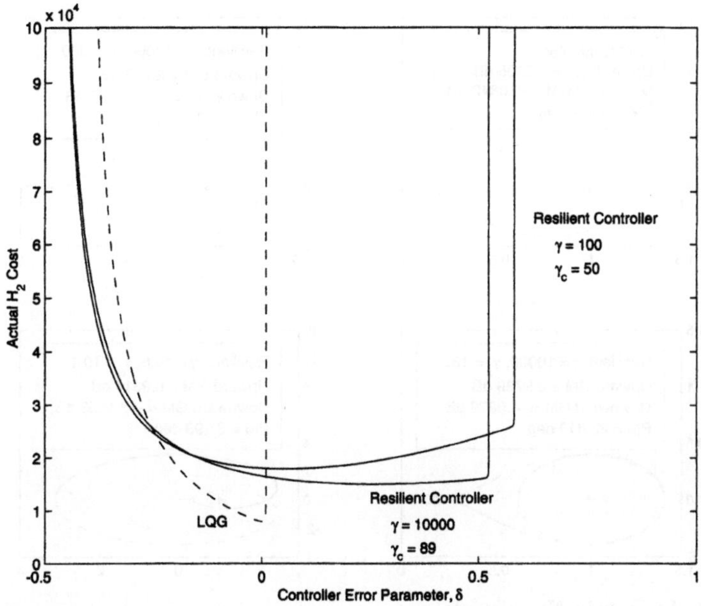

Fig. 11.3. Dependence of the $\mathcal{H}_2$ cost on the controller error parameter

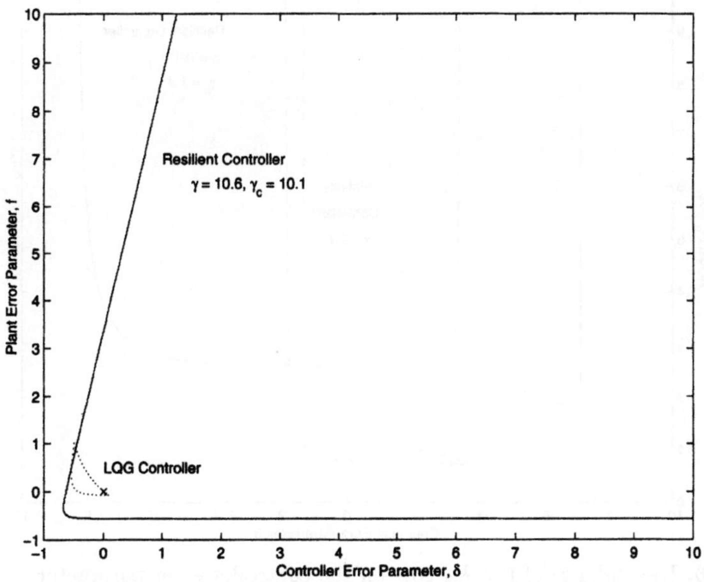

Fig. 11.4. Asymptotic stability regions

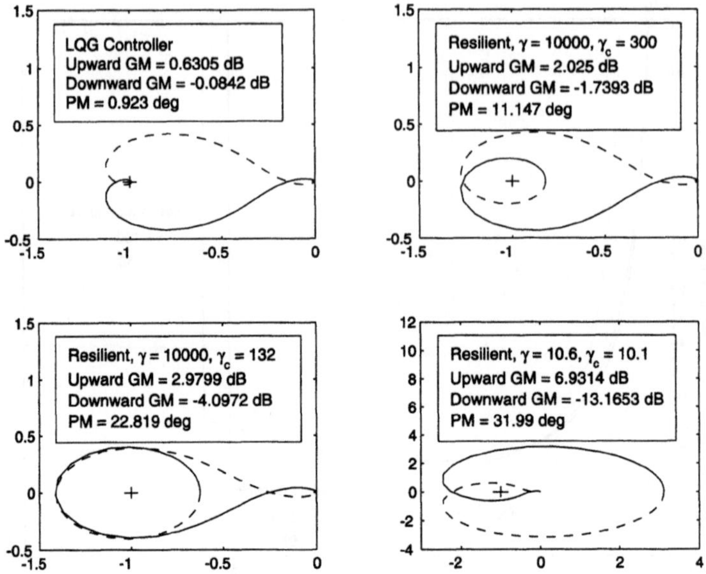

Fig. 11.5. Loop gain Nyquist plots

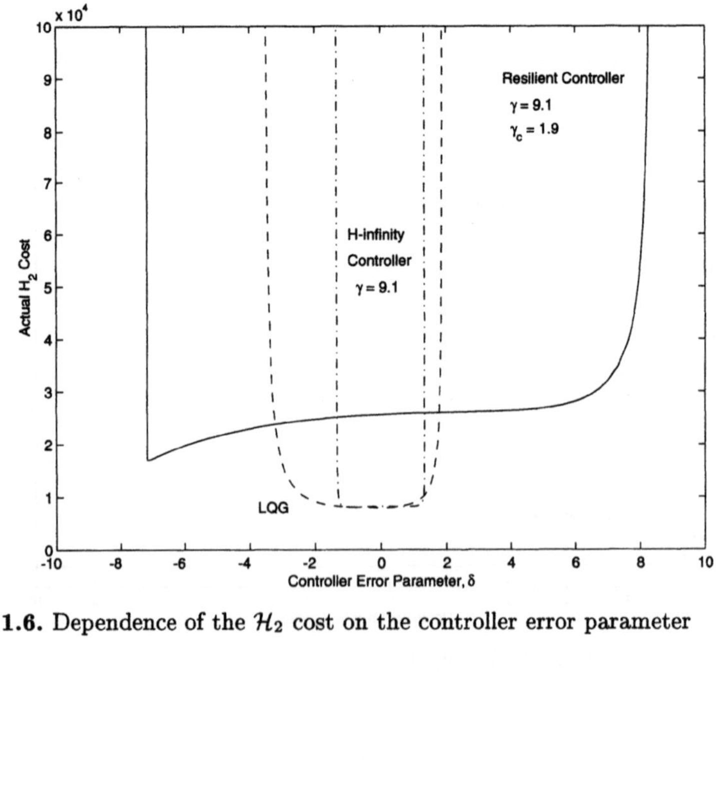

Fig. 11.6. Dependence of the $\mathcal{H}_2$ cost on the controller error parameter

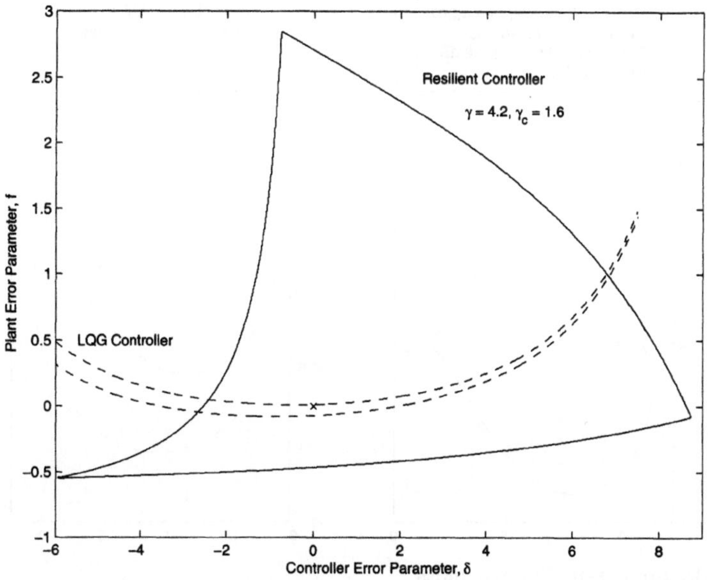

Fig. 11.7. Asymptotic stability regions

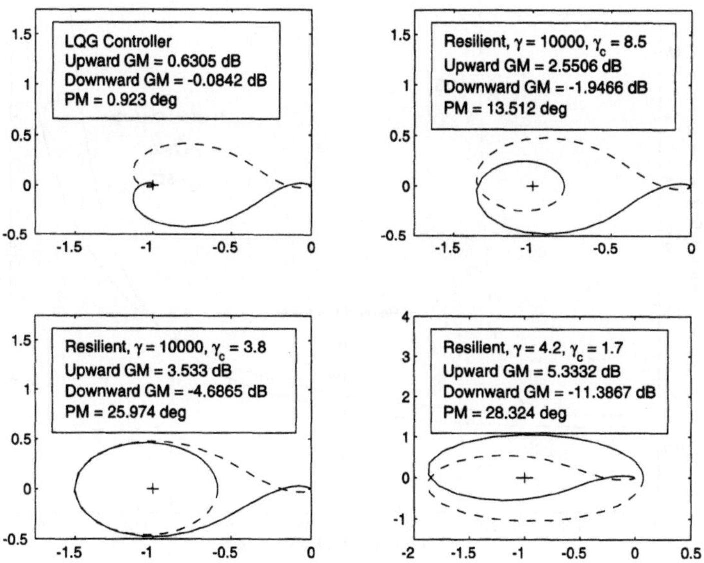

Fig. 11.8. Loop gain Nyquist plots

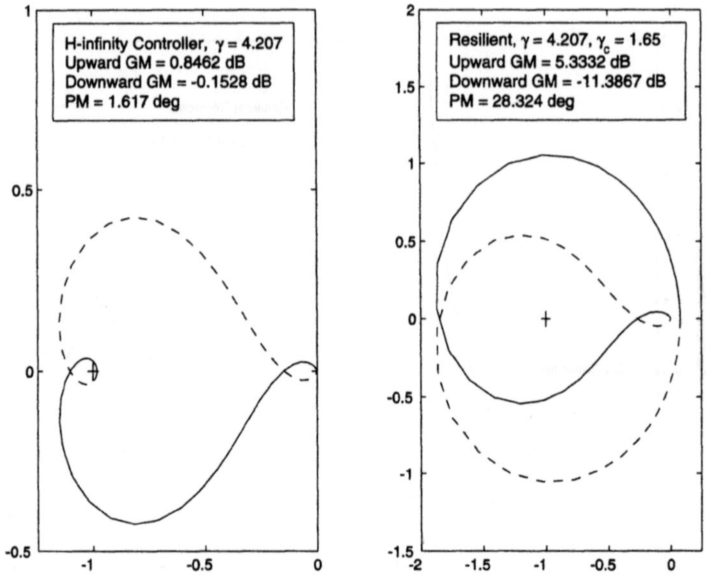

Fig. 11.9. Loop gain Nyquist plots

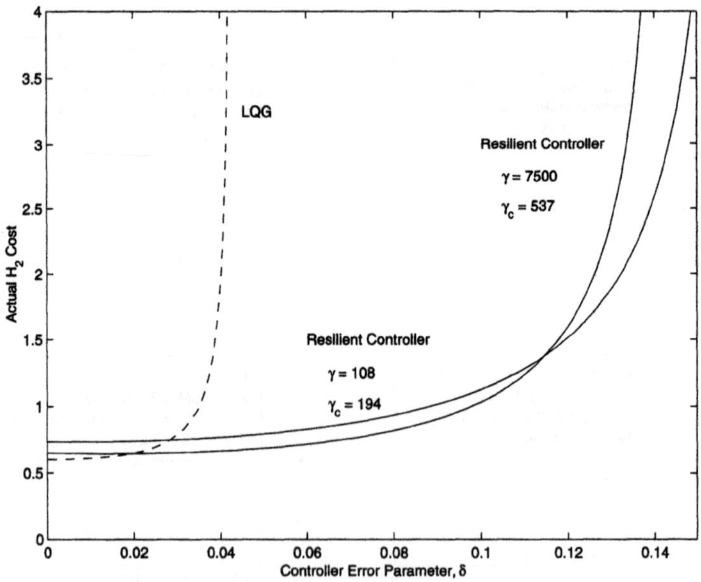

Fig. 11.11. Dependence of the $\mathcal{H}_2$ cost on the controller error parameter

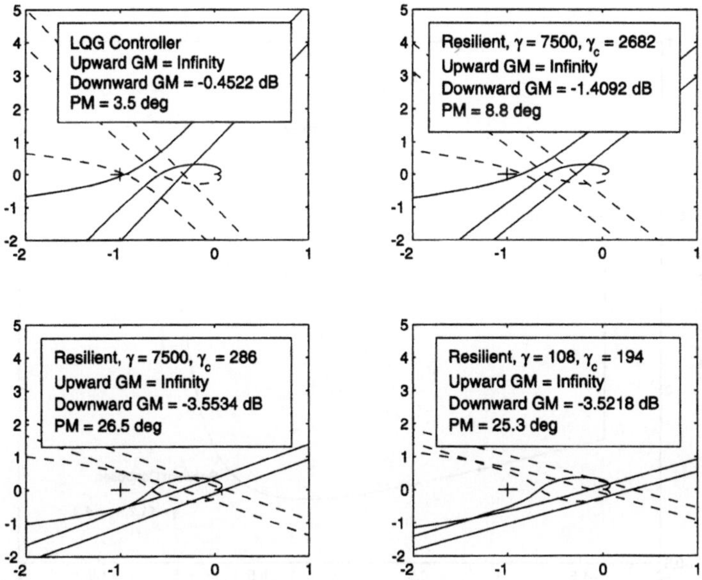

Fig. 11.12. Loop gain Nyquist plots

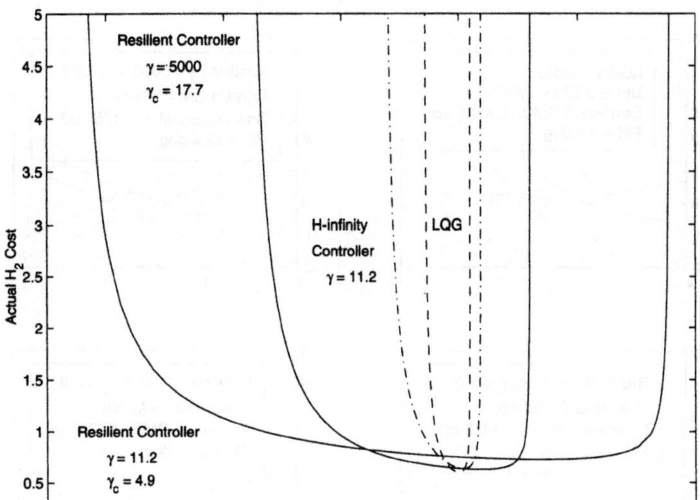

Fig. 11.13. Dependence of the $\mathcal{H}_2$ cost on the controller error parameter

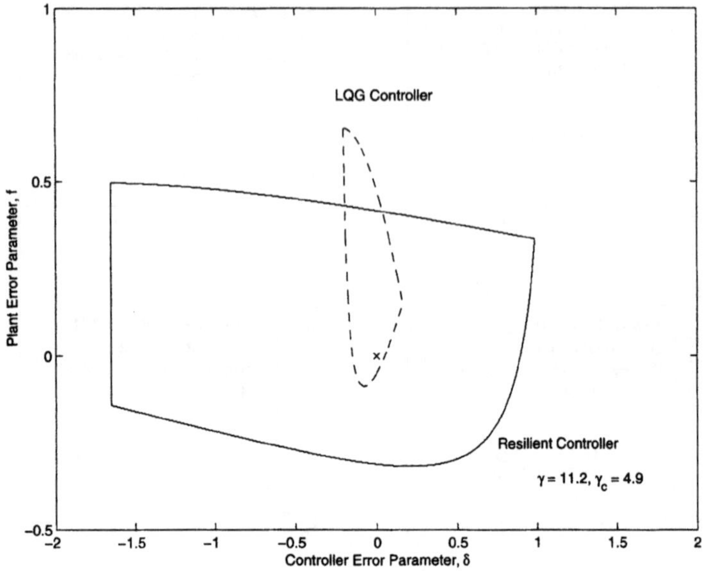

Fig. 11.14. Asymptotic stability regions

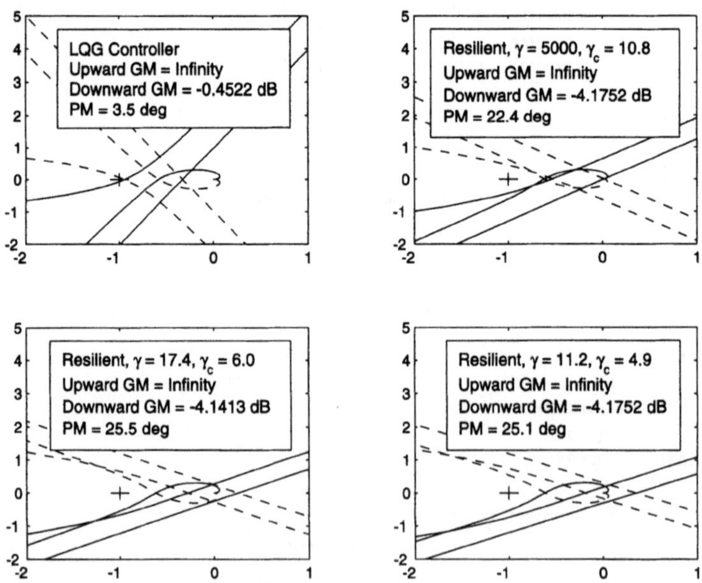

Fig. 11.15. Loop gain Nyquist plots

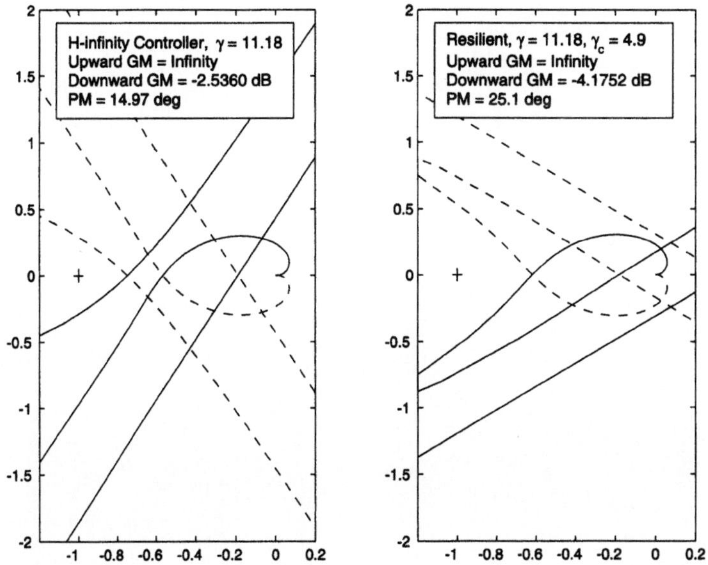

Fig. 11.16. Loop gain Nyquist plots

CHAPTER 12

ROBUST NON-FRAGILE CONTROLLER DESIGN FOR DISCRETE TIME SYSTEMS WITH FWL CONSIDERATION

Guang-Hong Yang, Jian Liang Wang, and Yeng Chai Soh

Abstract. This chapter is concerned with the problem of robust non-fragile $\mathcal{H}_2$ controller design for linear time-invariant discrete time systems with uncertainty. The controller to be designed is assumed to have additive gain variations or multiplicative gain variations, which are due to the finite word-length (FWL) effects when the controller is implemented. Sufficient conditions for the robust non-fragile control problem are given in terms of solutions to a set of matrix inequalities. The resulting design is such that the closed-loop system is robustly stable and has an $\mathcal{H}_2$ cost bound against plant uncertainty and controller gain variations. Iterative linear matrix inequalities (LMI) based algorithms are developed. Numerical examples are given to illustrate the design methods.

12.1 Introduction

In the area of robust control for linear systems, some important techniques for robust controller design have been developed [3,8,13,23,28]. An implicit assumption inherent in these design techniques is that the controller will be implemented exactly. However, in practice, controllers do have a certain degree of errors due to finite word-length (FWL) in digital systems, the imprecision inherent in analogue systems, and the need for additional tuning of parameters in the final controller implementation. In [12], it has been pointed out that a stable discrete time control system may become unstable when the digital controller is actually implemented with a digital control processor due to the FWL effects and that the stability of the closed-loop system depends on the structure of the digital controller. The optimal controller structure problem and stability issues due to the FWL effects have been investigated in [12,21,24] and the references therein. Moreover, in [19], Keel and Bhattacharyya have shown by a number of examples that the controllers designed by using weighted $\mathcal{H}_\infty$, μ and ℓ_1 synthesis techniques may be very sensitive, or fragile, with respect to errors in the controller coefficients. This brings a new issue at the stage of designing controllers: how to design a controller for a given plant such that the controller is insensitive to some amount of error with respect to its gains, *i.e.*, the controller is resilient or non-fragile.

Recently, there have been some results on how one can tackle the non-fragile controller design problem [4,7,10,14–18,22,25–27]. In particular, Dorato [7] gave an overview of non-fragile controller design for linear systems. Haddad and Corrado [14] extended the robust fixed-structure guaranteed cost controller synthesis framework to synthesise robust non-fragile controllers for multiplicative controller gain variations and system parametric uncertainty, the corresponding results for the case of additive controller gain variations are given in [15]. The problem of designing a robust resilient static output feedback controller for uncertain systems is addressed in [6]. In [25], a non-fragile $\mathcal{H}_\infty$ controller design method with respect to additive norm-bounded controller gain variations is given by using the Riccati inequality approach. The results for the case of multiplicative controller gain variations are given in [27]. A robust non-fragile Kalman filtering problem for uncertain linear systems with estimator gain uncertainty is addressed in [26].

This chapter will be concerned with the robust non-fragile $\mathcal{H}_2$ controller design problem for uncertain discrete time systems with FWL consideration. It is assumed that the controller to be designed has to tolerate additive controller gain variations or multiplicative controller gain variations, the gain variations are due to the FWL effects when the controller is implemented. The purpose of the chapter is to develop numerical algorithms to approach the robust non-fragile control problem. The chapter is organised as follows. Section 12.2 presents the problem under consideration and some preliminaries. In Section 12.3, a robust non-fragile $\mathcal{H}_2$ controller design method with respect to the additive controller gain variations is given in terms of solutions of matrix inequalities, and an LMI-based iterative algorithm to obtain solutions is developed. The design method for the case of the multiplicative controller gain variations is given in Section 12.4. Section 12.5 presents numerical examples to illustrate the design procedures and their effectiveness. Finally, some concluding remarks are given in Section 12.6.

12.2 Problem Statement and Preliminaries

Consider an uncertain linear time-invariant discrete time system $\boldsymbol{\Sigma}$ described by equations of the form:

$$\boldsymbol{\Sigma}: \quad x_{k+1} = Ax_k + B_1 w_k + B_2 u_k \tag{12.1}$$

$$z_k = C_1 x_k + D_{11} w_k + D_{12} u_k \tag{12.2}$$

$$y_k = C_2 x_k + D_{21} w_k + D_{22} u_k \tag{12.3}$$

where $x_k \in \mathbb{R}^n$ is the state vector, $u_k \in \mathbb{R}^m$ is the control-input vector, $w_k \in \mathbb{R}^r$ is the exogenous-input vector, $z_k \in \mathbb{R}^s$ the controlled-output vector and $y_k \in \mathbb{R}^p$ the measured-output vector. B_1, B_2, C_1, C_2, D_{11}, D_{12}, D_{21} and D_{22} are real constant matrices of appropriate dimensions. Without loss of generality, we assume $D_{22} = 0$. The matrix A satisfies:

$$A \in A_0 + \delta_P Co\{A_1, \cdots, A_L\} \tag{12.4}$$

where Co denotes the convex hull [5] of the vertexes $A_1, \cdots, A_L$, and A_0, $A_1, \cdots, A_L$ are given constant matrices.

Assume that the controller K to be designed has the following realisation with uncertainty to be tolerated:

$$\xi_{k+1} = (A_c + \Delta A_c)\xi_k + (B_c + \Delta B_c)y_k \tag{12.5}$$

$$u_k = (C_c + \Delta C_c)\xi_k + (D_c + \Delta D_c)y_k \tag{12.6}$$

where $\xi_k \in \mathbb{R}^{n_c}$ is the state vector of the controller. Denote:

$$X = [x_{ij}]_{(m+n_c)\times(p+n_c)} = \begin{bmatrix} D_c & C_c \\ B_c & A_c \end{bmatrix} \tag{12.7}$$

The uncertainty:

$$\Delta X = \begin{bmatrix} \Delta D_c & \Delta C_c \\ \Delta B_c & \Delta A_c \end{bmatrix} \tag{12.8}$$

due to the FWL effects, is in one of the following two forms:
1. Additive case:

$$\begin{aligned} \Delta X = \Delta X_a \triangleq [\Delta x_{ij}]_{(m+n_c)\times(p+n_c)}, \quad | \Delta x_{ij} | \leq \delta_a \\ i = 1, \cdots, m+n_c; \; j = 1, \cdots, p+n_c \end{aligned} \tag{12.9}$$

2. Multiplicative case:

$$\begin{aligned} \Delta X = \Delta X_m \triangleq [\delta_{ij} x_{ij}]_{(m+n_c)\times(p+n_c)}, \quad | \delta_{ij} | \leq \delta_m \\ i = 1, \cdots, m+n_c; \; j = 1, \cdots, p+n_c \end{aligned} \tag{12.10}$$

Remark 12.1. The additive gain variation model of form (12.9) is from [21] and [24], which has been extensively used to describe the FWL effects. The multiplicative gain variation model of form (12.10) is introduced in [10], which can describe the errors due to implementation of a controller with finite precision in the calculations. Examples of the errors include roundoff error, quantisation errors, and controller realisation errors, *etc.* [1,2,12].

Applying the controller K with the realisation described by (12.5) and (12.6) to the system Σ, the resulting closed-loop system is given by:

$$x^e_{k+1} = (A_e + \Delta A_e)x_k + (B_e + \Delta B_e)w_k \tag{12.11}$$

$$z_k = (C_e + \Delta C_e)x_e + (D_e + \Delta D_e)w_k \tag{12.12}$$

where $x^e_k = [x_k^T \;\; \xi_k^T]^T$:

$$A_e = \begin{bmatrix} A + B_2 D_c C_2 & B_2 C_c \\ B_c C_2 & A_c \end{bmatrix}, \quad \Delta A_e = \begin{bmatrix} B_2 \Delta D_c C_2 & B_2 \Delta C_c \\ \Delta B_c C_2 & \Delta A_c \end{bmatrix} \tag{12.13}$$

$$B_e = \begin{bmatrix} B_1 + B_2 D_c D_{21} \\ B_c D_{21} \end{bmatrix}, \quad \Delta B_e = \begin{bmatrix} B_2 \Delta D_c D_{21} \\ \Delta B_c D_{21} \end{bmatrix} \tag{12.14}$$

$$C_e = [C_1 + D_{12} D_c C_2 \quad D_{12} C_c], \quad \Delta C_e = [D_{12} \Delta D_c C_2 \quad D_{12} \Delta C_c] \tag{12.15}$$

$$D_e = D_{11} + D_{12} D_c D_{21}, \quad \Delta D_e = D_{12} \Delta D_c D_{21} \tag{12.16}$$

Denote:

$$G_e(z, K, \Delta X) = (C_e + \Delta C_e)(zI - A_e - \Delta A_e)^{-1}(B_e + \Delta B_e) + D_e + \Delta D_e \quad (12.17)$$

Then, the problem under consideration is as follows.

Robust non-fragile $\mathcal{H}_2$ control problem with controller gain variations. Consider the system Σ described by equations (12.1)-(12.3) with (12.4), and $\delta_P > 0$, $\gamma > 0$ and $\delta_a > 0$ (or $\delta_m > 0$) are constants. If possible, find a dynamic output feedback controller K having the realisation with uncertainty described by (12.5) and (12.6) such that $A_e + \Delta A_e$ is stable and $\| G_e(z, K, \Delta X) \|_2 < \gamma$ for all A satisfying (12.4) and ΔX satisfying (12.9) for the additive case (or ΔX satisfying (12.10) for the multiplicative case).

In Section 12.3 and Section 12.4, sufficient conditions for the solvability of the robust non-fragile $\mathcal{H}_2$ control problem will be presented for the additive case and the multiplicative case, respectively, and LMI-based iterative algorithms will be developed.

The following lemmas and preliminaries will be used in the sequel.

Notation: In this chapter, the symbol "*" within a matrix represents the symmetric term of the matrix, the operator Tr represents the trace operation as applied to a square matrix. I_i denotes the identity matrix of $i \times i$, and $0_{i\times j}$ denotes the zero matrix of $i \times j$. For a given transfer function matrix $G(z)$, the $\mathcal{H}_2$ norm of $G(z)$ is defined as:

$$\| G(z) \|_2^2 = \frac{1}{2\pi j} \oint_{|z|=1} \mathrm{Tr}\left\{G(z)(G(z))^H\right\} z^{-1} dz \quad (12.18)$$

Lemma 12.1. *([23]): Let $G(z) = C(zI-A)^{-1}B+D$ with A stable, and $\gamma > 0$ be a constant. Then $\| G(z) \|_2 < \gamma$ if and only if there exist positive-definite matrices P and R such that the following inequalities hold:*

$$\begin{bmatrix} -P^{-1} & A & B \\ A^T & -P & 0 \\ B^T & 0 & -I \end{bmatrix} < 0, \quad \begin{bmatrix} -R & C & D \\ C^T & -P & 0 \\ D^T & 0 & -I \end{bmatrix} < 0, \quad \mathrm{Tr}(R) < \gamma^2$$

Let:

$$\{E_k : k = 1, \cdots, \beta\} = \left\{ \sum_{i=1}^{m+n_c} \sum_{j=1}^{p+n_c} v_{aij} : v_{aij} = e_{1i}e_{2j}^T \text{ or } -e_{1i}e_{2j}^T \right\} \quad (12.19)$$

$$\{F_k(X) : k = 1, \cdots, \beta\} = \left\{ \sum_{i=1}^{m+n_c} \sum_{j=1}^{p+n_c} v_{mij} : v_{mij} = f_{1i}Xf_{2j}^T \text{ or } -f_{1i}Xf_{2j} \right\} \quad (12.20)$$

where X is defined by (12.7), $\beta = 2^{(m+n_c)\times(p+n_c)}$, $e_{1i} \in \mathbb{R}^{m+n_c}$, $e_{2i} \in \mathbb{R}^{p+n_c}$, $f_{1i} \in \mathbb{R}^{(m+n_c)\times(m+n_c)}$, $f_{2i} \in \mathbb{R}^{(p+n_c)\times(p+n_c)}$, and:

$$
\begin{aligned}
e_{1i} &= [\underbrace{0\cdots0}_{i-1}\ 1\ 0\cdots0]^T, \quad e_{2j} = [\underbrace{0\cdots0}_{j-1}\ 1\ 0\cdots0]^T \\
f_{1i} &= \mathrm{diag}[\underbrace{0\cdots0}_{i-1}\ 1\ 0\cdots0]^T, \quad f_{2j} = [\underbrace{0\cdots0}_{j-1}\ 1\ 0\cdots0]^T \\
&i = 1,\cdots,m+n_c;\ j = 1,\cdots,p+n_c
\end{aligned}
\tag{12.21}
$$

Then we have the following.

Lemma 12.2.

$$(i)\quad \{\Delta X_a : \Delta X_a \text{ satisfies (12.9) }\} = \delta_a Co\{E_k : k = 1,\cdots,\beta\} \tag{12.22}$$

$$(ii)\quad \{\Delta X_m : \Delta X_m \text{ satisfies (12.10) }\} = \delta_m Co\{F_k(X) : k = 1,\cdots,\beta\} \tag{12.23}$$

Proof. It is immediate from (12.8)-(12.10) and (12.19)-(12.21). □

In the sequel, we denote:

$$
\begin{aligned}
&\bar{A} = \begin{bmatrix} A & 0 \\ 0 & 0_{n_c\times n_c} \end{bmatrix}, \quad \bar{B}_2 = \begin{bmatrix} B_2 & 0 \\ 0 & I_{n_c\times n_c} \end{bmatrix}, \quad \bar{C}_2 = \begin{bmatrix} C_2 & 0 \\ 0 & I_{n_c\times n_c} \end{bmatrix} \\
&\bar{D}_{21} = \begin{bmatrix} D_{21} \\ 0_{n_c\times r} \end{bmatrix}, \quad \bar{B}_1 = \begin{bmatrix} B_1 \\ 0_{n_c\times r} \end{bmatrix}, \quad \bar{D}_{12} = [D_{12}\ \ 0_{s\times n_c}] \\
&\bar{C}_1 = [C_1\ \ 0_{s\times n_c}], \quad \bar{A}_i = \begin{bmatrix} A_i & 0 \\ 0 & 0_{n_c\times n_c} \end{bmatrix}, \quad i = 0,1,\cdots,L
\end{aligned}
\tag{12.24}
$$

12.3 Robust Non-fragile $\mathcal{H}_2$ Control with Additive Controller Uncertainty

The following theorem presents a sufficient condition under which the robust non-fragile $\mathcal{H}_2$ control problem with respect to additive controller uncertainty is solvable.

Theorem 12.1. *Consider the system Σ described by (12.1)-(12.4). $\gamma > 0$, $\delta_P > 0$ and $\delta_a > 0$ are constants. If there exist matrices $P > 0$, $P_0 > 0$, $R > 0$, and X such that the following inequalities hold:*

$$
\begin{aligned}
M_{a1jk} &= M_{a1jk}(P, P_0, X, \delta_P, \delta_a) \\
&\triangleq \begin{bmatrix} P - 2P_0 & P_0(\bar{A}_0 + \delta_P\bar{A}_j + \bar{B}_2X\bar{C}_2 + \delta_a\bar{B}_2E_k\bar{C}_2) & P_0(\bar{B}_1 + \bar{B}_2X\bar{D}_{21} + \delta_a\bar{B}_2E_k\bar{D}_{21}) \\ * & -P & 0 \\ * & * & -I_r \end{bmatrix} \\
&< 0, \qquad j = 1,\cdots,L;\ k = 1,\cdots,\beta
\end{aligned}
\tag{12.25}
$$

$$
\begin{aligned}
M_{a2k} &= M_{a2k}(R, P, X, \delta_a) \\
&\triangleq \begin{bmatrix} -R & \bar{C}_1 + \bar{D}_{12} X \bar{C}_2 + \delta_a \bar{D}_{12} E_k \bar{C}_2 & D_{11} + \bar{D}_{12} X \bar{D}_{21} + \delta_a \bar{D}_{12} E_k \bar{D}_{21} \\ * & -P & 0 \\ * & * & -I_r \end{bmatrix} \\
&< 0, \qquad k = 1, \cdots, \beta \qquad (12.26) \\
\mathrm{Tr}(R) &< \gamma^2 \qquad (12.27)
\end{aligned}
$$

then the controller K having the realisation with the additive uncertainty described by (12.5), (12.6) and (12.9), and with the controller parameter matrices A_c, B_c, C_c and D_c given by (12.7), solves the robust non-fragile $\mathcal{H}_2$ control problem.

The following lemmas will be used in the proof of Theorem 12.1.

Lemma 12.3. *Consider the system Σ described by (12.1)-(12.4), and the controller K described by (12.5) and (12.6). If there exist positive-definite matrices P and R such that the closed-loop system described by (12.11) and (12.12) satisfies (12.27) and:*

$$
M_1(P, \delta_P, X, \Delta X) \triangleq \begin{bmatrix} -P^{-1} & A_e + \Delta A_e & B_e + \Delta B_e \\ * & -P & 0 \\ * & * & -I \end{bmatrix} < 0 \qquad (12.28)
$$

$$
M_2(P, R, X, \Delta X) \triangleq \begin{bmatrix} -R & C_e + \Delta C_e & D_e + \Delta D_e \\ * & -P & 0 \\ * & * & -I \end{bmatrix} < 0 \qquad (12.29)
$$

for all A satisfying (12.4) and ΔX satisfying (12.9) for the additive case (or ΔX satisfying (12.10) for the multiplicative case). Then the controller K having the realisation described by (12.5) and (12.6) solves the robust non-fragile $\mathcal{H}_2$ control problem.

Proof. It is immediate from Lemma 12.1. □

Lemma 12.4. *(i) (12.28) holds for all A satisfying (12.4) and $\Delta X = \Delta X_a$ satisfying (12.9) if and only if there exists a matrix $P_0 > 0$ such that (12.25) holds.*
(ii) (12.29) holds for all $\Delta X = \Delta X_a$ satisfying (12.9) if and only if (12.26) holds.

Proof. (i) By (12.7), (12.8), (12.13)-(12.16) and (12.24), we have:

$$M_1(P,\delta_P,X,\Delta X_a) = \begin{bmatrix} -P^{-1} & \bar{A} & \bar{B}_1 \\ * & -P & 0 \\ * & * & -I \end{bmatrix} + \begin{bmatrix} 0 & \bar{B}_2(X+\Delta X_a)\bar{C}_2 & \bar{B}_2(X+\Delta X_a)\bar{D}_{21} \\ * & 0 & 0 \\ * & * & 0 \end{bmatrix} \tag{12.30}$$

From (12.4), (12.22) and (12.30), it follows that (12.28) holds for all A satisfying (12.4) and $\Delta X = \Delta X_a$ satisfying (12.9) if and only if:

$$M_{1a}(P,\delta_P,X,\delta_a) \triangleq \begin{bmatrix} -P^{-1} & \bar{A}_0+\delta_P\bar{A}_j+\bar{B}_2X\bar{C}_2+\delta_a\bar{B}_2E_k\bar{C}_2 & \bar{B}_1+\bar{B}_2X\bar{D}_{21}+\delta_a\bar{B}_2E_k\bar{D}_{21} \\ * & -P & 0 \\ * & * & -I \end{bmatrix} < 0 \quad j=1,\cdots,L;\ k=1,\cdots,\beta \tag{12.31}$$

In the following, we show that (12.31) holds if and only if there exists a matrix $P_0 > 0$ such that (12.25) holds. In fact, if (12.31) holds, then let $P = P_0$, we have:

$$\text{Block-diag}[P_0 \;\; I \;\; I]\, M_{1a}(P,\delta_P,X,\delta_a)\, \text{Block-diag}[P_0 \;\; I \;\; I] < 0$$

which implies that (12.25) holds for $P = P_0$. Conversely, if (12.25) holds, then from the definition of $M_{a1jk}(P,P_0,X,\delta_P,\delta_a)$ and:

$$P - 2P_0 = -P_0P^{-1}P_0 + (P-P_0)P^{-1}(P-P_0)$$

it follows that:

$$\begin{aligned} M_{a1jk}(P,P_0,X,\delta_P,\delta_a) &= \text{Block-diag}[P_0 \;\; I \;\; I]M_{1a}(P,\delta_P,X,\delta_a) \\ &\quad \times\text{Block-diag}[P_0 \;\; I \;\; I] \\ &\quad + \begin{bmatrix} (P-P_0)P^{-1}(P-P_0) & 0 & 0 \\ 0 & 0 & 0 \\ 0 & 0 & 0 \end{bmatrix} \end{aligned}$$

Thus:

$$M_{1a}(P,\delta_P,X,\delta_a) < 0$$

□

(ii) It is similar to that of (i), and the details are omitted. □

Proof of Theorem 12.1. It is immediate from Lemma 12.3 and Lemma 12.4. □

Remark 12.2. Theorem 12.1 presents a sufficient condition for the robust non-fragile $\mathcal{H}_2$ control problem with respect to the additive controller uncertainty, which is related to the solutions to a set of matrix inequalities (12.25)-(12.27). The condition may not be necessary because a common Lyapunov function matrix P is used in Lemma 12.3 for the presence of the plant uncertainty and controller uncertainty. The inequality (12.25) are not convex with respect to the variable P_0, it is difficult to solve the inequality directly. However, when P_0 is given, the inequalities (12.25)-(12.27) are convex with respect to P, X and R, and can be solved by using MATLAB® via the LMI Control Toolbox [11]. Based on the property, an iterative algorithm is developed in the following Algorithm 12.1.

Algorithm 12.1. Let $\delta_P > 0$ and $\delta_a > 0$ be given constants, and $\epsilon > 0$ be a small constant to denote a convergence rule.
Step 1. For the system Σ described by (12.1)-(12.4) with $\delta_P = 0$, find an optimal (or suboptimal) $\mathcal{H}_2$ controller K_0 (of n_c order) without controller uncertainty. Let $(A_{c0}, B_{c0}, C_{c0}, D_{c0})$ be a realisation of K_0, and denote:

$$\begin{aligned}
X_{K_0} &= \begin{bmatrix} D_{c0} & C_{c0} \\ B_{c0} & A_{c0} \end{bmatrix}, \quad A_{e0} = \begin{bmatrix} A_0 + B_2 D_{c0} C_2 & B_2 C_{c0} \\ B_{c0} C_2 & A_{c0} \end{bmatrix} \\
B_{e0} &= \begin{bmatrix} B_1 + B_2 D_{c0} D_{21} \\ B_{c0} D_{21} \end{bmatrix}, \quad D_{e0} = D_{11} + D_{12} D_{c0} D_{21} \\
C_{e0} &= [C_1 + D_{12} D_{c0} C_2 \quad D_{12} C_{c0}]
\end{aligned} \tag{12.32}$$

then minimise $\mathrm{Tr}(R_0)$ subject to the following LMI constraints:

$$\begin{bmatrix} -P_0 & P_0 A_{e0} & P_0 B_{e0} \\ * & -P_0 & 0 \\ * & * & -I \end{bmatrix} < 0 \tag{12.33}$$

and:

$$\begin{bmatrix} -R_0 & C_{e0} & D_{e0} \\ * & -P_0 & 0 \\ * & * & -I \end{bmatrix} < 0 \tag{12.34}$$

Let R_{0opt} and P_{0opt} be the optimal solutions to the optimisation, $P_0^1 = P_{0opt}$ and choose γ_1 such that $\gamma_1^2 \geq \mathrm{Tr}(R_{0opt})$.
Step 2. Maximise δ_P^i subject to the following LMI constraints:

$$M_{a1jk}(P^i, P_0^i, X^i, \delta_P^i, 0) < 0, \quad j = 1, \cdots, L \tag{12.35}$$

$$M_{a2k}(R^i, P^i, X^i, 0) < 0, \tag{12.36}$$

$$\mathrm{Tr}(R^i) < \gamma_1^2 \tag{12.37}$$

where M_{a1jk} and M_{a2k} are defined by (12.25) and (12.26), respectively, $P_0^i = P_{opt}^{i-1}$ $(i = 2, 3, \cdots)$, and P_{opt}^{i-1}, X_{opt}^{i-1} and δ_{Popt}^i denote the solutions of the $(i-1)$th optimisation. If $\delta_{Popt}^i > \delta_P$ for some i, then reduce γ_1 and repeat the above optimisation. If δ_{Popt}^i converges to $\delta_{P\gamma_1}$ and $\delta_{P\gamma_1} < \delta_P$, then increase γ_1 and repeat the above optimisation.
Step 3. (i) If $\delta_{Popt}^i < \delta_P - \epsilon$ for any i and $\gamma_1 > 0$, then stop. (ii) If δ_{Popt}^i converges to $\delta_{P\gamma_1}$ and $\delta_P < \delta_{P\gamma_1} \leq \delta_P + \epsilon$, then stop. Thus, one can choose an i such that $\delta_P < \delta_{Popt}^i \leq \delta_P + \epsilon$ (for the special case, $\delta_{Popt}^i \to \infty$ for $\gamma_1^2 = \mathrm{Tr}(R_{0opt})$, one can choose an i such that $\delta_P < \delta_{Popt}^i$). Denote $P_{01} = P_{opt}^i$, go to step 4.
Step 4. Choose $\gamma_2 \geq \gamma_1$, maximise δ_a^i subject to the following LMI constraints:

$$M_{a1jk}(P_i, P_{0i}, X_i, \delta_P, \delta_a^i) < 0, \quad j = 1, \cdots, L; \; k = 1, \cdots, \beta \tag{12.38}$$

$$M_{a2k}(R_i, P_i, X_i, \delta_a^i) < 0, \quad k = 1, \cdots, \beta \tag{12.39}$$

$$\mathrm{Tr}(R_i) < \gamma_2^2 \tag{12.40}$$

where $P_{0i} = P_{(i-1)\ opt}$ $(i = 2, 3, \cdots)$, and $P_{(i-1)\ opt}$, $X_{(i-1)\ opt}$ and δ_{aopt}^{i-1} denote the solutions of the $(i-1)$th optimisation. If $\delta_{aopt}^i > \delta_a$ for some i, then reduce γ_2 and repeat the above optimisation. If δ_{aopt}^i converges to $\delta_{a\gamma_2}$ and $\delta_{a\gamma_2} < \delta_a$, then increase γ_2 and repeat the above optimisation.
Step 5. (i) If $\delta_{aopt}^i < \delta_a - \epsilon$ for any i and $\gamma_2 \geq \gamma_1$, then stop. (ii) If δ_{aopt}^i converges to $\delta_{a\gamma_2}$ and $\delta_a < \delta_{a\gamma_2} \leq \delta_a + \epsilon$, then stop. Thus, one can choose an i such that $\delta_a < \delta_{aopt}^i \leq \delta_a + \epsilon$ (for the special case, $\delta_{aopt}^i \to \infty$ for $\gamma_2 = \gamma_1$, one can choose an i such that $\delta_a < \delta_{aopt}^i$). By using X_{iopt}, the controller parameter matrices A_c, B_c, C_c and D_c can be constructed by (12.7), the corresponding controller solves the robust non-fragile $\mathcal{H}_2$ control problem with $\gamma = \gamma_2$.

Remark 12.3. In Algorithm 12.1, Step 1 provides an initial solution for the iterative computation in Step 2, which is obtained by solving the standard fixed-order $\mathcal{H}_2$ control problem for the nominal system Σ with $\delta_P = 0$ (*i.e.*, without plant uncertainty). In Step 2 and Step 3, a robust $\mathcal{H}_2$ controller for the system Σ with the perturbation bound δ_P can be obtained by increasing the $\mathcal{H}_2$ cost γ of the resulting closed-loop system. After obtaining the robust $\mathcal{H}_2$ controller, a robust non-fragile $\mathcal{H}_2$ controller which tolerates the additive uncertainty with the bound δ_a is computed in Step 4 by further increasing the $\mathcal{H}_2$ cost of the closed-loop system. However, it should be pointed out that the number of LMI constraints involved in Step 4 will be very big ($\beta = 2^{(m+n_c \times (p+n_c))}$) if the designed controller is of high order, which may result in some numerical problem using MATLAB®.

Remark 12.4. For a nominal system without uncertainty, an optimal full-order $\mathcal{H}_2$ controller can be obtained by using the Riccati equation approach [28] or LMI approach [23]. However, the problem of designing an optimal fixed-order $\mathcal{H}_2$ controller has not been solved completely. For continuous time systems, a numerical method for designing suboptimal fixed-order stabilising controllers is given in [9]. For discrete time systems, if a stabilising controller K_0 of order n_c is known, then a suboptimal $\mathcal{H}_2$ controller can be obtained by iteratively minimising $\mathrm{Tr}(R^i)$ subject to the LMI constraints (12.35) and (12.36) with $\delta_P^i = 0$, where P_0^1 is the optimal solution to the optimisation problem: minimise $\mathrm{Tr}(R_0)$ subject to the LMI constraints (12.33) and (12.34).

Remark 12.5. It should be noted that the optimisation problems in Step 2 and Step 3 are generalised eigenvalue problems (GEVP) [5], which can be solved by using MATLAB® effectively. In fact, by (12.25) and (12.26), to maximise δ_P^i subject to (12.35)-(12.37) in Step 2 is equivalent to minimising $(\delta_P^i)^{-1}$ subject to (12.36), (12.37) and:

$$(\delta_P^i)^{-1} M_{a1jk}(P^i, P_0^i, X^i, 0, 0) < \begin{bmatrix} 0 & -P_0^i \bar{A}_j & 0 \\ * & 0 & 0 \\ * & * & 0 \end{bmatrix}$$
$$j = 1, \cdots, L \qquad (12.41)$$

which is a GEVP. For Step 3, maximising δ_a^i subject to (12.38)-(12.40) is equivalent to minimising $(\delta_a^i)^{-1}$ subject to (12.40):

$$(\delta_a^i)^{-1} M_{a1jk}(P_i, P_{0i}, X_i, \delta_P, 0) < \begin{bmatrix} 0 & -P_{0i}\bar{B}_2 E_k \bar{C}_2 & -P_{0i}\bar{B}_2 E_k \bar{D}_{21} \\ * & 0 & 0 \\ * & * & 0 \end{bmatrix}$$
$$j = 1, \cdots, L; \; k = 1, \cdots, \beta \qquad (12.42)$$

and:

$$(\delta_a^i)^{-1} M_{a2k}(R_i, P_i, X_i, 0) < \begin{bmatrix} 0 & -\bar{D}_{12} E_k \bar{C}_2 & -\bar{D}_{12} E_k \bar{D}_{21} \\ * & 0 & 0 \\ * & * & 0 \end{bmatrix}$$
$$j = 1, \cdots, L; \; k = 1, \cdots, \beta \qquad (12.43)$$

which also is a GEVP. Moreover, the sequences $\{\delta_{Popt}^i\}_0^\infty$ and $\{\delta_{aopt}^i\}_0^\infty$ given in Step 2 and Step 3 are monotonically increasing, and convergent if they are bounded.

12.4 Robust Non-fragile $\mathcal{H}_2$ control with Multiplicative Controller Uncertainty

The following theorem presents a sufficient condition under which the robust non-fragile $\mathcal{H}_2$ control problem with respect to multiplicative controller uncertainty is solvable.

Theorem 12.2. *Consider the system Σ described by (12.1)-(12.4). $\gamma > 0$, $\delta_P > 0$ and $\delta_m > 0$ are constants. If there exist matrices $P > 0$, $P_0 > 0$, $R > 0$, and X such that (12.27) and the following inequalities hold:*

$$\begin{aligned} M_{m1jk} &= M_{m1jk}(P, P_0, X, \delta_P, \delta_m) \\ &\triangleq \begin{bmatrix} P - 2P_0 & P_0(\bar{A}_0 + \delta_P \bar{A}_j + \bar{B}_2 X \bar{C}_2 + \delta_m \bar{B}_2 F_k(X) \bar{C}_2) & P_0(\bar{B}_1 + \bar{B}_2 X \bar{D}_{21} + \delta_a \bar{B}_2 F_k(X) \bar{D}_{21}) \\ * & -P & 0 \\ * & * & -I_r \end{bmatrix} < 0 \\ & j = 1, \cdots, L; \; k = 1, \cdots, \beta \end{aligned} \tag{12.44}$$

$$\begin{aligned} M_{m2k} &= M_{m2k}(R, P, X, \delta_m) \\ &\triangleq \begin{bmatrix} -R & \bar{C}_1 + \bar{D}_{12} X \bar{C}_2 + \delta_a \bar{D}_{12} F_k(X) \bar{C}_2 & D_{11} + \bar{D}_{12} X \bar{C}_2 + \delta_a \bar{D}_{12} F_k(X) \bar{D}_{21} \\ * & -P & 0 \\ * & * & -I_r \end{bmatrix} \\ &< 0, \qquad k = 1, \cdots, \beta \end{aligned} \tag{12.45}$$

where $F_k(X)$ is defined by (12.20). Then the controller K having the realisation with the multiplicative uncertainty described by (12.5), (12.6) and (12.10), and with the controller parameter matrices A_c, B_c, C_c and D_c given by (12.7), solves the robust non-fragile $\mathcal{H}_2$ control problem.

Proof. It is similar to that of Theorem 12.1, the details are omitted. □

Remark 12.6. Theorem 12.2 presents a sufficient condition for the robust non-fragile $\mathcal{H}_2$ control problem with respect to multiplicative controller uncertainty, which is related to the solutions to matrix inequalities (12.27), (12.44) and (12.45). When P_0 is given, the inequalities (12.27), (12.44) and (12.45) are convex with respect to P, X and R, and can be solved by using MATLAB® via the LMI Control Toolbox [11]. Similar to Algorithm 12.1 for the case of the additive controller uncertainty, we have the following algorithm for the case of the multiplicative controller uncertainty.

Algorithm 12.2. Let $\delta_P > 0$ and $\delta_m > 0$ be given constants, and $\epsilon > 0$ be a small constant to denote a convergence rule.
Step 1. Perform steps 1-3 in Algorithm 12.1, to obtain $\gamma_1 > 0$ and P^i_{opt} such that $\gamma = \gamma_1$ and $P_0 = P^i_{opt}$ are initial solutions to (12.27), (12.44) and (12.45) with $\delta_m = 0$. Denote $P_{01} = P^i_{opt}$, go to step 2.

Step 2. Choose $\gamma_2 \geq \gamma_1$, maximise δ_m^i subject to (12.40) and the following LMI constraints:

$$M_{m1jk}(P_i, P_{0i}, X_i, \delta_P, \delta_m^i) < 0, \quad j = 1, \cdots, L; \; k = 1, \cdots, \beta \tag{12.46}$$

$$M_{m2k}(R_i, P_i, X_i, \delta_m^i) < 0, \quad k = 1, \cdots, \beta \tag{12.47}$$

where $P_{0i} = P_{(i-1)\ opt}$ $(i = 2, 3, \cdots)$, and $P_{(i-1)\ opt}$, $X_{(i-1)\ opt}$ and δ_{mopt}^{i-1} denote the solutions of the $(i-1)$th optimisation. If $\delta_{mopt}^i > \delta_m$ for some i, then reduce γ_2 and repeat the above optimisation. If δ_{mopt}^i converges to $\delta_{m\gamma_2}$ and $\delta_{m\gamma_2} < \delta_m$, then increase γ_2 and repeat the above optimisation.
Step 3. (i) If $\delta_{mopt}^i < \delta_m - \epsilon$ for any i and $\gamma_2 \geq \gamma_1$, then stop. (ii) If δ_{mopt}^i converges to $\delta_{m\gamma_2}$ and $\delta_m < \delta_{m\gamma_2} \leq \delta_m + \epsilon$, then stop. Thus, one can choose an i such that $\delta_m < \delta_{mopt}^i \leq \delta_m + \epsilon$ (for the special case, $\delta_{mopt}^i \to \infty$ for $\gamma_2 = \gamma_1$, one can choose an i such that $\delta_m < \delta_{mopt}^i$). By using X_{iopt}, the controller parameter matrices A_c, B_c, C_c and D_c can be constructed by (12.7), the corresponding controller solves the robust non-fragile $\mathcal{H}_2$ control problem with $\gamma = \gamma_2$.

Remark 12.7. The above algorithm is similar to Algorithm 12.1, and the optimisation problem in Step 2 is a GEVP. In fact, from (12.44) and (12.45), maximising δ_m^i subject to (12.40), (12.46) and (12.47) is equivalent to minimising $(\delta_m^i)^{-1}$ subject to (12.40):

$$(\delta_m^i)^{-1} M_{m1jk}(P_i, P_{0i}, X_i, \delta_P, 0) < \begin{bmatrix} 0 & -P_{0i}\bar{B}_2 F_k(X)\bar{C}_2 & Z_0 \\ * & 0 & 0 \\ * & * & 0 \end{bmatrix}$$
$$j = 1, \cdots, L; \; k = 1, \cdots, \beta \tag{12.48}$$

and:

$$(\delta_m^i)^{-1} M_{m2k}(R_i, P_i, X_i, 0) < \begin{bmatrix} 0 & -\bar{D}_{12} F_k(X)\bar{C}_2 & -\bar{D}_{12} F_k(X)\bar{D}_{21} \\ * & 0 & 0 \\ * & * & 0 \end{bmatrix}$$
$$j = 1, \cdots, L; \; k = 1, \cdots, \beta \tag{12.49}$$

where $Z_0 = -P_{0i}\bar{B}_2 F_k(X)\bar{D}_{21}$. By [5], it is a GEVP.

12.5 Example

In this section, an example is presented to illustrate the proposed algorithms.

Example 12.1. The system under consideration is described by equations (12.1)-(12.4) with $L = 1$:

$$A_0 = \begin{bmatrix} 0.5 & 0.1 \\ 0.2 & 0 \end{bmatrix}, \quad A_1 = \begin{bmatrix} 1 & 0 \\ 1 & 1 \end{bmatrix}, \quad B_1 = \begin{bmatrix} 1 & 0 \\ 1 & 0 \end{bmatrix}$$

$$B_2 = \begin{bmatrix} 1 \\ 0 \end{bmatrix}, \quad C_1 = \begin{bmatrix} 1 & 1 \\ 0 & 1 \end{bmatrix}, \quad D_{11} = \begin{bmatrix} 1 & 0 \\ 0 & 1 \end{bmatrix}$$

$$D_{12} = \begin{bmatrix} 1 \\ 1 \end{bmatrix}, \quad C_2 = \begin{bmatrix} 0 & -1 \end{bmatrix}, \quad D_{21} = \begin{bmatrix} 1 & 1 \end{bmatrix}, \quad D_{22} = 0$$

In the following, we will consider the problem of designing first-order robust non-fragile $\mathcal{H}_2$ controller for the system. It is easy to check that the first-order controller with the parameters given by $X_0 = \begin{bmatrix} 0.1 & 0.1 \\ 0.1 & 0.1 \end{bmatrix}$ stabilises the system. Thus, by using the approach in Remark 12.4, a suboptimal $\mathcal{H}_2$ controller is given by:

$$X_{s0} = \begin{bmatrix} 0.0763 & -1.4432 \\ 0.2126 & 0.0294 \end{bmatrix}$$

and an $\mathcal{H}_2$ performance cost of 2.6951 is achieved for the closed-loop system. By performing the steps 1-3 in Algorithm 12.1, a robust $\mathcal{H}_2$ controller with $\delta_P = 0.1029$ is obtained by:

$$X_r = \begin{bmatrix} 0.2376 & -1.0789 \\ 0.1763 & 0.0831 \end{bmatrix} \tag{12.50}$$

with an $\mathcal{H}_2$ performance bound 2.9833. To guarantee the performance bound for the closed-loop system, the robust controller with the above realisation only can tolerate the additive controller uncertainty with a guaranteed bound $\delta_a = 0.0054$, and the multiplicative controller uncertainty with a guaranteed bound $\delta_m = 0.0249$, respectively.

For the robust non-fragile controller design with additive controller uncertainty, Algorithm 12.1 gives an $\mathcal{H}_2$ performance bound of 3.0822 for the closed-loop system with $\delta_P = 0.1$ and $\delta_a = 0.0961$, and the corresponding controller is given by:

$$X_a = \begin{bmatrix} 0.2191 & -0.4100 \\ 0.3629 & 0.0871 \end{bmatrix}$$

Under the same guaranteed performance bound for the closed-loop system, the robust controller given by (12.50) can tolerate the additive controller uncertainty with a guaranteed bound $\delta_a = 0.0733$, which is less than $\delta_a = 0.0961$ achieved by the robust non-fragile controller. Table 12.1 shows the trade-off between the plant uncertainty bound δ_P and the additive controller uncertainty bound δ_a for achieving a given $\mathcal{H}_2$ performance bound 3.0822 of the closed-loop system.

For the case of multiplicative controller uncertainty, Algorithm 12.2 gives an $\mathcal{H}_2$ performance bound of 3.0 for the closed-loop system with $\delta_P = 0.1$ and $\delta_m = 0.1328$, and the corresponding controller is given by:

$$X_a = \begin{bmatrix} 0.1823 & -1.0796 \\ 0.1423 & 0.0534 \end{bmatrix}$$

Table 12.1. Trade-off between δ_P and δ_a

δ_P	0.1	0.11	0.12	0.13	0.135	0.14
δ_a	0.0961	0.0745	0.0519	0.0275	0.0151	0.0025

In comparison with the robust controller given by (12.50), it can tolerate the multiplicative controller uncertainty with a guaranteed bound $\delta_m = 0.0887$ under the same guaranteed performance bound for the closed-loop system, which also is less than $\delta_m = 0.1328$ achieved by the robust non-fragile controller.

12.6 Conclusion

In this chapter, we have investigated the problem of robust non-fragile $\mathcal{H}_2$ controller design for linear time-invariant discrete time systems with uncertainty. The additive controller gain variations and multiplicative controller gain variations are considered. These can be due to the finite word-length (FWL) effects when the controller is implemented. Sufficient conditions for the robust non-fragile control problem are given in terms of solutions to a set of non-convex matrix inequalities. The resulting design is such that the closed-loop system is robustly stable and has an $\mathcal{H}_2$ cost bound against plant uncertainty and controller gain variations. Iterative LMI-based algorithms are developed. The examples have shown the effectiveness of the design methods.

Acknowledgements

This work is supported by the Academic Research Fund of the Ministry of Education, Singapore, under grant MID-ARC 3/97.

References

1. Ackerman, J., *Sampled-Data Control Systems*, Springer-Verlag: Berlin, 1985.
2. Astrom, K.J. and B. Wittenmark, *Computer Controlled Systems*, Prentice Hall: Upper Saddle River, N.J., 1997.
3. Basar, T. and P. Bernhard, *H_∞-Optimal Control and Related Minimax Design Problems: A Dynamic Game Approach*, Birkhauser: Boston, Massachusetts, 1991.
4. Blanchini, F., R.Lo Cigno and R. Tempo, "Control of ATM networks: fragility and robustness issues", *Proc. American Control Conference*, Philadephia, Pennsylvania, pp. 2947-2851, 1998.
5. Boyd, S., L. El Ghaoui, E. Feron and V. Balakrishan, *Linear Matrix Inequalities in Systems and Control Theory*, SIAM: Philadelphia, PA, 1994.

6. Corrado, J.R. and W.M. Haddad, "Static output feedback controllers for systems with parametric uncertainty and controller gain variation", *Proc. American Control Conference*, San Diego, California, pp. 915-919, 1999.
7. Dorato, P., "Non-fragile controller design, an overview", *Proc. American Control Conference*, Philadephia, Pennsylvania, pp. 2829-2831, 1998.
8. Doyle, J.C., K. Glover, P.P. Khargonekar and B.A. Francis, "State space solutions to standard H_2 and H_∞ control problems," *IEEE Trans. Automatic Contr.*, Vol. 34, no. 8, pp. 831 - 847, 1989.
9. El Ghaoui, L., F. Oustry, and M. AitRami, "A cone complementarity linearization algorithm for static output feedback and related problems," *IEEE Trans. Automatic Contr.*, Vol. 42, no. 8, pp. 1171-1176, 1997.
10. Famularo, D., C.T. Abdallah, A. Jadbabais, P. Dorato, and W.M. Haddad, "Robust non-fragile LQ controllers: the static state feedback case", *Proc. American Control Conference*, Philadephia, Pennsylvania, pp. 1109-1113, 1998.
11. Gahinet, P., A. Nemirovski, A.J. Laub, and M. Chilali, LMI Control Toolbox, Natick, MA: The MathWorks, 1995.
12. Gevers, M. and G. Li, *Parametrizations in Control, Estimation and Filtering Problem: Accuracy Aspects*, Springer-Verlag: Berlin, 1993.
13. Green, M. and D.J.N. Limebeer, *Linear Robust Control*, Prentice-Hall: Englewood Cliffs, New Jersey, 1995.
14. Haddad, W.M. and J.R. Corrado, "Resilient controller design via quadratic Lyapunov bounds", *Proc. IEEE Conf. Dec. Contr.*, San Diego, CA, pp. 2678-2683, 1997.
15. Haddad, W.M. and J.R. Corrado, "Robust resilient dynamic controllers for systems with parametric uncertainty and controller gain variations", *Proc. American Control Conference*, Philadephia, Pennsylvania, pp. 2837- 2841, 1998.
16. Jadbabaie, A., T. Chaouki, D. Famularo, and P. Dorato, "Robust, non-fragile and optimal controller design via linear matrix inequalities", *Proc. American Control Conference*, Philadephia, Pennsylvania, pp. 2842-2846, 1998.
17. Kaesbauer, D. and J. Ackermann, "How to escape from the fragility trap", *Proc. American Control Conference*, Philadephia, Pennsylvania, pp. 2832-2836, 1998.
18. Keel, L.H. and S.P. Bhattacharyya, "Stability margins and digital implementation of controllers", *Proc. American Control Conference*, Philadephia, Pennsylvania, pp. 2852-2856, 1998.
19. Keel, L.H. and S.P. Bhattacharyya, "Robust, fragile, or optimal?" *IEEE Trans. Automatic Contr.*, Vol. 42, no. 8, pp. 1098 - 1105, 1997.
20. Keel, L.H. and S.P. Bhattacharyya, "Authors' Reply to Comments on "Robust, Fragile, or Optimal?"", *IEEE Trans. Automatic Contr.*, Vol. 43, no. 9, pp. 1268, 1998.
21. Li, G., "On the structure of digital controllers with finite word length consideration", *IEEE Trans. Automatic Contr.*, Vol. 43, pp. 689-693, 1998.
22. Makila P.M., "Comments on "Robust, Fragile, or Optimal?", *IEEE Trans. Automatic Contr.*, Vol. 43, no. 9, pp. 1265-1268, 1998.
23. Masubuchi, I., A. Ohara and N. Suda, "LMI-based controller synthesis: a unified formulation and solution", *Int. J. Robust Nonlinear Control*, vol. 8, pp. 669-686, 1998.
24. Whidborne, J. F., J. Wu and R. S. H. Istepanian, "Finite word length stability issues in an ℓ_1 framework", *Int. J. Control*, vol. 73, no. 2, pp. 166-176, 2000.

25. Yang, G.-H., J.L. Wang, and C. Lin, "H_∞ control for linear systmes with additive controller gain variations", *Int. J. Control*, vol. 73, no. 16, pp. 1500-1506, 2000.
26. Yang, G.-H. and J.L. Wang, "Robust non-fragile Kalman filtering for uncertain linear systems with estimator gain uncertainty", *IEEE Trans. Automatic Contr.*, Vol. 46, no. 2, pp. 343-348, 2001.
27. Yang, G.-H. and J.L. Wang, "Non-fragile H_∞ control for linear systems with multiplicative controller gain variations", *Automatica* Vol. 37; no. 5 ; pp. 727-737 , 2001.
28. Zhou, K., J.C. Doyle, and K. Glover, *Robust Optimal Control*, Prentice Hall: Englewood Cliffs, NJ, 1996.

CHAPTER 13

SYNTHESIS OF CONTROLLERS WITH FINITE-PRECISION CONSIDERATIONS

Mauricio C. de Oliveira and Robert E. Skelton

Abstract. Truncation and rounding appear in digital systems due to the need for representing and storing information with a finite number of bits. In addition, some delays are introduced between signals that are subject to transport or arithmetic calculations. When a computer is used to implement a controller, these *finite-precision* phenomena can significantly affect the behaviour of the closed loop system. This paper addresses the problem of designing a digital controller that behaves as close as possible to its predicted performance in such a finite precision environment. Some classical results are revisited and procedures for the simultaneous design of the controller transfer function and its optimal realisation with respect to roundoff noise are investigated.

13.1 Introduction

Most controller design procedures available in the literature implicitly assume that the obtained discrete time controller will be implemented in a digital computer exactly, *i.e.*, that the behaviour of the digital computer representation of the designed controller will be exactly as prescribed by the designer. However, the implementation of digital controller imposes some fundamental limitations on the performance of the controller, and hence on the closed-loop performance, that are usually ignored in this first design phase. Despite the amazing speed at which computer processing speed and storage capabilities evolve, some issues related to the digital computer *architecture* are still relevant for the control engineer. For instance, arithmetic calculations will *always* be carried by an arithmetic processor with a finite number of digits (word-length). Consequently, digital arithmetic is necessarily corrupted by, at least, rounding or truncation effects. In control and filtering applications, also the inputs of the filter or controller (plant output measurements) are subject to sampling by an A/D (analogue-to-digital) converter that also assigns a finite number of bits to represent the current value of an analogue input signal. This process is known as *quantisation*. Therefore, virtually all quantities involved in the control law computation will necessarily be corrupted, at some point, by finite-precision effects.

The fact that the computer industry currently provides processors with longer word-lengths every few years is used by some to justify the design of controllers with little or no regard to finite-precision effects. In fact, for many simple control loops, this increase in word-length practically means that the quantisation effects can be ignored. However, faster and more precise computers provide also the opportunity to increase the complexity of a controlled system, both in terms of more sophisticated control algorithms and of the number of controlled devices. As observed in [1], this increased complexity will eventually face limitations in bandwidth, that is, the speed at which the devices communicate, pushing down the sampling rates (relative to the available processor word-length). In this scenario, a careful analysis of finite-precision effects will be certainly required.

An extra complication is that finite-precision effects, such as rounding, truncation and quantisation, are nonlinear phenomena. Even when the controller or filter is designed as a linear system, its "real-world" finite-precision implementation may present nonlinear behaviours, such as limit cycles and oscillations [11].

A fundamental observation is that contrary to (infinite-precision) linear systems theory, which suggests that input/output properties are independent of the state space realisation, the amount of error generated by the finite-precision implementation of filters and controllers is highly dependent on its realisation [10]. In other words, given a controller specified by its input-output behaviour, two distinct implementations of this controller can show radically different finite-precision performance. Therefore the design of the realisation of the controller *and* its transfer matrix can both influence the performance of a computer controlled system.

A second aspect of the implementation of digital filters and controllers considered in this chapter is the so-called *skewed sampling*. In a digital controller the modulation of the control sequence and the measurement samples are necessarily separated by some amount of time. This delay is, among other factors, introduced due to the time required by the controller processor to compute the next control value based on the previous measurement. Other sources of delay are the A/D and D/A (digital-to-analogue) converters. The consequence of this is that control modulation and measurement sampling appear "skewed" in time.

Despite the importance of these facts, it is unfortunate that the disciplines that take care of the control design (systems and control) and its implementation (signal processing) have been traditionally separated. By including models of the effects of finite-precision in the control design procedure, this chapter tries to reduce this gap. The provided results complement some previous efforts [8,19] by looking at the design problems in the light of recent advances in the synthesis of controllers via Linear Matrix Inequalities (LMI) [9,13].

13.2 A Model for Finite-precision Controller Design

Consider the discrete time time-invariant linear system described by the state space difference equations:

$$\begin{aligned} x(k+1) &= Ax(k) + B_w w(k) + B_u u(k) \\ z(k) &= C_z x(k) + D_{zw} w(k) + D_{zu} u(k) \\ y(k) &= C_y x(k) + D_{yw} w(k) \end{aligned} \tag{13.1}$$

where:

$x(k) \in \mathbb{R}^n$ is the state vector,
$z(k) \in \mathbb{R}^r$ is the controller output,
$y(k) \in \mathbb{R}^q$ is the measured output,
$u(k) \in \mathbb{R}^m$ is the control input,
$w(k) \in \mathbb{R}^p$ is an independent zero mean white noise source.

The zero mean white noise sequence $w(k)$ is assumed to be independent of the state and the control signal and is characterised by the covariance matrix E_w. Consider also the *dynamic linear controller* given by difference equation:

$$\begin{aligned} x_c(k+1) &= A_c x_c(k) + B_c y(k) \\ u(k) &= C_c x_c(k) + D_c y(k) \end{aligned} \tag{13.2}$$

where:

$x_c(k) \in \mathbb{R}^{n_c}$ is the controller state.

This discrete time linear system should be seen as the *ideal* control structure that the engineer wishes to implement.

A digital computer implementation of the ideal controller (13.2) typically performs the following operations: 1) quantisation and storage of the measurement signal $y(k)$; 2) summation of the products of the stored values of the controller state $x_c(k)$ and the measurements by the controller parameters; 3) storage of the results into the state and production of the output signal $u(k)$. Assuming that the controller will be implemented in a fixed-point architecture with a RBM (Roundoff Before Multiplication) model for the computations, a model for the finite-precision implementation of the ideal controller (13.2) is given by by [5,8]:

$$\begin{aligned} x_c(k+1) &= A_c \mathcal{Q}[x_c(k)] + B_c \mathcal{Q}[y(k)] \\ u(k) &= C_c \mathcal{Q}[x_c(k)] + D_c \mathcal{Q}[y(k)] \end{aligned} \tag{13.3}$$

where the operator $\mathcal{Q}[\cdot]$ denotes the occurrence of a quantisation process.

Although most modern digital filters and controller are implemented in a floating-point architecture, this chapter considers the analysis of finite-precision effects with respect to fixed-point arithmetic. The fixed-precision arithmetic structure provides simpler finite-precision models, which lead to stronger synthesis results. Furthermore, it is known that filters and controller whose structures have being optimised for implementation in a fixed-point architecture can also be optimal for some floating-point architectures [2].

The effects of quantisation of the control signal appear only in (13.1), and can be modelled by introducing the perturbed plant:

$$\begin{aligned} x(k+1) &= Ax(k) + B_w w(k) + B_u \mathcal{Q}[u(k)] \\ z(k) &= C_z x(k) + D_{zw} w(k) + D_{zu} \mathcal{Q}[u(k)] \\ y(k) &= C_y x(k) + D_{yw} w(k) \end{aligned} \tag{13.4}$$

The above models have served as a starting point for many works dealing with finite-precision design of filters and controllers [7,8,10,17,19]. Notice that (13.3) ignores any coefficient errors, that is, it assumes that the stored coefficients of the controller (13.3) are the same as the ones given in (13.2). There are several reasons why not to include coefficient errors in (13.3). The first is that coefficient errors interact with the controller input and state signals in a multiplicative way, a fact that significantly complicates the analysis. Secondly, and more importantly, the coefficient errors can be easily estimated from the designed controller coefficients *prior* to the hardware implementation of the controller. Hence, the effects of coefficient errors can be minimised by small parameter adjustments. Finally, the results of [8,18] suggest that a controller that has been designed for good performance with respect to signal quantisation also tends to be good with respect to coefficient errors.

A more tractable and convenient representation for the above systems can be obtained by defining the *quantisation errors*:

$$\begin{aligned} e_u(k) &:= \mathcal{Q}[u(k)] - u(k) \\ e_y(k) &:= \mathcal{Q}[y(k)] - y(k) \\ e_x(k) &:= \mathcal{Q}[x_c(k)] - x_c(k) \end{aligned} \tag{13.5}$$

Substituting these quantities into (13.3) and (13.4), one obtains the perturbed linear plant:

$$\begin{aligned} x(k+1) &= Ax(k) + B_w w(k) + B_u \left(u(k) + e_u(k)\right) \\ z(k) &= C_z x(k) + D_{zw} w(k) + D_{zu} \left(u(k) + e_u(k)\right) \\ y(k) &= C_y x(k) + D_{yw} w(k) \end{aligned} \tag{13.6}$$

and the perturbed linear controller:

$$\begin{aligned} x_c(k+1) &= A_c \left(x_c(k) + e_x(k)\right) + B_c \left(y(k) + e_y(k)\right) \\ u(k) &= C_c \left(x_c(k) + e_x(k)\right) + D_c \left(y(k) + e_y(k)\right) \end{aligned} \tag{13.7}$$

The advantage of considering the linear systems (13.6) and (13.7) instead of (13.3) and (13.4) is that the nonlinear character of the quantisation process is conveniently *hidden* in the perturbation signals $e_u(k)$, $e_y(k)$ and $e_x(k)$.

Defining the augmented state and noise vector:

$$\tilde{x}(k) := \begin{pmatrix} x(k) \\ x_c(k) \end{pmatrix}, \quad \tilde{w}(k) := \begin{pmatrix} w(k) \\ e_u(k) \\ e_y(k) \end{pmatrix} \tag{13.8}$$

and the augmented noise matrices:

$$B_{\tilde{w}} := \begin{bmatrix} B_w \; B_u \; \mathbf{0} \end{bmatrix}, \quad D_{y\tilde{w}} := \begin{bmatrix} D_{yw} \; \mathbf{0} \; \mathbf{I} \end{bmatrix}, \quad D_{z\tilde{w}} := \begin{bmatrix} D_{zw} \; D_{zu} \; \mathbf{0} \end{bmatrix} \tag{13.9}$$

the closed-loop connection of plant (13.6) and controller (13.7) produces the linear system:

$$\begin{aligned} \tilde{x}(k+1) &= (\mathcal{A} + \mathcal{B}\mathcal{K}\mathcal{C})\,\tilde{x}(k) + (\mathcal{G} + \mathcal{B}\mathcal{K}\mathcal{H})\,\tilde{w}(k) + \mathcal{B}\mathcal{K}\mathcal{I}e_x(k) \\ z(k) &= (\mathcal{E} + \mathcal{D}\mathcal{K}\mathcal{C})\,\tilde{x}(k) + (\mathcal{F} + \mathcal{D}\mathcal{K}\mathcal{H})\,\tilde{w}(k) + \mathcal{D}\mathcal{K}\mathcal{I}e_x(k) \end{aligned} \tag{13.10}$$

where the closed-loop matrices can be calculated as:

$$\begin{aligned} &\mathcal{A} := \begin{bmatrix} A & \mathbf{0} \\ \mathbf{0} & \mathbf{0} \end{bmatrix}, && \mathcal{B} := \begin{bmatrix} B_u & \mathbf{0} \\ \mathbf{0} & \mathbf{I} \end{bmatrix}, && \mathcal{C} := \begin{bmatrix} C_y & \mathbf{0} \\ \mathbf{0} & \mathbf{I} \end{bmatrix} \\ &\mathcal{D} := \begin{bmatrix} D_{zu} \; \mathbf{0} \end{bmatrix}, && \mathcal{E} := \begin{bmatrix} C_z \; \mathbf{0} \end{bmatrix}, && \mathcal{F} := \begin{bmatrix} D_{z\tilde{w}} \end{bmatrix} \\ &\mathcal{G} := \begin{bmatrix} B_{\tilde{w}} \\ \mathbf{0} \end{bmatrix}, && \mathcal{H} := \begin{bmatrix} D_{y\tilde{w}} \\ \mathbf{0} \end{bmatrix}, && \mathcal{I} := \begin{bmatrix} \mathbf{0} \\ \mathbf{I} \end{bmatrix} \end{aligned} \tag{13.11}$$

In (13.10) all controller coefficients have been gathered in the matrix:

$$\mathcal{K} := \begin{bmatrix} D_c & C_c \\ B_c & A_c \end{bmatrix} \tag{13.12}$$

The linear model (13.10) is subject to two distinct types of additive perturbations. This first is the combined effect of $w(k)$ and the input/output quantisation errors $e_u(k)$ and $e_y(k)$, which are represented by the augmented noise vector $\tilde{w}(k)$. The second perturbation is the state quantisation $e_x(k)$.

When the noisy inputs $e_u(k)$, $e_y(k)$ and $e_x(k)$ are set to zero, the system (13.10) reduces to the feedback connection of the linear plant (13.1) and the ideal controller (13.2), a control design structure that has been extensively studied in the literature [14,15].

13.3 The Noise Model

The achievement of an underlying linear model for finite-precision controller design was the main goal of Section 13.2. However, this simplicity hides the fact that the success of design procedures based on (13.10) should rely on reasonable characterisations of the noise sources and their connections. This section is dedicated to this issue.

A first attempt to characterise these noise sources is to consider the sequences $e_u(k)$, $e_y(k)$ and $e_x(k)$ as white noises. The adequacy of this choice has been investigated by many researchers. Assuming a fixed-point arithmetic unit, it has been shown in [16] that under *sufficient excitation conditions*, the controller quantisation error $e_x(k)$ can be modelled as a zero mean, white noise sequence independent of $x(k)$ and $w(k)$. This white noise sequence has a diagonal covariance matrix E_x whose (i,i) element is:

$$[E_x]_{(i,i)} := q_{x_i}, \quad q_{x_i} := \frac{1}{12} 2^{-2\beta_i} \tag{13.13}$$

where:

β_i is the word-length (number of bits) used to store the ith component of the state vector of the controller $x_c(k)$.

For simplicity, it is assumed that all states are stored in registers with the same word-length, that is:

$$\beta_i = \beta_x, \quad i = 1, \ldots, n_c, \quad \Rightarrow \quad E_x = q_x \mathbf{I}, \quad q_x := \frac{1}{12} 2^{-2\beta_x} \tag{13.14}$$

Following [8], it will also be assumed that $e_u(k)$ and $e_y(k)$ are zero mean, mutually independent uniform white noise sources with covariance matrices E_u and E_y given by:

$$E_u := q_u \mathbf{I}, \quad q_u := \frac{1}{12} 2^{-2\beta_u}, \qquad E_y := q_y \mathbf{I}, \quad q_y := \frac{1}{12} 2^{-2\beta_y} \tag{13.15}$$

where:

β_u is the word-length used to store the control signal and
β_y is the word-length used in the A/D converters.

The above structure implicitly assumes that all controller input channels (plant measurements) are quantised with the same precision, which is reasonable when all input channels use similar hardware (A/D converters). This assumption could be relaxed by assigning different variances to each diagonal entry of E_y. The same holds for the controller output channels (control signal). Finally, the noises $e_u(k)$ and $e_y(k)$ are assumed to be independent of $x(k)$, $w(k)$ and $e_x(k)$. Notice that a major limitation of this white noise model is that the stability of the closed loop (13.10) is entirely dictated by the stabilisability properties of the ideal controller (13.2). That is, the noise model assumes that roundoff errors can deteriorate performance but never destabilise the loop. It is important to stress that this is a consequence of the fact that the white noise sources are assumed to be independent of other loop signals. More sophisticated white noise models would indeed be able to destabilise the closed-loop system [4]. While an example where a finite-precision implementation of a ideal controller can destabilise a system can be easily generated [8], the independence hypothesis greatly simplifies the analysis of the control design problem. Moreover, although the noise characteristics do not exactly reproduce all finite-precision phenomena, at least the *noise structure* of finite-precision computation, that is, the way finite-precision issues affect the closed-loop system, seems to be precisely reproduced in (13.10).

13.4 Finite-precision Effects on Closed-loop Performance

Finite-precision effects on the closed-loop performance can be evaluated by looking at the variance of the output $z(k)$ in (13.10):

$$J := \lim_{k\to\infty} \mathcal{E}\left[z^T(k)z(k)\right] \tag{13.16}$$

where $\mathcal{E}[\cdot]$ denotes the expectation operator. Utilising the independence of the noise sources $w(k)$, $e_x(k)$, $e_y(k)$, and $e_y(k)$ the *cost function* (13.16) can be written as [14]:

$$\begin{aligned} J = \operatorname{tr}&\left[(\mathcal{E}+\mathcal{DKC})\,\mathcal{X}\,(\mathcal{E}+\mathcal{DKC})^T + (\mathcal{DKI})\,E_x\,(\mathcal{DKI})^T\right] + \\ &\operatorname{tr}\left[(\mathcal{F}+\mathcal{DKH})\,E_{\tilde{w}}\,(\mathcal{F}+\mathcal{DKH})^T\right] \end{aligned} \tag{13.17}$$

where $E_{\tilde{w}} = \operatorname{diag}[E_w, E_u, E_y]$ and $\mathcal{X} \in \mathbb{S}^{n+n_c}$ is the augmented *state covariance* (controllability Grammian) satisfying the Lyapunov equation:

$$\begin{aligned} \mathcal{X} = &(\mathcal{A}+\mathcal{BKC})\,\mathcal{X}\,(\mathcal{A}+\mathcal{BKC})^T + (\mathcal{BKI})\,E_x\,(\mathcal{BKI})^T + \\ &(\mathcal{G}+\mathcal{BKH})\,E_{\tilde{w}}\,(\mathcal{G}+\mathcal{BKH})^T \end{aligned} \tag{13.18}$$

Under the assumption that the closed loop (13.10) is asymptotically stable, matrix $\mathcal{X}$ is positive-definite, that is, $\mathcal{X} > 0$.

A useful reformulation of the above is obtained from duality. The dual of the cost function (13.16) can be written as [14]:

$$\begin{aligned} J = \operatorname{tr}&\left[E_{\tilde{w}}\{(\mathcal{G}+\mathcal{BKH})^T\mathcal{Y}\,(\mathcal{G}+\mathcal{BKH}) + (\mathcal{F}+\mathcal{DKH})^T(\mathcal{F}+\mathcal{DKH})\}\right] + \\ &\operatorname{tr}\left[E_x\{(\mathcal{BKI})^T\,\mathcal{Y}\,(\mathcal{BKI}) + (\mathcal{DKI})^T\,(\mathcal{DKI})\right] \end{aligned} \tag{13.19}$$

where $\mathcal{Y} \in \mathbb{S}^{n+n_c}$ is the observability Grammian satisfying the Lyapunov equation:

$$\mathcal{Y} = (\mathcal{A}+\mathcal{BKC})^T\,\mathcal{Y}\,(\mathcal{A}+\mathcal{BKC}) + (\mathcal{E}+\mathcal{DKC})^T\,(\mathcal{E}+\mathcal{DKC}) \tag{13.20}$$

For convenience, the cost function (13.19) can be split into the two components:

$$J = J_{\tilde{w}} + J_x \tag{13.21}$$

where:

$$J_{\tilde{w}} := \text{tr}\left[E_{\tilde{w}}\{(\mathcal{G}+\mathcal{BKH})^T\mathcal{Y}(\mathcal{G}+\mathcal{BKH})+(\mathcal{F}+\mathcal{DKH})^T(\mathcal{F}+\mathcal{DKH})\}\right]$$

$$J_x := \text{tr}\left[E_x\{(\mathcal{BKI})^T\mathcal{Y}(\mathcal{BKI})+(\mathcal{DKI})^T(\mathcal{DKI})\}\right] \tag{13.22}$$

This dual partitioning should be compared with the primal partitioning analysed in [8], which requires the evaluation of two distinct Lyapunov equalities. This simpler structure of the dual partitioning will be useful when dealing with controller synthesis.

The two components in (13.21) represent contributions to the closed-loop performance from three different sources. The first one, $J_{\tilde{w}}$, consists of the closed-loop *ideal* variance, which is associated with the closed-loop connection of (13.1) and (13.2), plus the contribution from the noise generated by quantisation of the inputs and outputs, that is, the perturbation vectors $e_u(k)$ and $e_y(k)$. It can be verified that the optimal controller that minimises the cost $J_{\tilde{w}}$ is a standard LQG controller, and design methods that can obtain such a controller are available in the literature [14]. It is also known that a similarity transformation can be applied on the LQG controller without affecting its optimality or, in other words, the optimal controller that minimises the cost component $J_{\tilde{w}}$ is coordinate independent. The second term, J_x, represents the finite-precision effects coming from quantisation at the storage of the controller state vector, represented by the perturbation $e_x(k)$. On the contrary of $J_{\tilde{w}}$, the minimisation of this term turns out to be dependent on the controller state space realisation.

The dependence of J_x on the controller realisation can be explained in the light of the the dual Lyapunov equation (13.20) (see also the discussion on Section 13.5). The particular structure exhibited by matrices $\mathcal{A}$, $\mathcal{E}$ and $\mathcal{D}$ in (13.11):

$$\mathcal{A} = \begin{bmatrix} \star & \mathbf{0} \\ \mathbf{0} & \mathbf{0} \end{bmatrix}, \quad \mathcal{E} = \begin{bmatrix} \star & \mathbf{0} \end{bmatrix}, \quad \mathcal{C} = \begin{bmatrix} \star & \mathcal{I} \end{bmatrix}$$

can be used to show that the $(2,2)$ block of the observability Grammian $\mathcal{Y}$ satisfies:

$$\mathcal{Y}_{(2,2)} = (\mathcal{BKI})^T\mathcal{Y}(\mathcal{BKI})+(\mathcal{DKI})^T(\mathcal{DKI}) \tag{13.23}$$

Comparing (13.23) with (13.22), the dual component cost J_x can be rewritten simply as:

$$J_x = \text{tr}\left[E_x\mathcal{Y}_{(2,2)}\right] \tag{13.24}$$

Hence, if the controller realisation (13.7) is subject to a similarity transformation defined by a non-singular matrix T_c in the form:

$$\tilde{\mathcal{K}} = \tilde{\mathcal{T}}_c^{-1}\mathcal{K}\tilde{\mathcal{T}}_c, \quad \tilde{\mathcal{T}}_c := \begin{bmatrix} \mathbf{I} & \mathbf{0} \\ \mathbf{0} & T_c \end{bmatrix} \tag{13.25}$$

the component cost J_x changes into:

$$J_x(T_c) = \text{tr}\left[E_x T_c^T \mathcal{Y}_{(2,2)} T_c\right] \tag{13.26}$$

while $J_{\tilde{w}}$ remains the same. Notice that, contrary to (13.22), the expression (13.26) does not directly depend on the controller gains $\mathcal{K}$, a feature that will be explored in Section 13.6. Another conclusion that can be drawn is that (13.26) is trivially minimised by making $T_c \to \epsilon\mathbf{I}$, with $\epsilon > 0$ sufficiently small. Notice that this happens at the expense of increasing the gain of the transfer function from the controller input (measured plant output) to the controller state, which is proportional to T_c^{-1}. This should be avoided, at the risk of introducing overflow into the controller state computation [5,18]. As will be discussed later, a standard way to overcome this behaviour is to introduce extra *scaling* conditions.

13.5 Optimal Controller Coordinates

As seen in the previous section, a similarity transformation T_c applied on the controller matrices is able to change the cost function J. The objective of this section is to obtain the transformation, hence the controller coordinates, which yields the minimum possible cost. For that sake, it is assumed that some controller $\mathcal{K}$, which is able to stabilise the closed-loop connection (13.10), has been previously calculated. A simple calculation shows that the effect of the similarity transformation (13.25) on this controller changes (13.17) and (13.18) into:

$$\begin{aligned} J(T_c) = \text{tr}&\left[(\mathcal{E}+\mathcal{DKC})\,\mathcal{X}\,(\mathcal{E}+\mathcal{DKC})^T + (\mathcal{DKI})\,T_c E_x T_c^T\,(\mathcal{DKI})^T\right] + \\ &\text{tr}\left[(\mathcal{F}+\mathcal{DKH})\,E_{\tilde{w}}\,(\mathcal{F}+\mathcal{DKH})^T\right] \end{aligned} \tag{13.27}$$

where $\mathcal{X} = \mathcal{X}^T$ is the solution to the Lyapunov equation:

$$\begin{aligned} \mathcal{X} = {}&(\mathcal{A}+\mathcal{BKC})\,\mathcal{X}\,(\mathcal{A}+\mathcal{BKC})^T + (\mathcal{BKI})\,T_c E_x T_c^T\,(\mathcal{BKI})^T + \\ &(\mathcal{G}+\mathcal{BKH})\,E_{\tilde{w}}\,(\mathcal{G}+\mathcal{BKH})^T \end{aligned} \tag{13.28}$$

The covariance matrix associated with the transformed controller $\tilde{\mathcal{K}}$ is related to $\mathcal{X}$ through $\tilde{\mathcal{X}} = \tilde{T}_c^{-1}\mathcal{X}\tilde{T}_c^{-T}$. As in (13.26), the cost (13.27) can be trivially minimised by making T_c arbitrarily close to zero. A popular strategy to overcome this indetermination is to look for a minimiser of (13.27) subject to the l_2-norm scaling constraint on the controller component of the transformed covariance matrix, that is, to impose the additional scaling constraint [5,18]:

$$\left[\tilde{\mathcal{X}}_{(2,2)}\right]_{(i,i)} = \left[T_c^{-1}\mathcal{X}_{(2,2)}T_c^{-T}\right]_{(i,i)} \le s, \quad i = 1,\ldots,n_c \tag{13.29}$$

where $s \in \mathbb{R}$ is a positive scalar related to the available dynamic range. This scaling constraint prevents T_c from becoming arbitrarily small. Notice that, under the assumption that the closed-loop connection (13.10) is asymptotically stable, $\mathcal{X}$ is bounded from below by the solution to the Lyapunov equation:

$$\bar{\mathcal{X}} = (\mathcal{A} + \mathcal{BKC})\,\bar{\mathcal{X}}\,(\mathcal{A} + \mathcal{BKC})^T + (\mathcal{G} + \mathcal{BKH})\,E_{\tilde{w}}\,(\mathcal{G} + \mathcal{BKH})^T \qquad (13.30)$$

that is, $\mathcal{X} \geq \bar{\mathcal{X}} \geq 0$. Conversely, notice that $J(T_c)$ is unbounded from above, even in the presence of the scaling constraints (13.29). In fact, the cost $J(T_c)$ can be made arbitrarily large still satisfying (13.29) by making T_c large. This fact indicates the existence of ***arbitrarily bad*** realisations for the controller with respect to roundoff effects. A simplified scaling constraint that is more tractable than the set of constraints (13.29) is:

$$T_c^{-1}\mathcal{X}_{(2,2)}T_c^{-T} \leq s\mathbf{I} \qquad (13.31)$$

which can be obtained from (13.29) by relaxation. It is clear that all inequalities in (13.29) hold whenever (13.31) holds. This provides the suboptimal controller realisation problem: *find T_c so as to minimise $J(T_c)$, defined in (13.27), subject to the constraints (13.28) and (13.31).* This problem can be solved with the help of the next lemma, which explores some ideas developed in [12].

Lemma 13.1. *Given a controller $\mathcal{K}$ that stabilises the closed-loop system (13.10), there exists a realisation for this controller parameterised by the similarity transformation matrix T_c such that the cost $J(T_c) < \mu$, defined in (13.27), subject to the scaling constraint (13.31) if, and only if, there exist matrices $\mathcal{X} \in \mathbb{S}^{n+n_c}$ and $Z \in \mathbb{S}^{n_c}$ such that the LMI:*

$$\begin{aligned}
&(\mathcal{A} + \mathcal{BKC})\,\mathcal{X}\,(\mathcal{A} + \mathcal{BKC})^T - \mathcal{X} + q_x\,(\mathcal{BKI})\,Z\,(\mathcal{BKI})^T + \\
&\qquad\qquad\qquad (\mathcal{G} + \mathcal{BKH})\,E_{\tilde{w}}\,(\mathcal{G} + \mathcal{BKH})^T < 0 \\
&W \geq (\mathcal{E} + \mathcal{DKC})\,\mathcal{X}\,(\mathcal{E} + \mathcal{DKC})^T + q_x\,(\mathcal{DKI})\,Z\,(\mathcal{DKI})^T \\
&\mathcal{X}_{(2,2)} \leq sZ, \quad \mathcal{X} > 0, \quad \operatorname{tr}[W] \geq \mu
\end{aligned} \qquad (13.32)$$

hold. If the LMI (13.32) have a feasible solution, such a similarity transformation matrix is given by:

$$T_c = V, \quad VV^T = Z \qquad (13.33)$$

where $V \in \mathbb{R}^{n_c \times n_c}$ is any non-singular factor of the symmetric and positive-definite matrix Z.

Proof. See Appendix 13.A.

Since μ appears affinely in the inequalities (13.32) the optimal controller realisation can be obtained by solving the following problem:

Problem 13.1. Solve

$$\min\{\mu \,:\, (13.32)\} \qquad (13.34)$$

To the authors' knowledge, there is no solution available for the optimal controller realisation problem with the original scaling constraint (13.29). However, a simplification of this problem does have a solution. The key to this is to replace the constraint (13.29) by the approximation:

$$\left[T_c^{-1}\bar{\mathcal{X}}_{(2,2)}T_c^{-T}\right]_{(i,i)} \leq s, \quad i = 1,\ldots,n_c \tag{13.35}$$

where the whole matrix $\bar{\mathcal{X}}$ (including of course its $(2,2)$ partition) does not depend on T_c and can be pre-computed from (13.30). Remembering that the dual expression for the cost $J_x(T_c)$ given in (13.26) is a function of the matrix $\mathcal{Y}_{(2,2)}$, that also does not depend on T_c and can be pre-computed from (13.20), this approximated optimal controller realisation problem can be stated as:

$$\min_{T_c}\left\{\operatorname{tr}\left[E_x T_c^T \mathcal{Y}_{(2,2)} T_c\right] : \left[T_c^{-1}\bar{\mathcal{X}}_{(2,2)}T_c^{-T}\right]_{(i,i)} = s,\ i = 1,\ldots,n_c,\right\} \tag{13.36}$$

The solution to this problem is available in the literature [7,8,10,17,19] and is given in the next lemma.

Lemma 13.2. *Given a minimal realisation of a controller $\mathcal{K}$ that stabilises the closed-loop system (13.10) the similarity transformation defined with the matrix:*

$$T_c = U_x \Sigma_x U_y \Pi V^T \tag{13.37}$$

where U_x, Σ_x, U_y and Σ_y are computed from the singular value decomposition of matrices:

$$U_x \Sigma_x^2 U_x^T = \mathcal{X}_{(2,2)}, \quad U_y \Sigma_y^2 U_y^T = \Sigma_x U_x^T \mathcal{Y}_{(2,2)} U_x \Sigma_x \tag{13.38}$$

and the diagonal matrix Π and the unitary matrix V are computed such that:

$$\Pi^{-2} = \frac{s n_c}{\operatorname{tr}\left[\Sigma_y\right]}\Sigma_y, \quad \left[V\Pi^{-2}V^T\right]_{(i,i)} = s, \quad i = 1,\ldots,n_c \tag{13.39}$$

solves problem (13.36). This controller realisation achieves the minimum cost:

$$J_x(T_e^*) = \frac{q_x}{s}\operatorname{tr}\left[\Sigma_y\right]^2 \tag{13.40}$$

Proof. See Appendix 13.B.

13.6 Optimal Controller Design

The optimal controller realisation obtained through Lemma 13.2 constitutes the first step towards a finite-precision controller design. Given a stabilising controller designed to meet some criterion, Lemmas 13.1 and 13.2 provide optimal realisations in which this controller could be implemented. When the cost to be minimised is the output variance defined in equation (13.16) this idea provides the two-step procedure outlined in the following algorithm [19].

Algorithm 13.1.

1. Design a controller to minimise the *ideal* output variance $J_{\tilde{w}}$ defined in (13.22).
2. Design an optimal finite-precision realisation for the controller obtained in Step 1 based on Lemma 13.1 or Lemma 13.2.

The main advantage of the above procedure is its simplicity. The controller in Step 1 can be obtained by solving a standard LQG control problem and solutions to the problems in Lemmas 13.1 and 13.2 can be easily computed. The main disadvantage is that there is no guarantee that it will provide a *globally* optimal finite-precision controller. In other words, the solution obtained by Algorithm 13.1 is usually suboptimal with respect to the total cost function $J = J_{\tilde{w}} + J_x$, defined in (13.16).

Unfortunately, computing this optimal controller that minimises $J = J_{\tilde{w}} + J_x$ does not seem to be an easy task. In [8], the authors derive the necessary optimality conditions with respect to the global cost function J and pursue the computation of a controller satisfying these conditions using some iterative algorithm, such as gradient or Newton's method. In the absence of strong properties on the formulation of this optimisation problem, only convergence to a local minimiser can be guaranteed. Nonetheless, the work [8] has provided some examples where the closed-loop performance can be somewhat improved. The rest of this chapter is dedicated to analysing the problem of jointly minimising the cost function $J = J_{\tilde{w}} + J_x$ in the light of recent advances in the synthesis of controllers via LMI [9,13].

Defining:

$$\mathcal{M} := \begin{bmatrix} \mathbf{0} \\ \mathbf{I} \end{bmatrix} \quad \text{so that} \quad \mathcal{Y}_{(2,2)} = \mathcal{M}^T \mathcal{Y} \mathcal{M}$$

and following the reasoning in Section 13.4, the total closed-loop cost function $J = J_{\tilde{w}} + J_x$ can be rewritten with the help of (13.23) as:

$$J = \mathrm{tr}\left[E_{\tilde{w}}\{(\mathcal{G} + \mathcal{B}\mathcal{K}\mathcal{H})^T \mathcal{Y} (\mathcal{G} + \mathcal{B}\mathcal{K}\mathcal{H}) + (\mathcal{F} + \mathcal{D}\mathcal{K}\mathcal{H})^T (\mathcal{F} + \mathcal{D}\mathcal{K}\mathcal{H})\}\right] + \mathrm{tr}\left[E_x \mathcal{M}^T \mathcal{Y} \mathcal{M}\right] \tag{13.41}$$

where $\mathcal{Y}$ is the observability Grammian (13.20). For synthesis purposes, and using standard arguments [15], the Lyapunov equation (13.20) can be relaxed into a strict Lyapunov inequality which, after applying Schur complement [15], can be equivalently restated as:

$$\begin{bmatrix} \mathcal{Y} & (\mathcal{A} + \mathcal{B}\mathcal{K}\mathcal{C})^T \mathcal{Y} & (\mathcal{E} + \mathcal{D}\mathcal{K}\mathcal{C})^T \\ \mathcal{Y}(\mathcal{A} + \mathcal{B}\mathcal{K}\mathcal{C}) & \mathcal{Y} & \mathbf{0} \\ (\mathcal{E} + \mathcal{D}\mathcal{K}\mathcal{C}) & \mathbf{0} & \mathbf{I} \end{bmatrix} > 0 \tag{13.42}$$

Defining the following partition structure for matrix $\mathcal{Y}$ and its inverse:

$$\mathcal{Y} := \begin{bmatrix} Y & V \\ V^T & \hat{Y} \end{bmatrix}, \quad \mathcal{Y}^{-1} := \begin{bmatrix} X & U^T \\ U & \hat{X} \end{bmatrix} \tag{13.43}$$

the change of variables [9,13]

$$\begin{bmatrix} \mathbf{D} & \mathbf{C} \\ \mathbf{B} & \mathbf{A} \end{bmatrix} := \begin{bmatrix} \mathbf{I} & \mathbf{0} \\ YB_u & V \end{bmatrix} \mathcal{K} \begin{bmatrix} \mathbf{I} & C_y X \\ \mathbf{0} & U \end{bmatrix} + \begin{bmatrix} \mathbf{0} \\ Y \end{bmatrix} A \, [\mathbf{0} \; X] \tag{13.44}$$

can be used to provide LMI conditions for the synthesis of controllers that guarantees some performance index (13.41) subject to the inequality constraint (13.42). This result is given in the following lemma.

Lemma 13.3. *There is a dynamic output feedback controller (13.7) of order $n_c = n$ that stabilises the closed-loop system (13.10) and guarantees that $J < \mu$, where J is the cost function (13.16), if, and only if, there exist a scalar μ and matrices $W \in \mathbb{S}^p$, $X, Y \in \mathbb{S}^n$, $\mathbf{A} \in \mathbb{R}^{n\times n}$, $\mathbf{B} \in \mathbb{R}^{n\times q}$, $\mathbf{C} \in \mathbb{R}^{m\times n}$, $\mathbf{D} \in \mathbb{R}^{m\times q}$ and a non-singular matrix $V \in \mathbb{R}^{n\times n}$ such that the LMI:*

$$\begin{bmatrix} X & (\star)^T & (\star)^T & (\star)^T & (\star)^T \\ \mathbf{I} & Y & (\star)^T & (\star)^T & (\star)^T \\ AX + B_u\mathbf{C} & A + B_u\mathbf{D}C_y & X & (\star)^T & (\star)^T \\ \mathbf{A} & YA + \mathbf{B}C_y & \mathbf{I} & Y & (\star)^T \\ C_zX + D_{zu}\mathbf{C} & C_z + D_{zu}\mathbf{D}C_y & \mathbf{0} & \mathbf{0} & \mathbf{I} \end{bmatrix} > 0$$

$$\begin{bmatrix} W_{\tilde{w}} & (\star)^T & (\star)^T & (\star)^T \\ B_{\tilde{w}} + B_u\mathbf{D}D_{y\tilde{w}} & X & (\star)^T & (\star)^T \\ YB_{\tilde{w}} + \mathbf{B}D_{y\tilde{w}} & \mathbf{I} & Y & (\star)^T \\ D_{z\tilde{w}} + D_{zu}\mathbf{D}D_{y\tilde{w}} & \mathbf{0} & \mathbf{0} & \mathbf{I} \end{bmatrix} \geq 0, \quad \begin{bmatrix} W_x & (\star)^T & (\star)^T \\ \mathbf{0} & X & (\star)^T \\ V & \mathbf{I} & Y \end{bmatrix} \geq 0$$

$$\mu_{\tilde{w}} \geq \operatorname{tr}[E_{\tilde{w}}W_{\tilde{w}}], \quad \mu_x \geq \operatorname{tr}[E_xW_x], \quad \mu \geq \mu_x + \mu_{\tilde{w}} \tag{13.45}$$

hold. If the LMI (13.45) have a feasible solution, such a controller is given by:

$$\mathcal{K} = \begin{bmatrix} \mathbf{I} & \mathbf{0} \\ -V^{-1}YB_u & V^{-1} \end{bmatrix} \begin{bmatrix} \mathbf{D} & \mathbf{C} \\ \mathbf{B} & \mathbf{A} - YAX \end{bmatrix} \begin{bmatrix} \mathbf{I} & -C_yXU^{-1} \\ \mathbf{0} & U^{-1} \end{bmatrix} \tag{13.46}$$

where $U = V^{-1}(\mathbf{I} - YX)$.

Proof. See Appendix 13.C.

In the above lemma, the compact notation $(\star)^T$ indicates a block that is the transpose of the corresponding symmetric block in the same matrix. The characterisation of this set as an LMI means that the set of feasible controllers is convex and that a controller satisfying the given conditions can be computed in polynomial time [3]. A major difference between Lemma 13.3

and other results that use the change-of-variables [9,13] is the fact that the LMI involve the matrix V as an optimisation variable. The practical consequence of this fact is that the controller realisation *cannot be arbitrarily chosen*, that is, given matrices that solve the LMI in the problem, the controller realisation (13.46) is the only one which is guaranteed to satisfy the design objectives. This is in accordance with the previous observations that the optimality of a finite-precision controller should be realisation dependent.

Although Lemma 13.3 characterises the set of feasible controllers, it cannot be used in the present form for optimal synthesis. That is because the third inequality in (13.45) implies that optimality is achieved with $V \to \epsilon \mathbf{I}$, for an arbitrarily small $\epsilon > 0$. Indeed, the Schur complement of (13.45) produces:

$$W_x > V^T \left(Y - X^{-1}\right)^{-1} V$$

and, since V does not appear in any other inequality, the component cost μ_x can be made arbitrarily close to zero by setting:

$$V \leftarrow \epsilon \mathbf{I} \Rightarrow W_x > \epsilon^2 \left(Y - X^{-1}\right)^{-1} \Rightarrow \mu_x > \epsilon^2 \operatorname{tr}\left[\left(Y - X^{-1}\right)^{-1}\right]$$

As in Section 13.5, this behaviour asks for the introduction of scaling conditions. Unfortunately, since the inequalities obtained in Lemma 13.3 have been obtained from a dual formulation, the standard scaling conditions used in Section 13.5, which impose constraints on the covariance of the controller, cannot be used here. Notice that only the partitions of matrix $\mathcal{Y}$ (and its inverse) are available as optimisation variables. The covariance matrix $\mathcal{X}$ is not explicitly computed in Lemma 13.3.

It is also hard to incorporate "dual scalings" on $\hat{Y}$ without destroying convexity of the inequalities in Lemma 13.3[1]. Alternatively, consider the scaling condition:

$$\hat{X} \leq s\mathbf{I} \tag{13.47}$$

which is related to the relaxation of the l_2-norm scaling constraint (13.31) considered in the previous section, this time applied to the block $\hat{X}$. Following the lines of the proof of Lemma 13.3, given in Appendix 13.C, one can show that the inequality (13.47) can be transformed[2] into the following LMI:

$$\begin{bmatrix} s\mathbf{I} & (\star)^T & (\star)^T \\ U^T & X & (\star)^T \\ \mathbf{0} & \mathbf{I} & Y \end{bmatrix} \geq 0$$

[1] Notice that the cost function penalises $\hat{Y}$ and hence a constraint in the form $\hat{Y} \leq s\mathbf{I}$ does not prevent V from approaching zero. Since $\mathcal{Y} > 0$, the constraint $s\mathbf{I} \geq \hat{Y} > V^T Y V$ is in fact easily satisfied by making V small. Instead, the non-convex (in the variables of Lemma 13.3) constraint $\hat{Y} \geq s\mathbf{I}$ could be considered!

[2] This LMI comes by applying Schur complement on $\hat{X} = \mathcal{M}^T \mathcal{Y}^{-1} \mathcal{M} = \mathcal{M}^T \mathcal{T} \left(\mathcal{T}^T \mathcal{Y} \mathcal{T}\right)^{-1} \mathcal{T}^T \mathcal{M} \leq s\mathbf{I}$.

which translates (13.47) as a function of the optimisation variables considered in Lemma 13.3. Although the above inequality is an LMI in the optimisation variables, it cannot be directly incorporated into (13.45). The reason for that is that X, Y, V and U are not truly independent variables: they have to be determined in order to satisfy the partitioning defined in (13.43). This implies that these variables must additionally satisfy the algebraic relation

$$YX + VU = \mathbf{I} \tag{13.48}$$

Notice that they can be indeed treated as independent variables as long as not all of them appear in the inequalities. For instance, given X, Y, and V satisfying (13.45), there always exists a matrix U (and also $\hat{X}$ and $\hat{Y}$) such that (13.48) holds. The situation changes when all four should be determined simultaneously, and (13.48) can no longer be ignored. Equation (13.48) cannot be added to (13.45) in Lemma 13.3 without destroying convexity.

These difficulties to deal with scalings on $\hat{Y}$ and $\hat{X}$ motivate the search for an alternative scaling procedure. From (13.26) and (13.29), the J_x component of the cost can be properly scaled by either scaling the controller Grammians $\mathcal{Y}_{(2,2)}$ and $\mathcal{X}_{(2,2)}$, or by scaling the similarity transformation T_c. The former seems to be the preferred choice in the literature. However, in the context of Lemma 13.3, an application of a similarity transformation defined with T_c on the controller (13.46) produces matrices that are given exactly by (13.46), provided V is replaced by VT_c. In other words, matrix V is able to generate all similar controller realisations, playing exactly the same role played by the similarity transformation T_c throughout Section 13.5. In this sense, scaling V can be seen as conceptually equivalent to scaling T_c. This interrelationship between T_c and V can be explored to provide an alternative scaling condition. One of the simplest scaling procedures available is to hold V constant at a given value and solve the inequalities (13.45). A choice of V that tries to compromise this form of scaling with the Grammian scaling is the one given in the following algorithm.

Algorithm 13.2.

1. Design a controller to minimise the ***ideal*** output variance $J_{\bar{w}}$ defined in (13.22).
2. Design an optimal finite-precision realisation for the controller obtained in Step 1 based on Lemma 13.1 or Lemma 13.2.
3. For the controller obtained in the previous item, compute matrix $\mathcal{Y}$ and extract matrix $\bar{V} = V$ from the partition of $\mathcal{Y}$ defined in (13.43). If $\bar{V}$ is singular, apply a small perturbation so as to make $\bar{V}$ non-singular[3].
4. Minimise μ subject to the inequalities (13.45) with $V = \bar{V}$ held constant.

---

[3] Notice that this can be done without perturbing the cost function [13].

The rationale behind this choice is that the cost function will necessarily decrease from the one obtained by Algorithm 13.1. More sophisticated scaling conditions on V with interesting system interpretation might also be available. This issue is currently under investigation. The main advantage of Algorithm 13.2 over [8] is that its solution can be completely computed using standard LMI solvers. The improvement in the cost function is obtained without having to iteratively compute a controller satisfying hard to compute necessary optimality conditions.

13.7 Skewed Sampling

In a digital controller, the computation of the control law depends on a sample of the measured output. After this sample is obtained, a finite amount of processing time is required to calculate the amplitude of the control signal. Consequently output sampling and control modulation occurs in distinct instants of time, that is, they are *skewed*, as illustrated in Figure 13.1. In this figure, the *skew time* is denoted by $\delta > 0$, which is chosen to be equal to or greater than all computational or transport delays. The interval δ accounts for the time elapsed between the control modulation and the next output sampling. Equivalently, $(T - \delta)$ is the time required for the computation of the control law. The skew time δ can also accommodate other sources of delay as, for instance, the conversion time of the A/D and D/A converters.

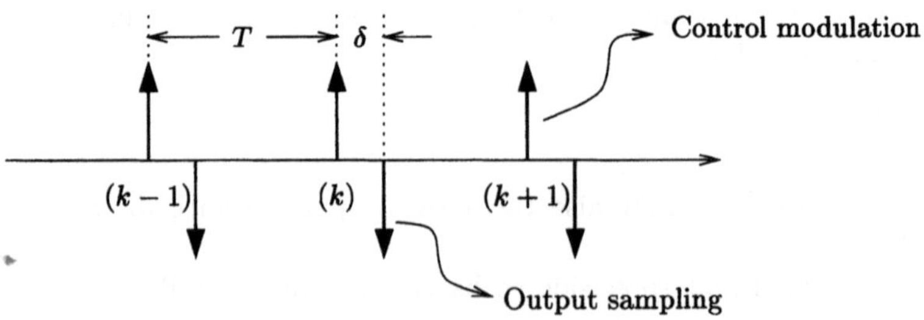

Fig. 13.1. Skewed Sampling.

When the skew time is such that $\delta = \kappa T$, $\kappa \in \mathbb{N}$, that is, when δ is a positive integer multiple of the sampling period, a standard discrete time plant (13.1) augmented with the necessary delay elements can be used to describe the behaviour of the discrete time system. In the more interesting situation[4] when $0 < \delta < T$, the following modified plant:

$$\begin{aligned} x(k+1) &= Ax(k) + B_w w(k) + B_u\left(u(k) + e_u(k)\right) \\ z(k) &= C_z x(k) + D_{zw} w(k) + D_{zu}\left(u(k) + e_u(k)\right) \\ y(k+\delta/T) &= C_y x(k) + D_{yw} w(k) + D_{yu}\left(u(k) + e_u(k)\right) \end{aligned} \tag{13.49}$$

should be considered. Notice that when the above discrete time model is obtained by discretisation of a continuous time plant assuming that the inputs are held constant between consecutive samples, then the matrices C_y, and D_{yu} can be computed[5] as a function of the continuous time process matrices[6] (F, G, H) and the skew time δ as:

$$C_y = He^{F\delta}, \quad D_{yu} = H\int_0^\delta e^{F\tau} G d\tau \tag{13.50}$$

The presence of skewed sampling also calls for a modified controller structure:

$$\begin{aligned} x_c(k+1) &= A_c\left(x_c(k) + e_x(k)\right) + B_c\left(y(k+\delta/T) + e_y(k+\delta/T)\right) \\ u(k+1) &= C_c\left(x_c(k) + e_x(k)\right) + D_c\left(y(k+\delta/T) + e_y(k+\delta/T)\right) \end{aligned} \tag{13.51}$$

which reflects the delayed use of the measurements $y(k+\delta/T)$ at instant $k+1$. Notice that both plant and controller (13.49) and (13.51) already contain the appropriate noise sources that take care of roundoff effects.

The main goal of this section is to show that with the introduction of the auxiliary control signal $\tilde{u}(k) := u(k+1)$, the design of finite-precision controllers with skewed sampling can be be handled by the same techniques discussed so far. Indeed, incorporating the signal $u(k)$ in the state vector and considering the augmented noise vector and its associated matrices defined in (13.8) and (13.9) with $D_{y\tilde{w}}$ replaced by:

$$D_{y\tilde{w}} := \left[D_{yw}\ D_{yu}\ \mathbf{0}\right] \tag{13.52}$$

the plant (13.49) can be rewritten as

$$\begin{aligned} \begin{pmatrix} x(k+1) \\ u(k+1) \end{pmatrix} &= \begin{bmatrix} A & B_u \\ \mathbf{0} & \mathbf{0} \end{bmatrix} \begin{pmatrix} x(k) \\ u(k) \end{pmatrix} + \begin{bmatrix} B_{\tilde{w}} \\ \mathbf{0} \end{bmatrix} \tilde{w}(k) + \begin{bmatrix} \mathbf{0} \\ \mathbf{I} \end{bmatrix} \tilde{u}(k) \\ z(k) &= \left[C_z\ D_{zu}\right] \begin{pmatrix} x(k) \\ u(k) \end{pmatrix} + D_{z\tilde{w}} \tilde{w}(k) \\ y(k+\delta/T) &= \left[C_y\ D_{yu}\right] \begin{pmatrix} x(k) \\ u(k) \end{pmatrix} + D_{y\tilde{w}} \tilde{w}(k) \end{aligned} \tag{13.53}$$

---

[4] A skew time δ greater than but not a positive integer multiple of the period T can be considered by decomposing δ as $\delta = \kappa T + \delta_T$, $\kappa \in \mathbb{N}$, $0 < \delta_T < T$. In this case, κ unity delays are introduced in the plant and δ_T serves as a *normalised* skew time.

[5] See [18] for detail on discretisation of noise processes.

[6] $\dot{x}(t) = Fx(t) + Gu(t), \quad y(t) = Hx(t)$.

and the controller (13.51) as:

$$\begin{aligned} x_c(k+1) &= A_c\,(x_c(k) + e_x(k)) + B_c\,(y(k+\delta/T) + e_y(k+\delta/T)) \\ \tilde{u}(k) &= C_c\,(x_c(k) + e_x(k)) + D_c\,(y(k+\delta/T) + e_y(k+\delta/T)) \end{aligned} \quad (13.54)$$

Plant (13.53) and controller (13.54) are now very similar to the standard configuration (13.6)-(13.7). In fact, the connection of (13.53) and (13.54) produces a closed-loop system in the form (13.10) where:

$$\begin{aligned} &\mathcal{A} := \left[\begin{array}{cc|c} A & B_u & 0 \\ \mathbf{0} & \mathbf{0} & \mathbf{0} \\ \hline \mathbf{0} & \mathbf{0} & \mathbf{0} \end{array}\right], && \mathcal{B} := \left[\begin{array}{c|c} \mathbf{0} & \mathbf{0} \\ \mathbf{I} & \mathbf{0} \\ \hline \mathbf{0} & \mathbf{I} \end{array}\right], && \mathcal{C} := \left[\begin{array}{cc|c} C_y & D_{yu} & \mathbf{0} \\ \mathbf{0} & \mathbf{0} & \mathbf{0} \\ \hline \mathbf{0} & \mathbf{0} & \mathbf{I} \end{array}\right] \\ &\mathcal{D} := \left[\begin{array}{cc|c} \mathbf{0} & \mathbf{0} & \mathbf{0} \end{array}\right], && \mathcal{E} := \left[\begin{array}{cc|c} C_z & D_{zu} & \mathbf{0} \end{array}\right], && \mathcal{F} := \left[D_{z\tilde{w}} \right] \\ &\mathcal{G} := \left[\begin{array}{c} B_{\tilde{w}} \\ \mathbf{0} \\ \hline \mathbf{0} \end{array}\right], && \mathcal{H} := \left[\begin{array}{c} D_{y\tilde{w}} \\ \mathbf{0} \\ \hline \mathbf{0} \end{array}\right], && \mathcal{I} := \left[\begin{array}{c} \mathbf{0} \\ \mathbf{0} \\ \hline \mathbf{I} \end{array}\right] \end{aligned} \quad (13.55)$$

The techniques for optimal realisation and controller designed discussed in the previous sections can be modified with not much effort to cope with the new definitions given in (13.55).

13.8 A Numerical Example

A very simple numerical example is given in this section to illustrate the design procedures discussed throughout the text. The plant is the read/write head of a disk drive that have been considered as a demo (`diskdemo.m`) for the MATLAB® Control Toolbox. It consists of the second-order plant:

$$J\ddot{\theta} + C\dot{\theta} + K\theta = K_i i$$

where

θ is the (angular) position of the head in radians,
i is input current in Amperes,
$J = 0.01$ Kg m^2 is the inertia of the head assembly,
$C = 0.004$ Nm/(rad s^{-1}) is the damping coefficient of the bearings,
$K = 10$ Nm/rad is the return spring constant,
$K_i = 0.5$ Nm/A is the motor torque constant.

This plant has been discretised assuming a sample period of $5 \times 10^{-3}s$ and assuming a zero-order hold on the current input. The complete model considered for design (including the controlled output and the noise inputs) is given by (13.1) with matrices:

$$A = \begin{bmatrix} 1.9731 & -0.9980 \\ 1 & 0 \end{bmatrix}, \qquad B_w = \begin{bmatrix} \mathbf{I} \ \mathbf{0} \end{bmatrix}, \qquad B_u = \begin{bmatrix} 1.5625 \times 10^{-2} \\ 0 \end{bmatrix}$$

$$C_z = \begin{bmatrix} C_y \\ \mathbf{0} \end{bmatrix}, \qquad D_{zw} = \mathbf{0}, \qquad D_{zu} = \sqrt{0.01} \begin{bmatrix} \mathbf{0} \\ \mathbf{I} \end{bmatrix}$$

$$C_y = 10^{-3} \times \begin{bmatrix} 3.9890 \ 3.9864 \end{bmatrix}, \quad D_{yw} = \begin{bmatrix} 0 \ 0 \ 10 \end{bmatrix}$$

Controllers based on the four design methods below have been calculated.

Design I: Compute the controller that minimises $J_{\tilde{w}}$ followed by a diagonal scaling,
Design II: Run Algorithm 13.1 using Lemma 13.1,
Design III: Run Algorithm 13.1 using Lemma 13.2,
Design IV: Run Algorithm 13.2 with $\bar{V}$ computed from Lemma 13.2.

The word-length of the registers used to store the controller state, the control input and the control output have been assumed to be the same ($\beta_x = \beta_u = \beta_y = \beta$). A controller has been designed by each design method for values of β varying from 4 to 16. Table 13.1 compares the variance of the output z achieved by these different design for the different choices of word-length.

Table 13.1. Design Comparison

Wordlength	Design I	Design II	Design III	Design IV
16	8.3102×10^{-1}	8.3102×10^{-1}	8.3102×10^{-1}	8.3102×10^{-1}
12	8.3103×10^{-1}	8.3103×10^{-1}	8.3102×10^{-1}	8.3102×10^{-1}
8	8.3178×10^{-1}	8.3104×10^{-1}	8.3104×10^{-1}	8.3104×10^{-1}
6	8.4310×10^{-1}	8.3133×10^{-1}	8.3133×10^{-1}	8.3133×10^{-1}
4	10.2419×10^{-1}	8.3599×10^{-1}	8.3587×10^{-1}	8.3585×10^{-1}

Notice that in this simple example, the controllers produced by Algorithm 13.1 and Algorithm 13.2 practically attain the same performance level. A small improvement in the performance of the controller computed with Algorithm 13.2 can be seen for $\beta = 4$. It is expected that this improvement shows up at larger values of β for more complex plants. Notice also the significant degradation in performance from the (almost) ideal cost $J = 0.83$ computed with $\beta = 16$ to $J = 1.02$ with $\beta = 4$ on the simply scaled Design I, which represents 23% of the ideal cost. This should be compared with the degradation of only 0.6% achieved by the Design II from $\beta = 16$ to $\beta = 4$.

Acknowledgements

Maurício C. de Oliveira is supported by a grant from FAPESP, "Fundação de Amparo à Pesquisa do Estado de São Paulo", Brazil.

References

1. G. Amit and U. Shaked. Minimization of roundoff errors in digital realizations of Kalman filters. *IEEE Transactions on Acoustics, Speech and Signal Processing*, 37(12):1980–1982, 1989.
2. B. W. Bomar, L. M. Smith, and R. D. Joseph. Roundoff noise analysis of state-space digital filters implemented on floating-point digital signal processors. *IEEE Transactions on Circuits and Systems – II: analog and digital signal processing*, 44(11):952–955, 1997.
3. S. P. Boyd, L. El Ghaoui, E. Feron, and V. Balakrishnan. *Linear Matrix Inequalities in System and Control Theory.* SIAM, Philadelphia, PA, 1994.
4. M. C. de Oliveira and R. E. Skelton. State feedback control of systems with finite signal-to-noise models. In *Proceedings of the 39th Proceedings of the IEEE Conference on Decision and Control*, Sydney, Australia, 2000.
5. M. Gevers and G. Li. *Parametrizations in Control, Estimation and Filtering Problems.* Springer-Verlag, London, UK, 1993.
6. R. A. Horn and C. R. Johnson. *Matrix Analysis.* Cambridge University Press, Cambridge, UK, 1985.
7. S. Hwang. Minimum uncorrelated unit noise in state-space digital filtering. *IEEE Transactions on Acoustics, Speech and Signal Processing*, 25(8):273–281, 1977.
8. K. Liu, R. E. Skelton, and K. Grigoriadis. Optimal controllers for finite wordlength implementation. *IEEE Transactions on Automatic Control*, 37(9):1294–1304, 1992.
9. I. Masubuchi, A. Ohara, and N. Suda. LMI-based controller synthesis: A unified formulation and solution. *International Journal of Robust and Nonlinear Control*, 8(8):669–686, 1998.
10. C. Mullis and R. Roberts. Synthesis of minimun round-off noise fixed-point digital filters. *IEEE Transactions on Circuits and Systems*, 23(9):551–562, 1976.
11. R. A. Roberts and C. T. Mullis. *Digital Signal Processing.* Addison-Wesley, Reading, MA, 1987.
12. M. A. Rotea and D. Williamson. Optimal realizations of finite wordlength digital filters and controllers. *IEEE Transactions on Circuits and Systems – I: fundamental theory and applications*, 42(2):61–72, 1995.
13. C. W. Scherer, P. Gahinet, and M. Chilali. Multiobjective output-feedback control via LMI optimization. *IEEE Transactions on Automatic Control*, 42(7):896–911, 1997.
14. R. E. Skelton. *Dynamics Systems Control: linear systems analysis and synthesis.* John Wiley & Sons, Inc, New York, NY, 1988.
15. R. E. Skelton, T. Iwasaki, and K. Grigoriadis. *A Unified Algebraic Approach to Control Design.* Taylor & Francis, London, UK, 1997.
16. A. Sripad and D. Snyder. A necessary and sufficient condition for quantization error to be uniform and white. *IEEE Transactions on Acoustics, Speech and Signal Processing*, 5(10):442–448, 1977.
17. D. Williamson. Finite wordlength design of digital Kalman filters for state estimation. *IEEE Transactions on Automatic Control*, 30(10):30–39, 1985.
18. D. Williamson. *Digital Control and Implementation: a signal processing viewpoint.* Prentice Hall, Sydney, AUS, 1991.
19. D. Williamson and K. Kadiman. Optimal finite wordlength linear quadratic regulators. *IEEE Transactions on Automatic Control*, 34(12):1218–1228, 1989.

13.A Proof of Lemma 13.1

Use (13.14) and relax the equality in (13.28) so as to obtain the Lyapunov inequality:

$$(\mathcal{A}+\mathcal{BKC})\,\mathcal{X}\,(\mathcal{A}+\mathcal{BKC})^T - \mathcal{X} + q_x\,(\mathcal{BKI})\,T_cT_c^T\,(\mathcal{BKI})^T + \\ (\mathcal{G}+\mathcal{BKH})\,E_{\tilde{w}}\,(\mathcal{G}+\mathcal{BKH})^T < 0$$

It is a standard result that any solution to the above inequality serves as an upper bound to the solution to the equality [15]. Multiply the scaling constraint (13.31) on the left by T_c and on the right by T_c^T to get:

$$\mathcal{X}_{(2,2)} \leq sT_cT_c^T$$

Notice that the stability of the closed-loop ensures that $\mathcal{X} > 0$, which implies that $T_cT_c^T > 0$. Implicitly define the variable W such that:

$$W \geq (\mathcal{E}+\mathcal{DKC})\,\mathcal{X}\,(\mathcal{E}+\mathcal{DKC})^T + q_x\,(\mathcal{DKI})\,T_cT_c^T\,(\mathcal{DKI})^T$$

and $\mu \geq \operatorname{tr}[W] > J_x(T_c)$. Finally, since $T_cT_c^T$ is positive-definite, the one-to-one change-of-variable:

$$Z := T_cT_c^T$$

provides the equivalence with the inequalities (13.32) and guarantees the existence of the factorisation (13.33). □

13.B Proof of Lemma 13.2

The idea is to use the non-singular similarity transformation matrix $T_1 := U_x\Sigma_xU_y$, where U_x, U_y and Σ_x are computed from (13.38), which is able to simultaneously diagonalise the matrices appearing in problem (13.32), bringing them to the form:

$$T_1^T\mathcal{Y}_{(2,2)}T_1 = \Sigma_y^2, \quad T_1^{-1}\bar{\mathcal{X}}_{(2,2)}T_1^{-T} = \mathbf{I} \tag{13.56}$$

From (13.56) a second similarity transformation defined by matrix $T_2 := U\Pi V^T$, which appears here factored in the unitary matrices U and V and diagonal $\Pi > 0$, can then be used to define the total coordinate change:

$$T_c := T_1T_2 = U_x\Sigma_xU_yU\Pi V^T \tag{13.57}$$

The substitution of (13.57) into (13.36) provides the equivalent optimisation problem:

$$\min_{U,V,\Pi}\left\{n_cq_x\operatorname{tr}\left[\Pi^2U^T\Sigma_y^2U\right] : \left[V\Pi^{-2}V^T\right]_{(i,i)} \leq s,\, i=1,\ldots,n_c,\right\} \tag{13.58}$$

The following lemma provide the key to the solution of this problem.

Lemma 13.4. *Given $M \in \mathbb{R}^{n\times n}$, $M > 0$ and diagonal, there exists a unitary matrix $V \in \mathbb{R}^{n\times n}$ such that $\left[VMV^T\right]_{(i,i)} = s$, $i = 1,\ldots,n$, if, and only if,* $\mathrm{tr}\,[M] = sn$.

Proof. See [7] for a proof. This work also provides a constructive algorithm based on Givens' rotations to obtain such unitary matrix V. □

Using Lemma 13.4, problem (13.58) can be restated as

$$\min_{U,V,\Pi} \left\{n_c q_x \,\mathrm{tr}\left[\Pi^2 U^T \Sigma_y^2 U\right] : \mathrm{tr}\left[\Pi^{-2}\right] = s n_c\right\} \tag{13.59}$$

The minimiser of (13.59) can be obtained from Cauchy-Schwartz inequality.

Lemma 13.5 (Cauchy-Schwartz). *For any two given matrices $A, B \in \mathbb{R}^{n\times n}$ it is true that* $\mathrm{tr}\left[A^T A\right]\mathrm{tr}\left[B^T B\right] \geq |\,\mathrm{tr}\left[A^T B\right]|^2$ *where equality holds if, and only if, $A = \mu B$ for some $\mu \in \mathbb{R}$.*

Proof. See [6] page 261 for $\langle x, y\rangle := \mathrm{tr}\left[x^T y\right]$. □

Set $A = \Pi^{-1}$, $B = \Pi U^T \Sigma_y$ in the Cauchy-Schwartz inequality to obtain:

$$\mathrm{tr}\left[\Pi^2 U^T \Sigma_y^2 U\right]\mathrm{tr}\left[\Pi^{-2}\right] = s n_c \,\mathrm{tr}\left[\Pi^2 U^T \Sigma_y^2 U\right] \geq |\,\mathrm{tr}\left[U^T \Sigma_y\right]|^2$$

Optimality (lower bound of the Cauchy-Schwartz inequality) is attained when $\Pi^{-1} = \mu \Pi U^T \Sigma_y$ or, equivalently, when:

$$\Pi^{-2} = \mu U^T \Sigma_y \quad \Rightarrow \quad U = \mathbf{I} \text{ and } \mu = \frac{\mathrm{tr}\left[\Pi^{-2}\right]}{\mathrm{tr}\left[\Sigma_y\right]} = \frac{s n_c}{\mathrm{tr}\left[\Sigma_y\right]}$$

This completes the proof. □

13.C Proof of Lemma 13.3

Define the transformation matrix:

$$\mathcal{T} = \begin{bmatrix} X & \mathbf{I} \\ U & \mathbf{0} \end{bmatrix}, \quad \mathcal{Y}\mathcal{T} := \begin{bmatrix} \mathbf{I} & Y \\ \mathbf{0} & V^T \end{bmatrix}$$

Following [13,9], matrix V can be assumed to be non-singular without loss of generality so that the transformation matrix $\mathcal{T}$ is non-singular. Apply a congruence transformation on (13.42) by multiplying it on the right by $T = \mathrm{diag}[\mathcal{T}, \mathcal{T}, \mathbf{I}]$ and on the left by T^T so as to obtain the equivalent inequality:

$$\begin{bmatrix} \mathcal{T}^T \mathcal{Y} \mathcal{T} & \mathcal{T}^T (\mathcal{A} + \mathcal{B}\mathcal{K}\mathcal{C})^T \mathcal{Y}\mathcal{T} & \mathcal{T}^T (\mathcal{E} + \mathcal{D}\mathcal{K}\mathcal{C})^T \\ \mathcal{T}^T \mathcal{Y} (\mathcal{A} + \mathcal{B}\mathcal{K}\mathcal{C})\mathcal{T} & \mathcal{T}^T \mathcal{Y} \mathcal{T} & \mathbf{0} \\ (\mathcal{E} + \mathcal{D}\mathcal{K}\mathcal{C})\mathcal{T} & \mathbf{0} & \mathbf{I} \end{bmatrix} > 0$$

Implicitly define matrices $W_{\tilde{w}} \in \mathbb{S}^{p+m+q}$, $W_x \in \mathbb{S}^n$ such that:

$$\begin{aligned} W_{\tilde{w}} &\geq (\mathcal{G} + \mathcal{BKH})^T \mathcal{Y} (\mathcal{G} + \mathcal{BKH}) + (\mathcal{F} + \mathcal{DKH})^T (\mathcal{F} + \mathcal{DKH}) \\ W_x &\geq \mathcal{M}^T \mathcal{Y} \mathcal{M} \end{aligned}$$

An application of a Schur complement on these provides the equivalent inequalities:

$$\begin{bmatrix} W_{\tilde{w}} & (\mathcal{G} + \mathcal{BKH})^T \mathcal{Y} & (\mathcal{F} + \mathcal{DKH})^T \\ \mathcal{Y}(\mathcal{G} + \mathcal{BKH}) & \mathcal{Y} & \mathbf{0} \\ (\mathcal{F} + \mathcal{DKH}) & \mathbf{0} & \mathbf{I} \end{bmatrix} \geq 0, \quad \begin{bmatrix} W_x & \mathcal{M}^T \mathcal{Y} \\ \mathcal{Y}\mathcal{M} & \mathcal{Y} \end{bmatrix} \geq 0$$

Apply the congruence transformation $T = \text{diag}[\mathbf{I}, \mathcal{T}, \mathbf{I}]$ on the first two inequalities above and the congruence transformation $T = \text{diag}[\mathbf{I}, \mathcal{T}]$ on the third inequality so as to obtain:

$$\begin{bmatrix} W_{\tilde{w}} & (\mathcal{G} + \mathcal{BKH})^T \mathcal{Y}\mathcal{T} & (\mathcal{F} + \mathcal{DKH})^T \\ \mathcal{T}^T\mathcal{Y}(\mathcal{G} + \mathcal{BKH}) & \mathcal{T}^T\mathcal{Y}\mathcal{T} & \mathbf{0} \\ (\mathcal{F} + \mathcal{DKH}) & \mathbf{0} & \mathbf{I} \end{bmatrix} \geq 0$$

$$\begin{bmatrix} W_x & \mathcal{M}^T \mathcal{Y}\mathcal{T} \\ \mathcal{T}^T\mathcal{Y}\mathcal{M} & \mathcal{T}^T\mathcal{Y}\mathcal{T} \end{bmatrix} \geq 0$$

Finally, the change of variable (13.44) and the following identities:

$$\begin{aligned} \mathcal{T}^T\mathcal{Y}(\mathcal{A} + \mathcal{BKC})\mathcal{T} &= \begin{bmatrix} AX + B_u\mathbf{C} & A + B_u\mathbf{D}C_y \\ \mathbf{A} & YA + \mathbf{B}C_y \end{bmatrix}, \quad \mathcal{T}^T\mathcal{Y}\mathcal{T} = \begin{bmatrix} X & \mathbf{I} \\ \mathbf{I} & Y \end{bmatrix} \\ (\mathcal{E} + \mathcal{DKC})\mathcal{T} &= \begin{bmatrix} C_zX + D_{zu}\mathbf{C} & C_z + D_{zu}\mathbf{D}C_y \end{bmatrix} \\ \mathcal{T}^T\mathcal{Y}(\mathcal{G} + \mathcal{BKH}) &= \begin{bmatrix} B_{\tilde{w}} + B_u\mathbf{D}D_{y\tilde{w}} \\ YB_{\tilde{w}} + \mathbf{B}D_{y\tilde{w}} \end{bmatrix}, \quad \mathcal{T}^T\mathcal{Y}\mathcal{M} = \begin{bmatrix} \mathbf{0} \\ V \end{bmatrix} \end{aligned}$$

recover all inequalities in Lemma 13.3.

The controller matrices (13.46) are recovered by inverting the change-of-variables (13.44). Notice that due to the particular partitioning (13.43) the matrix U is determined by solving the identity (see also discussion in Section 13.6):

$$YX + VU = \mathbf{I} \Rightarrow U = V^{-1}(\mathbf{I} - YX)$$

This completes the proof of Lemma 13.3. □

CHAPTER 14

QUANTISATION ERRORS IN DIGITAL IMPLEMENTATIONS OF FUZZY CONTROLLERS

Inés del Campo, José M. Tarela, and Koldo Basterretxea

Abstract. Fuzzy Logic Controllers (FLCs) have proven useful in the control of complex and nonlinear processes. Unlike conventional control, which is based on a precise model of a process, fuzzy control is able to handle linguistic information in the form of IF–THEN rules. These rules usually encapsulate the experience of human operators and engineers. At present, most FLCs are implemented digitally. Microprocessors, digital signal processors (DSPs), and application specific integrated circuits (ASICs) are used to cope with real time fuzzy control. Therefore, the quantisation noise due to the finite length of digital words is to be taken into account in designing fuzzy systems. Digital implementations of FLCs involve three main types of quantisation errors: the analogue-to-digital (A/D) errors, the membership function errors, and the arithmetic errors. The consequences of these errors on the behaviour of a typical FLC are analysed and the problem of the selection of a digital format for fuzzy information is addressed.

14.1 Introduction

Fuzzy logic control has been successfully applied to a wide variety of industrial applications and consumer products [1–7]. The main feature of fuzzy control techniques is their ability to handle imprecise information. Unlike conventional control, which usually requires a precise model of a process, fuzzy control is able to perform the automatic control of a process on the basis of linguistic IF–THEN rules. These rules make use of imprecise or uncertain information generally derived from the experience of human operators, engineers, and designers.

Nowadays, most of the FLCs are implemented on standard microprocessors, microcontrollers, or digital signals processors (DSPs). Furthermore, a variety of application specific integrated circuits (ASICs) for fuzzy control have been developed to cope with the problem of high-speed real time applications [8–10]. Although both digital and analogue ASICs have been reported in the literature, the former is nowadays the most extensive [11–17]. As in other industrial sectors, where embedded electronics systems are used, it is likely that the expansion of digital techniques in fuzzy control will continue.

Although important advances in FLC theory have been reported in the past few years [18–21], the issue of FLC design methods that ensure the stability and the performance of the systems requires further investigation. This limitation of fuzzy technology is even more important when a digital implementation of the design is to be made. In this case, the quantisation effects, inherent to discrete systems, are also to be taken into account. In this situation, the consequences of the discretisation on the behaviour of a closed-loop fuzzy system are very difficult to predict, mainly due to the nonlinear nature of both the fuzzy controller and the quantisation function.

This work analyses the consequences of the finiteness of the word-length in digital implementations of FLCs. In a digitised fuzzy system, signals and fuzzy information must be represented using a finite number of bits. Therefore, quantisation errors are introduced due to the finiteness of binary words. The finite nature of digital information is not a serious problem if a floating-point microprocessor or DSP can be used. However, in fixed-point implementations, the consequences of the finiteness of the word-length are to be taken into account to avoid undesired effects. As in the case of conventional control, bias effects and limit cycles can degrade the response of discrete fuzzy systems [22].

In the next section, a brief introduction to the main concepts in fuzzy set theory and fuzzy control are included for the convenience of the reader. Design aspects concerning digital implementation of FLCs are also introduced. Section 14.3 is dedicated to the different sources of quantisation errors that arise when an analogue process is to be controlled using a digitised FLC. The consequences of the quantisation errors as a function of the digital word-length are analysed in Section 14.4. Next, Section 14.5 addresses the problem of the digitisation from the viewpoint of the dynamic of a feedback fuzzy system. Finally, some concluding remarks regarding the design of digital FLCs are presented.

14.2 Fuzzy Systems

Fuzzy sets, linguistic rules, and fuzzy inferences are basic tools in fuzzy system design. Let us briefly introduce these concepts.

14.2.1 Basic Concepts

The fuzzy set is the basic data type in fuzzy systems. The idea of fuzzy set, as introduced by Zadeh [23,24] is a generalisation of the ordinary concept of a set. In the same way as classical set theory establishes a sharp distinction between members and nonmembers of a set, fuzzy sets eliminate this dichotomy, allowing a partial degree of membership of an element to a class. Many of the classes we commonly use in natural language, such as the classes of "middle age", "high temperature", or "small error", can be conveniently characterised by fuzzy sets. Formally, a fuzzy set A in a universe of discourse

U is represented as a set of ordered pairs of a generic element u and its degree of membership:

$$A = \{(u, \mu_A(u)) | u \in U\} \tag{14.1}$$

where $0 \leq \mu_A(u) \leq 1$. The value of the membership function μ_A at u is the degree to which the object u belongs to A; $\mu_A(u) = 0$ represents no membership and $\mu_A(u) = 1$ represents full membership. Although membership functions are arbitrary functions, they become useful in system modelling only when they represent some semantic property of an element in the universe of discourse. Figure 14.1 depicts possible membership functions for the concepts "low", "medium" and "high" in a temperature universe. Readers interested in fuzzy set theory and related topics may consult [25–28].

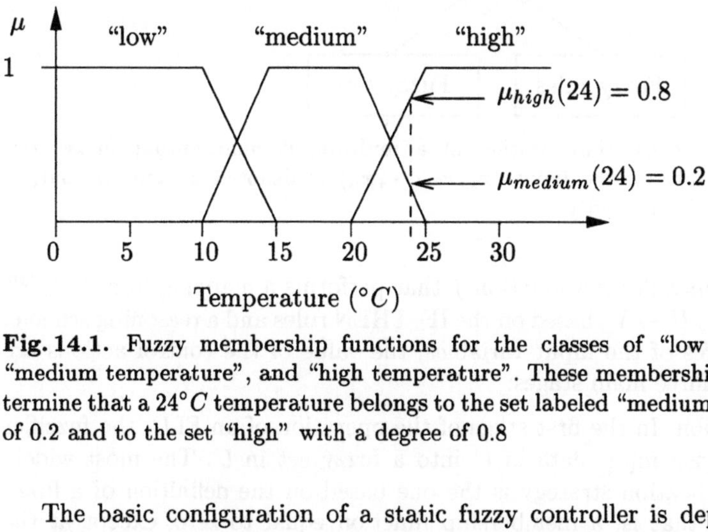

Fig. 14.1. Fuzzy membership functions for the classes of "low temperature", "medium temperature", and "high temperature". These membership functions determine that a 24°C temperature belongs to the set labeled "medium" with a degree of 0.2 and to the set "high" with a degree of 0.8

The basic configuration of a static fuzzy controller is depicted in Figure 14.2. Fuzzy controllers are constructed of a fuzzy rule base, an inference mechanism, a fuzzification interface, and a defuzzification interface. Many different algorithms are available to implement each block [29–33].

The fuzzy rule base is the collection of rules involved in an application. A typical fuzzy rule base consists of IF–THEN rules with the form:

$R^{(l)}$: IF x_1 is A_1^l and x_2 is A_2^l and $\cdots$ and x_n is A_n^l
THEN y_1 is B_1^l and y_2 is B_2^l and $\ldots$ and y_m is B_m^l

where $R^{(l)}$ is the lth rule with $1 \leq l \leq r$, $(x_1, \cdots, x_n) \in U_1 \times \cdots \times U_n \subset \mathbb{R}^n$ are the input variables, $(y_1, \cdots, y_m) \in V_1 \times \cdots \times V_m \subset \mathbb{R}^m$ are the output variables, and A_i^l and B_j^l are linguistic values defined as fuzzy sets in U_i and V_j, respectively. For instance, the conditional statement "IF temperature is low THEN heating is big", is a typical fuzzy rule in a temperature control problem. In this application "temperature" and "heating" are the input and output variables, respectively, and "low" and "big" are possible linguistic values numerically characterised as fuzzy sets like those depicted in Figure 14.1. The rule base provides a qualitative model, which expresses a control strategy using linguistic terms [34]. The rules are usually derived from the experience and knowledge of human operators and system engineers. However, for some complex systems, the fuzzy model is very difficult to develop due to the large number of variables involved in the design. In this sense, learning algorithms have been proposed to help in tuning both the fuzzy rules and the membership functions [35].

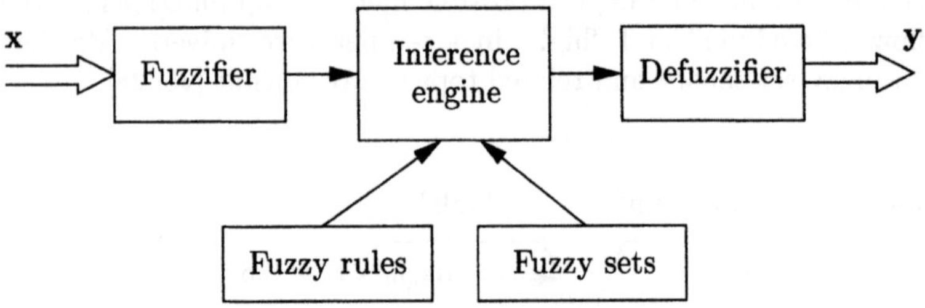

Fig. 14.2. The basic configuration of a multi-input multi-output fuzzy controller. For simplicity, the input $(x_1, x_2, \cdots, x_n)$ is denoted **x**, and the output $(y_1, y_2, \cdots, y_m)$ is denoted **y**

A fuzzy controller is a function f that performs a mapping from $U \subset \mathbb{R}^n$ to $V \subset \mathbb{R}^m$, $f : U \to V$, based on the IF–THEN rules and a reasoning scheme. Given measures of the input variables, the value of the control actions are computed in three main stages:

1. Fuzzification. In the first stage of the operation of an FLC, the fuzzifier maps a crisp input data in U into a fuzzy set in U. The most widely used fuzzification strategy is the one based on the definition of a fuzzy singleton, that is, a membership function equal to zero, except at the measured point, at which the membership degree equals one.
2. Inferences. Once the input data have been fuzzified, the inference mechanism evaluates the rule base and derives fuzzy control actions. This block performs a mapping from fuzzy sets in U to fuzzy sets in V. In a large number of FLCs, the inference mechanism is based on the compositional rule of inference. Particularly, the max-min method and the max-prod method are two of the most popular. The motivation is often that both methods lead to intuitively correct reasoning and are suitable for simple and extensible implementations.
3. Defuzzification. In the last stage of the FLC operation, the defuzzification interface maps fuzzy control actions in V into crisp control actions in V. In control applications the most commonly used defuzzification strategy is the centre of gravity. This method produces a control action equal to the centre of gravity of the membership function.

The above introduced concepts are the keys that distinguish fuzzy approximate reasoning from other symbolic frameworks. In a fuzzy system, the knowledge is linguistically encoded as structured rules, but in a numerical framework. As a consequence, mathematical algorithms can be applied and hardware implemented.

14.2.2 FLC Development

The development process of an FLC involves four main stages, which are not completely disjointed:

1. Analysis and specification
2. Design
3. Simulation and tuning
4. Implementation

In the first stage, an exhaustive analysis of the problem is performed and the specifications are clearly stated. After that, a suitable control strategy is to be selected; fuzzy technology is an alternative solution in the case of nonlinear processes that are difficult to model in a precise way, otherwise conventional control strategies would be more adequate. An important task that takes place in this stage is the identification of the controller interfaces. It is necessary to know the number of inputs and outputs, as well as their nature: data format, range of definition, variation frequency, and precision.

The object of the second stage is to design a controller capable of achieving the desired specifications. The design stage can be divided into three main steps: knowledge acquisition, knowledge representation and reasoning scheme selection. The knowledge, provided by human operators and engineers, is to be coded in the form of IF–THEN rules and fuzzy sets. The nature of the problem determines the number of rules as well as the kind and number of membership functions associated with each variable. A suitable reasoning scheme is also to be selected in this stage; the main algorithms to be selected are fuzzification, defuzzification and inferences.

The third stage, simulation and tuning, is greatly simplified if CAD tools for fuzzy control are available. Simulation and tuning steps are to be alternated until the desired response is obtained. This stage is one of the most time consuming due to the lack of a systematic design methodology. In this sense, a number of software tools incorporate design methods, based on the learning capability of neural networks, which minimise the effort required to obtain a satisfactory prototype [35]. The result of this stage is a software design, or prototype, which verifies the specifications given in the first stage.

Finally, in the fourth stage, the FLC prototype is to be implemented using a suitable technology. The implementation alternatives range from complete flexibility of software solutions to high speed ASICs [8]. Most of the available CAD tools automatically generate code for implementing the design, usually

C code, or assembly code for a microprocessor or DSP. Besides this, a number of tools have been reported in the bibliography that generate a representation of the design in a hardware description language like VHDL [36]. After programming the design into the selected hardware and interfacing it with the process, real time verifications are to be performed.

Let us briefly summarise the main design parameters, concerning the information format and the storage ability of the FLC, which will be used in the following sections:

- number of input variables (n)
- number of output variables (m)
- number of fuzzy rules (r)
- number of fuzzy sets in the partition of the input universe
- number of fuzzy sets in the partition of the output universe
- internal word-length (L)
- number of elements of the universe (C)
- membership degree word-length (N)

The choice of the above design parameters, which takes place at different stages of the FLC development cycle, is determined by application requirements such as desired precision, processing speed, or hardware dimensions, among others. Special care is to be taken in selecting those parameters that cannot be easily modified once the FLC development process enters an advanced stage. Proper examples of this are parameters L and N in digital ASIC approaches. In the following section, we will focus our attention on digital implementation technologies and on the problems that arise when the software prototype is to be translated into a finite-precision representation.

14.3 Sources of Quantisation Errors

Signal processing based on digital technology involves both amplitude discretisation and time discretisation. The former is referred to as quantisation and the latter as sampling. The precision of quantisation depends on the number of bits dedicated to represent each word of information. This design parameter, known as word-length, is a key parameter in the implementation of digital hardware. The word-length must be large enough to achieve the desired performance. On the other hand, large word-lengths penalise parameters such as speed, complexity, and cost of the circuits.

We will concentrate our attention on quantisation rather than sampling. The motivation is that word-length is one of the main design parameters in digital systems. Note that the word-length conditions the circuit architecture as well as the interfaces of the digital circuit with its environment. From the viewpoint of hardware designers, the addition of a single bit to the word-length can suppose important design modifications. On the other hand, the operation frequency could be changed without major circuit modifications.

Obviously, frequency limitations due to technological aspects and system performance are to be taken into account. To overcome the consequences of time discretisation, sampling rates verifying the Nyquist sampling theorem will be used.

Quantisation affects discretised fuzzy systems in several ways. As an illustration, consider the systems depicted in Figure 14.3, which shows the block-diagram for a typical open-loop fuzzy system using digital technology. First, the A/D interface converts the continuous signals $x_1(t)$ and $x_2(t)$ into sequences of digital samples $\{\hat{x}_1(t_k)\}$ and $\{\hat{x}_2(t_k)\}$. Next, the digital FLC produces the control sequence $\{\hat{y}(t_k)\}$ based on the input sequences. Finally, the digital-to-analogue (D/A) interface converts the discrete control sequence $\{\hat{y}(t_k)\}$ into an analogue signal $y(t)$ that is then applied to the process. Quantisation errors are introduced at different points in the system due

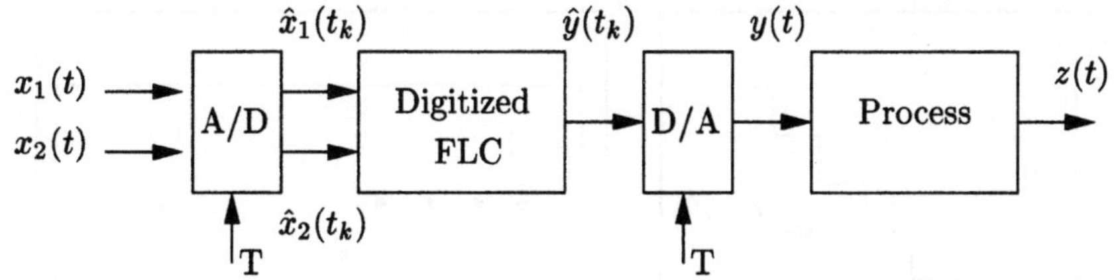

Fig. 14.3. Block-diagram of a typical open-loop fuzzy control system where an analogue process is controlled by using digital technology. For simplicity, a two-input single-output fuzzy controller has been considered

to the quantisation of both signals and coefficients. The first type of errors are the A/D errors where the samples of the analogue input signals are to be represented using a finite word-length. The second type of errors we will consider are the membership function errors that result from the storage of rules and membership degrees in a memory. Finally, the third type of errors are the arithmetic errors in the digital implementation of fuzzy algorithms. Although a wide variety of digital implementations of fuzzy systems can be found in the literature, all of them involving some kind of quantisation error, we will focus on digital implementations based on standard algorithms. The next sections give an insight into the above errors.

14.3.1 A/D Errors

In practice, the A/D conversion process is undertaken by means of electronic circuits that perform the following operations: sampling and hold, quantisation, and binary coding. The sample-and-hold circuit samples the analogue signals $x_1(t)$ and $x_2(t)$ and holds them constant until the next samples are taken. While the sampled values are held constant, the quantiser circuit generates a finite-precision representation of the samples:

$$\hat{x}_1(t_k) = Q[x_1(t_k)] \quad (14.2a)$$

$$\hat{x}_2(t_k) = Q[x_2(t_k)] \quad (14.2b)$$

In the last stage of the A/D conversion process, a binary number is assigned to each quantisation level according to a prescribed code. In most cases the quantisation function $Q[.]$ is performed by rounding or truncation. Figure 14.4(a) depicts the transfer function defining a 3-bit quantiser where quantisation has been performed by uniform rounding. Note that a 3-bit quantiser is able to represent eight different values, which are known as quantisation levels. In any B-bit implementation, independently of the selected code, 2^B quantisation levels are allowed.

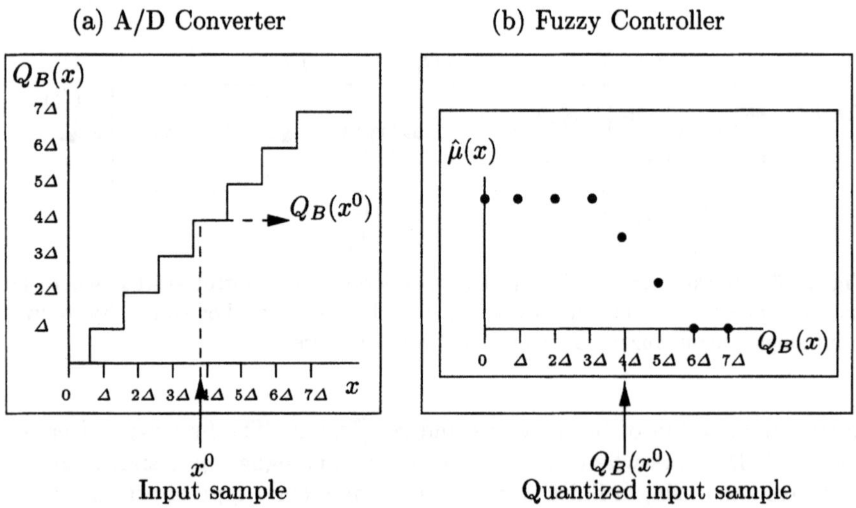

Fig. 14.4. Wordlength and quantisation. (a) Rounding quantisation in a 3-bit A/D converter. (b) The A/D word-length (external word-length) is selected to match the controller word-length (internal word-length)

The difference between the quantised sample and its true value is the quantisation error:

$$e_{x_1} = Q_B[x_1] - x_1 \quad (14.3a)$$

$$e_{x_2} = Q_B[x_2] - x_2 \quad (14.3b)$$

In the case of uniform rounding $-\Delta/2 < e_{x_1}, e_{x_2} \leq \Delta/2$, and for truncation $-\Delta < e_{x_1}, e_{x_2} \leq 0$, where Δ is the difference between consecutive quantisation levels. If a sample is out of range (overflow) the quantised result is usually obtained by saturation. Although the A/D circuitry is not a subject of digital fuzzy designers, its precision constrains the selection of the FLC word-length. In this sense, we will assume that the A/D word-length (external word-length) exactly matches the FLC word-length (internal word-length). Any other configuration generates additional quantisation errors [16]. As an example of this design constraint, Figure 14.4b) depicts a typical membership function in a 3-bit FLC.

14.3.2 Membership Function Errors

Another source of error in digital implementations of FLCs is the quantisation of the membership functions. Each antecedent and consequent of the rules is associated with a membership function whose definition in a digital format involves quantisation errors. The standard representation of a membership function in a digital format is composed of C binary words, each one composed of N bits; C being the number of elements of the discretised universe of definition, and N being the number of bits dedicated to represent each membership degree (membership degree word-length). Figure 14.5 shows an arbitrary membership function $\hat{\mu}(x)$ defined over an 8-element ($C = 8$) universe with 2 bits per membership degree ($N = 2$). The distance from black circles to the continuous function gives a measure of the membership function error.

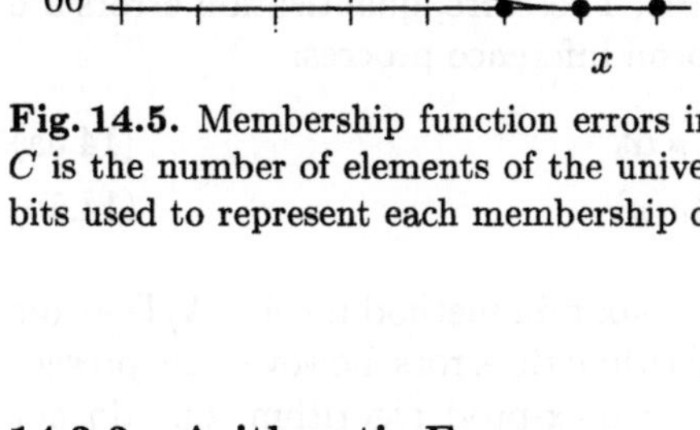

Fig. 14.5. Membership function errors in a digital FLC using $C = 8$ and $N = 2$. C is the number of elements of the universe of definition, and N is the number of bits used to represent each membership degree

14.3.3 Arithmetic Errors

The effects of quantisation depend on the inference algorithms used. Since the max-min method and the max-prod method are the kernel of most ASICs for fuzzy control, let us analyse these algorithms. In the following we will consider a typical two-input single-output FLC with rules:

$$R^{(l)} : \text{IF } x_1 \text{ is } A_1^l \text{ and } x_2 \text{ is } A_2^l \text{ THEN } y \text{ is } B^l$$

where x_1 and x_2 are input variables, and y is an output variable. The membership functions of the fuzzy subsets $A_1^l \subseteq U_1$, $A_2^l \subseteq U_2$, and $B^l \subseteq V$ are denoted $\mu_{A_1^l}(x_1)$, $\mu_{A_2^l}(x_2)$, and $\mu_{B^l}(y)$, respectively. Given crisp input data x_1^0 and x_2^0 which are fuzzified as fuzzy singletons, the control action, $\mu_B^0(y)$, is obtained as follows:

$$\alpha_1^l = \mu_{A_1^l}(x_1^0) \tag{14.4a}$$

$$\alpha_2^l = \mu_{A_2^l}(x_2^0) \tag{14.4b}$$

$$\omega^l = \alpha_1^l * \alpha_2^l \tag{14.4c}$$

$$\mu_{B^l}^0(y) = \omega^l * \mu_{B^l}(y) \tag{14.4d}$$

$$\mu_B^0(y) = \max \mu_{B^l}^0(y) \tag{14.4e}$$

where l is the number of the rule and $*$ is the product operation. The above introduced equations (14.4a) to (14.4e) define the max-prod inference method. A similar set of equations can be obtained using the minimum operation instead of the product. This alternative inference method is known as the max-min method. In a large number of digital hardware implementations, the inference processes given above are implemented as a parallel two stage architecture. In the first stage, the activation levels of the rules, ω^l, are evaluated, while in the second stage, the overall conclusion is obtained as an aggregation of partial rule contributions.

Consider the system characterised by a fuzzy rule base like the one above, and equations (14.4a) to (14.4e). Let us assume that the membership functions are stored using N-bit length registers, and the inputs are quantised into a L-bit digital format. While the minimum operation in a max-min inference does not introduce arithmetic errors, the product in equations (14.4c) and (14.4d) must be rounded or truncated to N bits before the maximum evaluation in equation (14.4e) takes place. Therefore, quantisation errors are introduced in both stages of the max-prod inference process:

$$e_{\omega^l} = Q_N[\alpha_1^l * \alpha_2^l] - \alpha_1^l * \alpha_2^l \tag{14.5a}$$

$$e_{\mu_{B^l}^0(y)} = Q_N[\omega^l * \mu_{B^l}(y)] - \omega^l * \mu_{B^l}(y) \tag{14.5b}$$

Both the max-prod method and the max-min method involve A/D errors as well as membership function errors. Arithmetic errors, however, are present in digital implementations based on the max-prod algorithm, but do not appear if a max-min algorithm is used [22].

14.4 Digitised FLCs

The block-scheme we propose for the analysis of the quantisation errors is reported in Figure 14.6. We will simulate the fuzzy system and measure the quantisation noise as a function of the digital word-length.

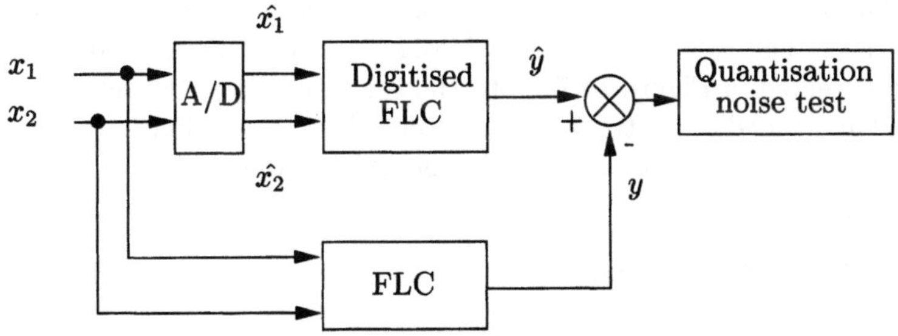

Fig. 14.6. Block-scheme for the evaluation of the overall quantisation noise

14.4.1 FLC Under Test

A typical two-input single-output FLC will be used to evaluate the sensitivity of both max-min and max-prod algorithms to the digitisation. A singleton fuzzifier and a centre of gravity defuzzifier will be used. The input variables are the "error" (e) and the "change of error" (Δe), and the output variable is the "action of control" (y). The fuzzy rules are as follows:

$$R^{(l)} : \text{IF } e \text{ is } A_1^l \text{ and } \Delta e \text{ is } A_2^l \text{ THEN } y \text{ is } B^l$$

Although arbitrary shaped membership functions can be defined, in spite of standardisation we use piecewise linear trapezoids and triangles in both the antecedent and the consequent of the rules. The input universes U_1 and U_2 have been partitioned into five fuzzy sets, and the output universe V has been partitioned into seven fuzzy sets. The membership functions of these fuzzy sets are depicted in Figure 14.7.

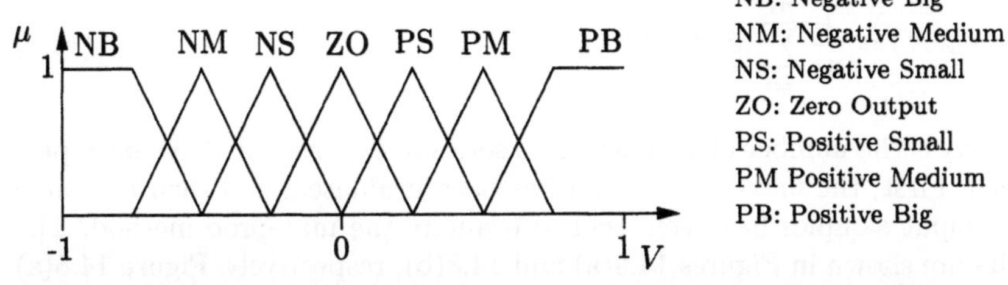

Fig. 14.7. Fuzzy partition of the input universes (U_1, U_2) and the output universe (V)

Table 14.1 shows a synthetic representation of the rule base, which is composed of 25 rules. The rule base has been designed in conjunction with the fuzzy sets represented in Figure 14.7. Both the rules and the fuzzy sets have have designed and tuned to control a second-order plant in a feedback system. This closed-loop system will be analysed in the next section.

Table 14.1. Rule base for computation of y

e \ Δe	NH	NL	Z	PL	PH
NH	NB	NB	NM	NS	ZO
NL	NB	NM	NS	ZO	PS
Z	NM	NS	ZO	PS	PM
PL	NS	ZO	PS	PM	PB
PH	ZO	PS	PM	PB	PB

The digitisation of the system has been carried out by discretising signals, membership degrees and fuzzy algorithms. Quantisation has been performed by uniform rounding and all digitised signals and coefficients have been represented by fixed-point binary numbers. Two independent digitisation parameters have been considered, the internal word-length (L), and the membership degree word-length (N). Each L with $3 \leq L \leq 10$ different values of N, into a realisable range ($2 \leq N \leq 16$), have been considered. Uniform random numbers have been fed to the controller inputs to evaluate the quantisation noise accordingly to the scheme proposed in Figure 14.6.

14.4.2 Consequences of the Digitisation on the FLC

In order to evaluate the controller response using different word-lengths, the mean error (e_y) of the continuous FLC control action (y) and the digitised FLC control action ($\hat{y}$) has been computed for each L and N:

$$e_y(L,N) = \frac{1}{n_s} \sum_{i=1}^{n_s} |\hat{y}_i(L,N) - y_i| \tag{14.6}$$

where n_s is the number of samples. Two series of experiments have been performed. First, the max-min method has been evaluated, and then the same set of input samples has been used to evaluate the max-prod method. The results are shown in Figures 14.8(a) and 14.8(b), respectively. Figure 14.8(a) shows the mean error as a function of the membership degree word-length for different internal word-lengths. It can be seen that independently of the internal word-length, the mean error is not significantly reduced by increasing the membership degree word-length over 6 bits. Note that curves with low precision ($L = 3$ and $L = 4$) exhibit this behaviour even from 4 bits. The saturation observed in the error curves indicates that there is a maximum number of significant bits in the representation of the membership degrees. The number of significant bits depends not only on L but also on the type and number of fuzzy sets in the partition of the universes [37]. Most fuzzy control applications use symmetric membership functions, uniformly distributed along the universe of definition in a similar way as appears in Figure 14.7. These kind of partitions require a reduced number of membership degrees because the same pattern is repeated several times. Therefore, the number of significant bits of N is lower than L. Moreover, the number of significant bits of N is reduced as the number of fuzzy sets in the partition is increased.

On the other hand, the addition of bits to the internal word-length reduces the mean error in all the curves under study. Figure 14.8(b) depicts the mean error as a function of the membership degree word-length obtained with the max-prod method. As can be seen, the results are qualitatively similar to those obtained using the max-min method. The main noticeable difference between both sets of curves is that the mean errors are slightly greater in Figure 14.8(b) than in Figure 14.8(a). The reason could be found in the presence of arithmetic errors (see equations 14.5a and 14.5b) in the digitisation of the max-prod algorithm that do not appear in a max-min inference algorithm. Both methods can be implemented with mean errors less than 1% selecting $N \geq 4$ and $L \geq 6$.

In Figure 14.9 the mean error has been represented, for both the max-min and the max-prod method, as a function of the internal word-length (L); the data have been extracted from Figures 14.8(a) and 14.8(b) in the saturation region ($N \geq 6$). Note that using a logarithmic scale in the ordinate axis, a linear relation between the mean error and the internal word-length has been obtained. This means that:

$$\log(e_y) = -a \cdot L + b \tag{14.7}$$

with a and b real constants. A linear curve fit in the case of the max-min method gives $a_1 = 0.303$ and $b_1 = 1.763$, and for the max-prod method $a_2 = 0.306$ and $b_2 = 1.796$. Note that $a_1 \simeq a_2 \simeq \log(2)$ with an error lower than 2%. Using this fact in equation (14.7) results in:

$$\log(e_y) = -\log(2) \cdot L + b \tag{14.8}$$

and, therefore:

$$e_y = \frac{c}{2^L} \tag{14.9}$$

where $c = 10^b$, and 2^L is the number of quantisation levels in a L-bit digital format. Therefore, in agreement to the experimental data, the mean error

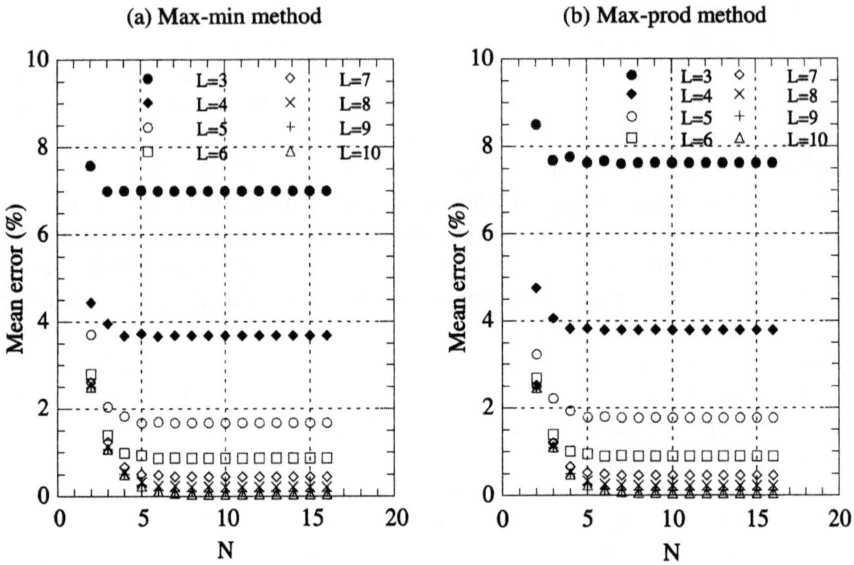

Fig. 14.8. Mean error as a function of the membership degree word-length (N), and the internal word-length (L), (a) using a max-min method and (b) using a max-prod method

is inversely proportional to the number of quantisation levels of the internal word-length. This relation provides a useful tool in the selection of the FLC internal word-length. In fact, given a maximum mean error, the internal word-length can be easily estimated from equation (14.9) between an error close to 2% in both the max-min and the max-prod methods.

In view of the results discussed above, and taking into account other important design parameters like implementation cost, hardware dimension, or speed requirements, a tentative digital format for fuzzy processing can be selected. However, in most practical applications, the FLC is embedded in a closed-loop control system and the effects of quantisation due to signals feedback should also be considered. In the next section, the quantisation noise in feedback fuzzy systems will be analysed, and the problem of the selection of the word-length in digital hardware for fuzzy control will be addressed.

14.5 Consequences of the Digitisation in Feedback Fuzzy Systems

To continue, the FLC outlined in the previous section will be used to control a second-order plant in a feedback system. The controller has been designed and tuned using continuous signals and "infinite-precision" arithmetic. If a

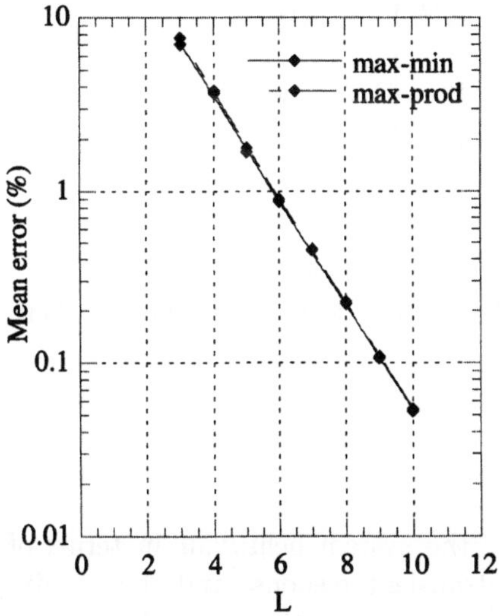

Fig. 14.9. Mean error as a function of the internal word-length (L) for a max-min algorithm and a max-prod algorithm in the saturation region ($N \geq 6$)

constant reference signal is applied to the continuous system, the output becomes equal to the reference and remains constant after a certain time. The same FLC, but digitally implemented, may lead to oscillations or steady state errors as a consequence of the quantisation errors. Let us investigate the quantisation distortions in a typical feedback fuzzy system.

14.5.1 Digitised FLC in a Feedback System

Figure 14.10 depicts the block-diagram of a fuzzy system composed of the digitised FLC introduced in the previous section, a second-order plant, and the necessary interfaces. The plant has a single analogue input $y(t)$ and a single analogue output $z(t)$. The output $z(t)$ is fed back to generate the error signal $e(t) = z_R(t) - z(t)$, $z_R(t)$ being the reference signal. Next, the A/D interface converts $e(t)$ at the sampling instant t_k into a digital number $\hat{e}(t_k)$. The digitised FLC produces the control signal $\hat{y}(t_k)$ based on the input $\hat{e}(t_k)$ and $\hat{\Delta} e(t_k)$. The D/A interface converts $\hat{y}(t_k)$ into a continuous control action $y(t)$. In addition to the blocks depicted in Figure 14.10, the system also includes scaling factors that multiply the controller input and output.

The rule base and the membership functions of the FLC have been designed to improve the overall control performance in the continuous case. The main considerations that have been taken into account for tuning the FLC are the following:

- to reduce the settling time

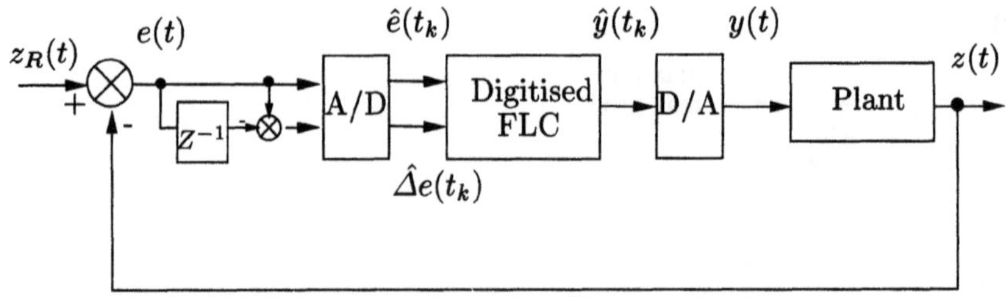

Fig. 14.10. Feedback system composed of a digitised FLC, a second-order plant, an A/D converter, and a D/A converter

- to avoid large overshoot
- to reduce the steady-state error

In this experiment, we will evaluate the system behaviour in terms of time domain characteristics. Both the transient response and the steady-state response will be evaluated in a qualitative rather than quantitative way. The continuous FLC will be digitised using different internal word-lengths, while the membership degrees will be assumed to take on "infinite-precision" values ($N \to \infty$). This strategy has been adopted in view of the results discussed in the previous section and taking into account some additional design considerations:

- The role of the internal word-length is more important than that of the membership degree word-length, as states equation (14.9). The effects of the membership degree word-length on the FLC response are to be taken into account only when small word-lengths are to be used (see Figure 14.8(a) and Figure 14.8(b)).
- The FLC internal word-length constrains the selection of the A/D and D/A circuits. It is desirable for the internal word-length to match the external word-length in order to avoid additional quantisation errors.
- In the case of software implementations, the membership functions are usually defined on discretised universes but using continuous membership degrees. In these cases the quantisation errors are also to be considered.

The discretisation of the universes is best analysed using the number of elements per universe (C) as a parameter instead of the internal word-length (L). In the case of hardware implementations only those values of C that verify $C = 2^L$ are suitable. Software implementations, however, allow arbitrary values of C. Therefore, in spite of the usefulness of the results, arbitrary values of C will be investigated.

14.5.2 Selection of the Wordlength in Digital Fuzzy Controllers

The set of curves depicted in Figure 14.11 accounts for the effects of quantisation errors on the feedback system of Figure 14.10. The curves depict the control action and the system output for different universe discretisations: C=7, 13, 25, and 385. Although these results have been obtained using a max-prod inference method, they do not significantly differ from those obtained by the authors using a max-min method in a similar experiment [22].

A discretisation of the input universes into seven elements ($C = 7$) accounts only for the most meaningful points of the fuzzy sets in the input partition, that is, those with full membership ($\mu = 1$) or null membership ($\mu = 0$) (see Figure 14.7). Therefore, α_1^l, α_2^l, and ω^l can take on only two values: 0 or 1 (see equations (14.4a) to (14.4c). Moreover, there is a unique rule in the rule base with $\omega^l = 1, 1 \leq l \leq r$; only this rule contributes to the conclusion. As a consequence, the control action is limited to a reduced set of values, each one associated with one of the seven fuzzy sets in the output partition. The system loses its ability to perform fuzzy reasoning and behaves like a conventional multirelay expert system. A discretisation of $C = 13$ leads to a quite different behaviour, close to that obtained in the continuous case. The process output reaches the reference with neither overshoot nor steady-state error. It is remarkable that this discretisation, though rather coarse, is indeed able to control the process efficiently. A new refinement of the discretisation up to $C = 25$ does not improve the system response. On the contrary, the system output oscillates indefinitely around the reference. This effect, referred to as limit cycle behaviour, is a typical consequence of quantisation in feedback systems. In the last case ($C = 385$), the system performance does not differ significantly from that achieved in the continuous case. These experimental results indicate that greater word-lengths do not always lead to an improvement in the system behaviour.

The above effects are best understood if we take a look at Figure 14.12 where the system trajectory in the $(e, \Delta e)$ plane has been represented. The figure grid represents the discretisation used in each case. Note that due to the signal quantisation, the trajectory points are to be rounded to allowed values (points of the grid). In the case $C = 13$ (Figure 14.12(a)), the system output reaches a stationary state with $e(t) = 0$ and $\Delta e = 0$. However, for $C = 25$ (Figure 14.12(b)), the system trajectory arrives to a stationary state characterised by two dark regions in the $(e, \Delta e)$ plane. Points in both regions are rounded to $\hat{e} = 0$ until a certain value of e is reached. This value of e is rounded to $\hat{e} \neq 0$ and activates the control action (see the pulses in Figure 14.11 with $C = 25$). As a consequence, the sign of Δe changes and $\hat{e} = 0$ again (the point jumps back to the other region). This repetitive process is the origin of the oscillations in the stationary state.

The experimental results presented in the previous section can be used to select a tentative digital format. Taking into account the application requirements and the precision of the peripheral circuits, a suitable internal word-length can be selected. To select the membership degree word-length, the number of significant bits in the representation of the membership degrees

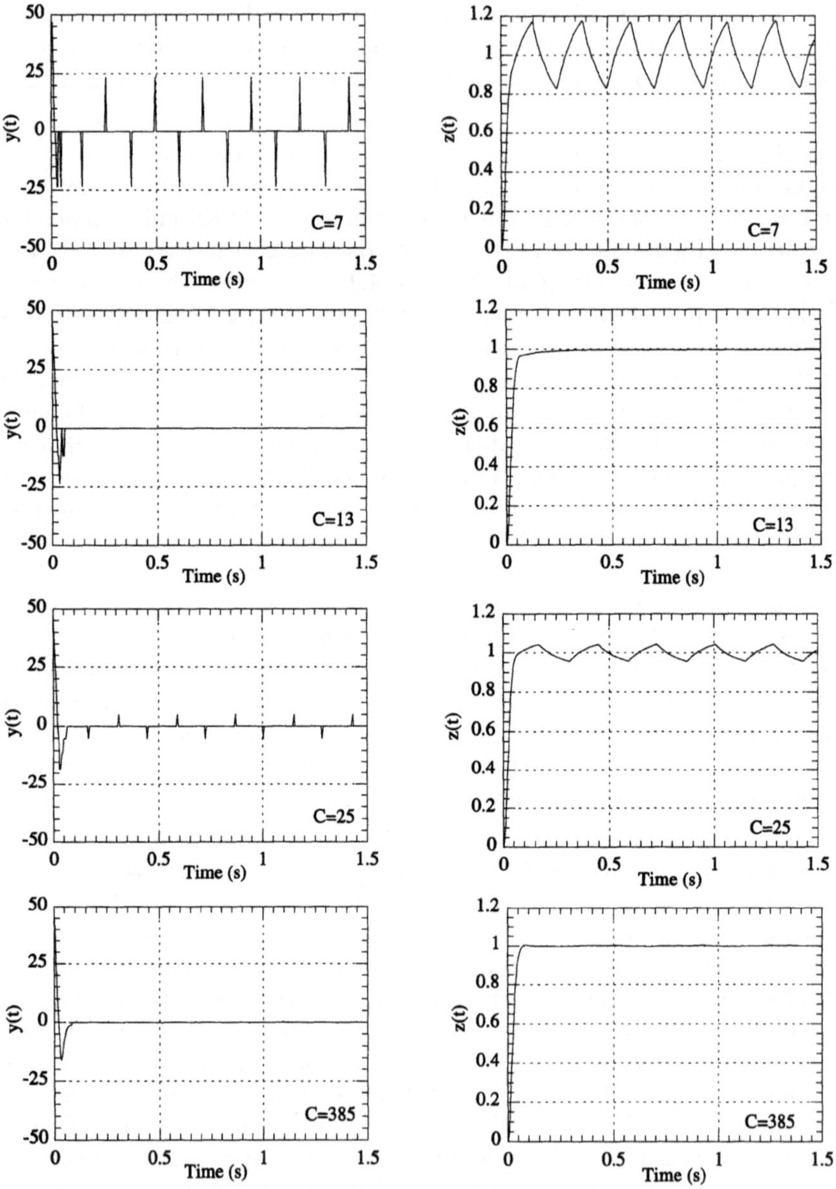

Fig. 14.11. Control actions and system outputs for different discretisations of the universes of definition; (C) is the number of elements of the universes

is to be estimated. This parameter is easily obtained from the partition of the universes. It is important to perform simulations of the digitised system to validate the selected digital format. The digitised FLC should be represented by implementing the N-bit membership degrees and the C-element universes, while the process should be represented with longer word-length to approximate the continuous process.

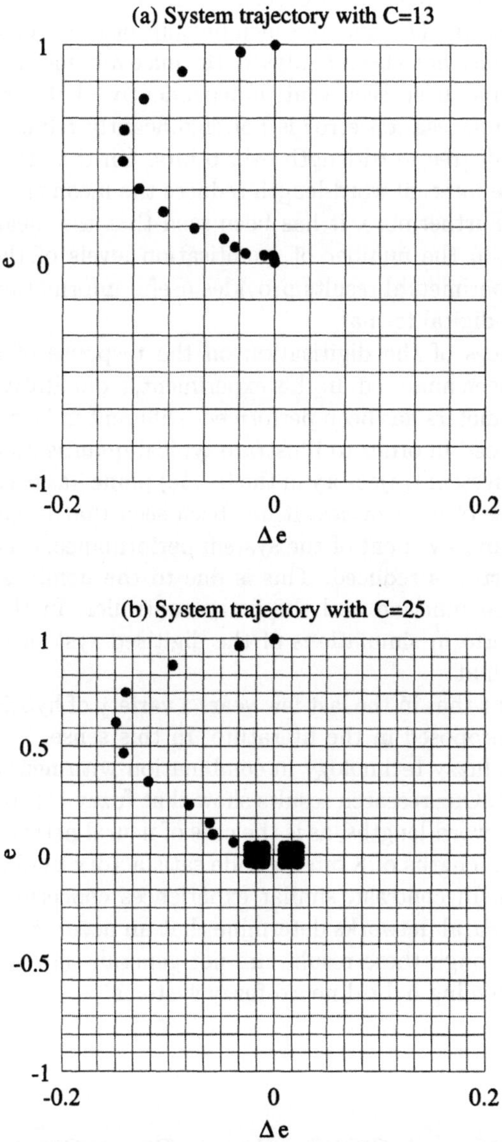

Fig. 14.12. System trajectory in the $(e, \Delta e)$ plane, (a) with $C = 13$ and (b) with $C = 25$. The figure grid represents the discretisation used in each case.

14.6 Conclusions

Quantisation noise is introduced at different points in a fuzzy system due to the quantisation of both signals and membership degrees. First, the main sources of quantisation noise have been introduced: the A/D errors, the membership function errors, and the arithmetic errors. Next, the quantisation errors in a typical two-input single output FLC have been evaluated as a function of the internal word-length (L) and the membership degree word-length (N). Two reasoning schemes have been analysed, the max-min method and the max-prod method. It has been seen that, independently of the internal word-length, the mean quantisation error is not significantly reduced by increasing the membership degree word-length over 6 bits. On the other hand, the addition of bits to the internal word-length reduces the mean error in all the curves under study. Furthermore, it has been seen that the mean error is inversely proportional to the number of quantisation levels of the internal word-length. These experimental results provides useful information in the selection of an adequate digital format.

In addition, the consequences of the digitisation on the response of a closed-loop fuzzy system have been analysed. In this experiment, a qualitative evaluation of time-domain parameters has been performed. Different universe discretisations have been proposed in order to illustrate typical quantisation effects. A representation of the system trajectory in the $(e, \Delta e)$ plane has been included to explain the presence of limit cycles. It has been seen that larger word-lengths do not imply an improvement of the system performance, even when the mean quantisation error is reduced. This is due to the nonlinear nature of both the quantisation function and the fuzzy controller. In this situation, it is important to perform simulations of the digitised system to validate the selected digital format.

Finally, it is to be pointed out that in the last few years a variety of hybrid intelligent systems have been proposed in the literature. In this sense, it is frequently advantageous to use fuzzy technology in combination with neural networks (neuro-fuzzy systems). Our research results show that fuzzy control hardware does not require large word-lengths, as is the case of neural network implementations. Internal word-lengths of 8 bits provide satisfactory performances in control applications. In contrast, similar experiences concerning the implementation of digital neural networks determine that an accuracy of 16 bits is usually required. Although these results are not general, they are to be taken into account in designing hybrid neuro-fuzzy systems.

Acknowledgements

The authors are grateful to the Spanish CICYT and to the Basque Country University for partial support of this work (projects TIC1999–0357 and UPV 224.310–EA105/99, respectively).

References

1. Mamdani, E. H., Assilian, S. (1975) A fuzzy logic controller for a dynamic plant. Int. J. Man-Machine Std. **7**, 1–13
2. Kickert, W. J. M, Van Nauta Lemke, H. R. (1976) Application of a fuzzy controller in a warm water plant. Automatica **12**, 301–308
3. Sugeno, M. (ed.) (1985) Industrial Applications of Fuzzy Control. Elsevier Science Publishers B. V. (North-Holland), The Netherlands
4. Hellendoorn, H., Palm, R. (1994) Fuzzy system technologies at Siemens R & D. Fuzzy Sets Syst. **63**, 245–269
5. Bonissone, P., Badami, V. et al. (1995) Industrial applications of fuzzy logic at General Electric. Proc. IEEE **83**, 450–465
6. van der Wal, A. J. (1995) Application of fuzzy logic control in industry. Fuzzy Sets Syst. **74**, 33–41
7. Bonissone, P., Chen, Y. et al. (1999) Hybrid soft computing systems: Industrial and commercial applications. Proc. IEEE **87**, 1641–1667
8. Jamshidi, M. (1994) On software and hardware applications of fuzzy logic, in Fuzzy Sets, Neural Networks and Soft Computing. Van Nostrand Reinhold, New York, 396–430
9. Costa, A., de Gloria, A. et al. (1995) Hardware solutions for fuzzy control. Proc. IEEE **83**, 422–434
10. Kandel, A., Langholz, G. (eds.) (1998) Fuzzy Hardware: Architectures and Applications. Kluwer Academic Publishers, USA
11. Togai, M., Watanabe, H. (1986) Expert system on a chip: An engine for real-time approximate reasoning. IEEE Expert Syst. Mag. **1**, 55–62
12. Lim, M., Takefuji, Y. (1990) Implementing fuzzy rule-based systems on silicon chip. IEEE Expert Syst. Mag. **5**, 31–45
13. Samoladas, V., Petrou, L. (1994) Special-purpose architectures for fuzzy logic controllers. Microprocessimg and Microprogramming **40**, 275–289
14. Eichfeld, H., Klimke, M. (1995) A General-purpose fuzzy inference processor. IEEE Micro **15**, 12–17
15. Ascia, G., Catania, A. et al. (1996) A reconfigurable parallel architecture for fuzzy processor. Information Sciences **88**, 299–315
16. Patyra, M. J., Grantner, J. L., Koster, K. (1996) Digital fuzzy logic controller: Design and implementation. IEEE Trans. Fuzzy Syst. **4**, 439–459
17. del Campo, I., Callao, R., Tarela, J. M. (1998) Automatic implementation of different inference architectures for fuzzy control on PLDs. Computers and Electrical Engineering **24**, 113–121
18. Ying, H., Silver, W., Buckley, J. J. (1990) Fuzzy control theory: A nonlinear case. Automatica **26**, 513–520
19. Hwang, G. C., Lin, S. C. (1992) A stability approach to fuzzy control design for nonlinear systems. Fuzzy Sets Syst. **48**, 279–287
20. Wang, H. O., Tanaka, K., Griffin, M. F. (1996) An approach to fuzzy control of nonlinear systems: Stability and design issues. IEEE Trans. Fuzzy Syst. **4**, 14–23
21. Chen, B., Tseng, C, Uang, H. (2000) Mixed H_2/H_∞ fuzzy output feedback control design for nonlinear dynamic systems: An LMI approach. IEEE Trans. Fuzzy Syst. **8**, 249–265

22. del Campo, I., Tarela, J. M. (1999) Consequences of the digitization on the performance of a fuzzy logic controller. IEEE Trans. Fuzzy Syst. **7**, 85–92
23. Zadeh, L. A. (1965) Fuzzy sets. Inform. Control **8**, 338–353
24. Zadeh, L. A. (1968) Fuzzy algorithms. Inform. Control **12**, 94–102
25. Dubois, D., Prade, H. (1980) Fuzzy Sets and Systems. Theory and Applications. Academic Press, London
26. Klir, G., Folger, T. (1988) Fuzzy Sets, Uncertainty, and Information. Prentice Hall International, London
27. Zimmerman, H. (1990) Fuzzy Set Theory–and its Applications, 2nd edition. Kluwer Academic Pub., Boston
28. Kosko, B. (1992) Neural Networks and Fuzzy Systems. Prentice Hall, New Jersey
29. Pedrycz, (1989) Fuzzy control and fuzzy systems. John Wiley & Sons, Chichester
30. Lee, C. C. (1990) Fuzz logic in control systems: Fuzzy logic controller–Part I. IEEE Trans. Syst., Man, Cybern. **20**, 404–418
31. Lee, C. C. (1990) Fuzz logic in control systems: Fuzzy logic controller–Part II. IEEE Trans. Syst., Man, Cybern. **20**, 419–435
32. Terano, T., Asai, K., Sugeno, M. (1991) Fuzzy Systems Theory and its Applications. Academic Press, London
33. Mendel, J. M. (1995) Fuzzy logic systems for engineering: A tutorial. Proc. IEEE **83**, 345–377
34. Sugeno, M., Yasukawa, T. (1993) A Fuzzy-logic-based approach to qualitative modeling. IEEE Trans. Fuzzy Syst. **1**, 7–31
35. Jang, J. S., Sun, C. T., Mizutani, E. (1997) Neuro-Fuzzy and Soft Computing. A computational Approach to Learning and Machine Intelligence. Prentice Hall, USA
36. Hollstein, T., Halgamuge, S. K., Glesner, M. (1996) Computer-aided design of fuzzy system based on generic VHDL. IEEE Trans. Fuzzy Syst. **4**, 403–417
37. Surmann, H., Ungering, A. P. (1995) Fuzzy rule-based systems on general purpose processors. IEEE Micro **15**, 40–48

INDEX

MIX
Papier aus verantwortungsvollen Quellen
Paper from responsible sources
FSC® C105338

If you have any concerns about our products,
you can contact us on
ProductSafety@springernature.com

In case Publisher is established outside the EU,
the EU authorized representative is:
Springer Nature Customer Service Center GmbH
Europaplatz 3, 69115 Heidelberg, Germany

Printed by Libri Plureos GmbH
in Hamburg, Germany